Religious Tourism and Globalization

The Search for Identity and Transformative Experience

CABI Religious Tourism and Pilgrimage Series

General Editors:

Dr Razaq Raj, Leeds Business School, Leeds Beckett University, UK; Dr Kevin Griffin, School of Hospitality Management and Tourism, Dublin Institute of Technology, Ireland

This series examines the practical applications, models and illustrations of religious tourism and pilgrimage management from a variety of international perspectives. Pilgrimage is not only a widespread and important practice in Islam, Judaism and Christianity, but also in other major religious traditions such as Buddhism, Hinduism and Sikhism.

The series explores the emergence and trajectories of religious tourism and pilgrimage. Inclusive of all denominations, religions, faiths and spiritual practices, it covers evaluations of religious tourism and pilgrimage, management guides, economic reports and sets of represented actions and behaviours within various cultural, management and marketing contexts. A key strength of the series is the presentation of current and diverse empirical research insights on aspects of religious tourism and pilgrimage, juxtaposing this with state-of-the-art reflections on the emerging theoretical foundations of the subject matter.

The series illustrates the principles related to religion, pilgrimage and the management of tourist sites. It aims to provide a useful resource for researchers and students of the subject, and increase understanding of this vital aspect of tourism studies.

Titles Available

Pilgrimage and Tourism to Holy Cities: Ideological and Management Perspectives Edited by Maria Leppäkari and Kevin Griffin

Conflicts, Religion and Culture in Tourism Edited by Razaq Raj and Kevin Griffin

The Many Voices of Pilgrimage and Reconciliation Edited by Ian S. McIntosh and Lesley D. Harman

Local Identities and Transnational Cults within Europe Edited by Fiorella Giacalone and Kevin Griffin

Religious Pilgrimage Routes and Trails: Sustainable Development and Management Edited by Anna Trono and Daniel Olsen

Risk and Safety Challenges for Religious Tourism and Events Edited by Maximiliano Korstanje, Kevin Griffin and Razaq Raj

Pilgrimage in Practice: Narration, Reclamation and Healing Edited by Ian S. McIntosh, E. Moore Quinn and Vivienne Keely

Religious Tourism in Asia
Edited by Shin Yasuda, Razaq Raj and Kevin Griffin

Islamic Tourism: Management of Travel Destinations
Edited by Ahmad Jamal, Kevin Griffin and Razaq Raj

Managing Religious Tourism
Edited by Maureen Griffiths and Peter Wiltshier

Tourism, Pilgrimage and Intercultural Dialogue
Edited by Dolors Vidal-Casellas, Silvia Aulet and Neus Crous-Costa

Spiritual and Religious Tourism: Motivations and Management
Edited by Ruth Dowson, Jabar Yaqub and Razaq Raj

Dark Tourism and Pilgrimage
Edited by Daniel Olsen and Maximiliano Korstanje

Religious Tourism and the Environment
Edited by Kiran Shinde and Daniel Olsen

Pilgrims: Values and Identities
Edited by Darius Liutikas

The Politics of Religious Tourism
Edited by Dino Bozonelos and Polyxeni Moira

Religious Tourism and Globalization: The Search for Identity and Transformative Experience
Edited by Darius Liutikas and Razaq Raj

Religious Tourism and Globalization

The Search for Identity and Transformative Experience

Edited by

Darius Liutikas

Lithuanian Centre for Social Sciences, Vilnius, Lithuania

and

Razaq Raj

Leeds Beckett University, Leeds, UK

CABI

CABI is a trading name of CAB International

CABI
Nosworthy Way
Wallingford
Oxfordshire OX10 8DE
UK

Tel: +44 (0)1491 832111
E-mail: info@cabi.org
Website: www.cabi.org

CABI
200 Portland Street
Boston
MA 02114
USA

Tel: +1 (617)682-9015
E-mail: cabi-nao@cabi.org

A catalogue record for this book is available from the British Library, London, UK.

ISBN-13: 9781800623651 hardback)
9781800623668 (ePDF)
9781800623675 (ePub)

DOI: 10.1079/9781800623675.0000

Commissioning Editor: Claire Parfitt
Editorial Assistant: Emma McCann
Production Editor: James Bishop

Typeset by Straive, Pondicherry, India
Printed and bound in the UK by CPI Group (UK) Ltd, Croydon, CR0 4YY

Contents

Contributors

Vitor Ambrósio (PhD) has been teaching at Escola Superior de Hotelaria e Turismo do Estoril [Estoril Higher Institute for Tourism and Hotel Management] (Portugal) since 1991. He also teaches on several master's degree and doctoral programmes for other institutions. Prior to this he worked as a travel agent promoter and tourist guide. His main research interest is religious tourism. His PhD thesis is on religious tourism and the development of sanctuary towns. He is the author of the book *Fátima as a Religious Centre for Incoming Tourism* (2000, Instituto Nacional de Formação Turística) and is also the author of several book chapters and articles on religious tourism.
Escola Superior de Hotelaria e Turismo do Estoril, Estoril, Portugal; and CiTUR – Centre for Tourism Research, Development and Innovation, Av. Condes de Barcelona 808, 2769-510 Estoril, Portugal. E-mail: vitor.ambrosio@eshte.pt

Silvia Aulet (PhD) is professor at the Faculty of Tourism of the University of Girona. Her research on religious tourism and pilgrimages focuses on management of sacred sites and in related areas such as peacebuilding, intercultural dialogue and cultural accessibility. She has participated in several research projects both as an academic researcher (European projects) and as a consultant. She is a member of the Editorial Board of the *International Journal on Religious Tourism and Pilgrimage*, the association Future for Religious Heritage and the University Twinning and Networking Programme (UNITWIN)/United Nations Educational, Scientific and Cultural Organization (UNESCO) Chair 'Culture Tourism and Development'. She is the director of the UNESCO Chair on Cultural Policies and Cooperation at the University of Girona.
Faculty of Tourism, University of Girona, Girona, Spain. E-mail: silvia.aulet@udg.edu

Caglar Bideci (PhD) is a researcher and lecturer in marketing in the School of Business and Creative Industries at the University of the West of Scotland's London campus. He has teaching experience in design and marketing and is currently working on visitor experience and sustainability. He has published articles on the topic of visitor experiences in high-ranked journals, including *Tourism Management Perspectives*, *Qualitative Market Research*, *Journal on Tourism & Sustainability* and the *TQM Journal*. He has contributed to various book chapters on sustainability and visitor experience.
Marketing Department, School of Business and Creative Industries, University of the West of Scotland, London campus, Import Building, 2 Clove Crescent, London, E14 2BE, UK. E-mail: caglar.bideci@uws.ac.uk

Mujde Bideci (PhD) is a researcher and senior lecturer at Canterbury Christ Church University in the UK. Her research interests focus on pilgrimage and archaeological heritage sites, with a particular emphasis on consumer experiences and immersive technology. She has published articles on the topic of pilgrimage experiences in high-ranked journals, including *Tourism Management*. Mujde is actively involved in various editorial roles, such as serving as an editorial board member for *Transnational Marketing Journal*, a review editor for *Frontiers in Sustainable Tourism*, and an active reviewer for journals such as *Tourism Management*, *Tourism Management Perspectives*, *Journal of Hospitality and Tourism Management* and *International Journal of Culture, Tourism, and Hospitality Research*.
Marketing Department, Christ Church Business School, Canterbury Christ Church University, North Holmes Road, Canterbury, CT1 1QU, UK. E-mail: mujde.bideci@canterbury.ac.uk

Elżbieta Bilska-Wodecka (PhD) is an associate professor at the Institute of Geography and Spatial Management, Jagiellonian University (Kraków). Her research interests include cultural and religious heritage, history and dynamics of changes in Catholic pilgrimages, the devotion to the Passion of Jesus Christ, the origins and typology of calvaries, religious tourism, and geography of religions especially transformations of religious structures in the 20th century. She has published around 42 articles, 23 book chapters and six books.
Institute of Geography and Spatial Management, Jagiellonian University, Gronostajowa 7, 30-381 Kraków, Poland. E-mail: elzbieta.bilska-wodecka@uj.edu.pl

Dino Bozonelos earned his PhD in political science from the University of California, Riverside. Presently, he serves as a lecturer in the Departments of Political Science and Global Studies at California State University, San Marcos, as well as in the Department of International Business at Kedge Business School. His primary research interests revolve around various global issues, including geopolitics, religious tourism and pilgrimage, religion and politics, and comparative political economy. He has published numerous journal articles and book chapters and has co-authored three open-source textbooks. He recently co-edited an edited volume with Polyxeni Moira titled *The Politics of Religious Tourism* (2023, CAB International).
Departments of Political Science and Global Studies, California State University, 333 S Twin Oaks Valley Road, San Marcos, CA 92096, USA. E-mail: dbozonelos@csusm.edu

Nour Farra Haddad (PhD) is a senior scientific researcher at the Lebanese University (Beirut). Research interests include sustainable tourism development, pilgrimage and religious tourism, anthropology of tourism, sacred sites, religious figures, rituals and interreligious dialogue. She has published around 30 articles and book chapters in Lebanon and international academic journals. She is the author of two touristic guides 'Wiz Kids' and 'Eco Lebanon: Nature and Rural Tourism' and she developed the first mobile phone app promoting interreligious tourism in Lebanon: 'Holy Lebanon'. She is also managing her own travel consultancy company NEOS.
Faculty of Tourism & Hotel Management, Lebanese University, Borj El Barajneh, Beirut, Lebanon. E-mail: nour@neosl.com

Stephen F. Haller (PhD) is an associate professor of philosophy at Wilfrid Laurier University in Brantford, Ontario, Canada. His research interests include philosophical problems of tourism. He has a particular interest in discussing pilgrimage and religious experiences in ways consistent with a scientific worldview. He is the author of numerous articles on: (i) the ethics and philosophy of tourism; (ii) chemistry; (iii) science and ethics; and (iv) the ethics of decision making under uncertainty. He is the author of *Apocalypse Soon?: Wagering on Warnings of Global Catastrophe* (2002, McGill-Queens's University Press, Montreal/Kingston, Canada).
Faculty of Social and Environmental Justice, Wilfrid Laurier University, 73 George Street, Brantford, Ontario, Canada N3T 2Y3. E-mail: shaller@wlu.ca

Jaffer Idris is a senior lecturer, teaching on the events management courses at Sheffield Hallam University since 2012. Prior to his role in academia, he worked in the private, public and voluntary sectors in a variety of management roles. He has a degree in business engineering, a master's degree in human resource management as well as several teaching qualifications. He also holds several professional accreditations that include being a chartered member of the Chartered Institute of Personnel and Development (CIPD) as well as a fellow of the Higher Education Academy. Jaffer's research interests cover cultural events and religious events, including pilgrimage such as the Hajj, and he has authored several book chapters in these areas.
Sheffield Business School, Sheffield Hallam University, Stoddart Building, Sheffield S1 1WB, UK. E-mail: j.idris@shu.ac.uk

Antonietta Ivona (PhD) is an associate professor in economic and political geography in the Department of Economy and Finance at the University of Bari Aldo Moro (Italy). Her research interests include geography of pilgrimages, economic and political geography, geography of tourism, regional economics and transport. She has published many articles and book chapters in international academic journals and books. Antonietta Ivona is a member of several editorial boards of international journals including the *International Journal of Religious Tourism and Pilgrimage* and the *Central European Journal of Geography and Sustainable Development*. She is director of a peer-review editorial series and coordinator and member of both national and international research groups.
Department of Economy and Finance, University of Bari Aldo Moro, Largo Abbazia S. Scolastica n. 53, 70124 Bari, Italy. E-mail: antonietta.ivona@uniba.it

Isilda Leitão has a Filología Hispánica and a European PhD, both granted by the University of Barcelona. She is a senior scientific researcher at CiTUR-Estoril (Center for Research, Development and Innovation in Tourism) and in the Instituto de Estudos de Literatura e Tradição (IELT) at the Universidade NOVA de Lisboa. She taught at ESHTE (Estoril Higher Institute for Tourism and Hotel Studies) from 1996 until 2021. She is a member of the scientific board of the *Revista Internacional de Turismo Empresa y Territorio*. She has published national and international articles, book chapters and books (two co-authored) about Portuguese and foreign culture and history, literature (in particular travel literature), and religious and cultural tourism.
Instituto de Estudos de Literatura e Tradição (IELT), Universidade NOVA de Lisboa, Av. de Berna, 26, C, 1069-061, Lisboa, Portugal; and CiTUR – Centre for Tourism Research, Development and Innovation, Av. Condes de Barcelona 808, 2769-510 Estoril, Portugal. E-mail: isilda.leitao@eshte.pt

Darius Liutikas (PhD) is a senior scientific researcher at the Lithuanian Centre for Social Sciences (Vilnius). Research interests include modern and traditional pilgrimage, sociology of tourism, anthropology of tourism forms of mobility, sacred places and sacred spaces, and geography of religions. He has published around 50 articles and book chapters in Lithuanian and international academic journals. He has prepared and published six books. He participated in more than 50 international scientific conferences as guest speaker or speaker. Darius Liutikas is a member of Editorial Boards of the journal *Creativity Studies*, the *International Journal of Religious Tourism and Pilgrimage*, and the *International Journal of Research in Sociology and Anthropology*. He is a chairman of the Community of Lithuanian Pilgrims (www.piligrimai.lt (accessed 20 August 2023)). He initiated and coordinated different projects on pilgrimage and religious tourism. His most recent book, of which he is the editor, is *Pilgrims: Values and Identities* (2021, CAB International). He was the main organizer of the International Religious Tourism and Pilgrimage Conference in Vilnius in 2022.
Lithuanian Centre for Social Sciences, Bitininku 1C-6, LT-08417 Vilnius, Lithuania. E-mail: darius.liutikas@gmail.com

Polyxeni Moira is Professor in Sociology of Tourism in the Department of Tourism Management, University of West Attica in Greece. She was also a visiting professor at the University of Ioannina, Harokopeion University, the University of the Aegean, the National and Kapodistrian University of Athens and the Hellenic Open University. She holds a PhD in sociology from Panteion University of Political and Social Sciences, Greece. She has BA degrees in public administration, international and European studies, and communication and mass media. She also holds MScs in rural and regional development, pedagogy, and administrative law. She has been teaching in higher education for the last 30 years. She has published 20 books in the field of tourism (e.g. on religious tourism and pilgrimage, marine tourism, cruise tourism and industrial heritage tourism), and articles in more than 50 international and Greek journals. Her research interests include sociology of tourism, consumers' behaviour in tourism, and alternative forms of tourism (especially religious tourism, marine tourism, industrial heritage tourism, slow tourism, etc.).
Department of Tourism Management, University of West Attica, Egaleo, Greece. E-mail: polmoira@uniwa.gr

Dimitrios Mylonopoulos is Professor of Tourism Law in the Department of Tourism Management, University of West Attica (Greece). He is also a visiting professor at the University of Patras, the Hellenic Open University, the National and Kapodistrian University of Athens, Piraeus University, Hellenic Army Academy and Hellenic Naval Academy. He holds a PhD in public law from Panteion University of Political and Social Sciences, Greece. He has studied at Democriteion University of Thrace, Panteion University, Harokopieio University, the Economic University of Athens, National and Kapodistrian University of Athens and Hellenic Navy Academy. He holds BA degrees in law, public administration, international and European studies, and communication and mass media. He also holds MScs in applied geography and area management, pedagogy, European studies and maritime studies. He is involved in the promotion of mental health and prevention of mental disorders. He has been teaching in higher education for the last 30 years and is a retired senior officer in the Hellenic Coast Guard. He has published more than 30 scientific books in the field of law (tourism law, marine tourism and maritime issues) and articles in international and Greek journals.
Department of Tourism Management, University of West Attica, Egaleo, Greece. E-mail: dimilon@uniwa.gr

Eleanor O'Keeffe (PhD) is a historian and social researcher. She is interested in how collective ritual or cultural practices interface with wider social and political currents in modern and contemporary UK society. Her PhD thesis explored the synergies between urban politics and culture and the reinvigoration of the military and power and identity after World War I. Her postdoctoral research has examined a wide range of subjects, including contemporary grassroots memorialization, pilgrimage practices and the work of digital memorialization in the coronavirus disease 2019 (COVID-19) pandemic. Her work has been published in journals across various fields of study, including *Mortality*, the *International Journal of Religious Tourism and Pilgrimage*, the *Journal of Urban History* and *War & Society*. She is currently the principal investigator on a British Academy & Leverhulme-funded research project, seeking a 'data-collaborative' approach to understanding contemporary pilgrimage practices in the UK.
National Centre for Social Research, 35 Northampton Square, London, ECIV 0AX, UK. E-mail: eleanor.okeeffe@natcen.ac.uk

Spyridon Parthenis is a PhD candidate in tourism, politics and public policy (University of West Attica, Greece). He is an alumnus of the Greek National School of Public Administration, specializing in tourism. He holds a master's degree in applied geography and spatial planning (Harokopio University of Athens, Greece), a master's degree in public administration (MPA) (Neapolis University Pafos, Cyprus) and an MSc degree in operations management, with specialization in agrotourism, wine tourism and regional development (University of West Attica, Greece). He has been a tourism public administration practitioner for over 15 years. His research interests are tourism

public policy, policy analysis, politics of tourism, international relations and tourism, policy process theories, policy narratives, tourism mobilities, lifestyle migration, and tourism development-induced displacement and resettlement. He has presented papers at international academic conferences and has published several articles and edited chapters on special interest tourism, with an emphasis on cultural and industrial heritage tourism, religious tourism and pilgrimage as well as tourism education.
Department of Tourism Management, University of West Attica, Egaleo, Greece. E-mail: s.parthenis@uniwa.gr

Donatella Privitera is a full professor of geography in the Department of Educational Sciences at the University of Catania (Catania, Italy). She attended an Executive Master in Agri-Business at the Catholic University of Milan, Italy. At present, she is teaching economic geography and implications with tourism and regional development. Research and publication interests include the socio-economic implications of tourism geography; sustainable cities; economic agriculture and food tourism; transport geography. In these fields, she published many scientific contributions in international academic journals, also book chapters. She is involved in several national and international research projects in public and private sectors.
Department of Educational Sciences, University of Catania, via Biblioteca 4, 95124 Catania, Italy. E-mail: donatella.privitera@unict.it

Ricardo Nicolas Progano (PhD) is a research fellow at Wakayama University, Japan. His research interests include religious tourism, heritage management and cross-cultural studies. He has carried out fieldwork on the contemporary tourism development of Japanese pilgrimage sites, such as Kumano Kodo and Koyasan. His research publications include: (i) Progano, R.N., Kato, K. and Cheer, J.M. (2021) Visitor diversification in pilgrimage destinations: comparing national and international visitors through means-end. *Tourism Geographies* 23(4), 833–854; (ii) Progano, R.N. (2021) The impact of COVID-19 on temple stays: a case study from Koyasan, Japan. *International Journal of Religious Tourism and Pilgrimage* 9(5), article 12; and (iii) Progano, R.N., Cheer, J.M. and Santos, X.M. (eds) (2023) *Host Communities and Pilgrimage Tourism: Asia and Beyond.* Springer, Cham, Switzerland.
Centre for Tourism Research, Wakayama University, Wakayama, 640-8441, Japan. E-mail: nprogano@gmail.com

Razaq Raj (PhD) is an internationally renowned academic with extensive experience in teaching, research, leadership and management in the UK, Malta, Portugal, Spain, Italy, Germany, China, Japan and South Korea. He is Associate Professor in Business Strategy and Financial Management and is a visiting professor at the Kedge Business School, France and an honorary professor at the University of Vitez, Bosnia. He is founding co-editor of the *International Journal of Religious Tourism and Pilgrimage* and Series Editor of CABI Religious Tourism and Pilgrimage Series (with Dr Kevin Griffin). He has published work on strategic management, economic and financial impacts, cultural festivals and sustainable tourism and religious tourism. He has both organized international conferences and delivered keynote speeches. He has published in a number of peer-reviewed journals and conference proceedings.
Leeds Business School, Leeds Beckett University, 520 Rose Bowl, Portland Crescent, Leeds LS1 3HB, UK. E-mail: r.raj@leedsbeckett.ac.uk

Alison T. Smith is Adjunct Professor at the College of Charleston and retired Associate Professor at The Citadel (Charleston, South Carolina). Smith is also the former Associate Director of Women's and Gender Studies at the College of Charleston. Her publications are included in the following edited volumes: (i) *Pilgrim Paths: Journeys of Transformation* (Farrelly and Keely; 2015, Inter-Disciplinary Press); (ii) *Pilgrimage in Practice: Narration, Reclamation and Healing* (McIntosh, Quinn and Keely; 2018, CAB International); (iii) *Peace Journeys: a New Direction in Religious Tourism and*

Pilgrimage Research (McIntosh, Haddad and Munro; 2020, Cambridge Scholars); (iv) *Multidisciplinary Perspectives on Pilgrimage* (Warfield; 2023, Peter Lang); and (v) *Food and the Pilgrim* (Munro, Buttigieg and Olsen; 2023, Peter Lang). Smith is co-editor of the volume *Women and Pilgrimage* with E. Moore Quinn (2022, CAB International) and she serves on the Advisory Board of the Sacred Journeys Project.

Department of French, Francophone and Italian Studies, Affiliate Faculty, Women's and Gender Studies, College of Charleston, 66 George Street, Charleston, SC 29424, USA. E-mail: smitha@cofc.edu

1 Introduction to Religious Tourism and Globalization: Challenges of Religious Tourism in a Global World

Darius Liutikas[1]* and Razaq Raj[2]
[1]*Lithuanian Centre for Social Sciences, Vilnius, Lithuania;*
[2]*Leeds Beckett University, Leeds, UK*

Abstract

This introductory chapter embarks on a journey to explore the dynamic interplay of globalization and its effects on religious tourism. It endeavours to uncover both the positive and the negative impacts of this intersection. The chapter delves into the profound significance of the quest for identity and transformative experiences, shedding light on how globalization reshapes notions of personal and social identity, as well as the transformative essence of pilgrimage. The chapter covers a critical reassessment of globalization's influence on religious tourism, spanning shifts in values, the adoption of new rituals and practices among pilgrims, changes in traditions, the embrace of technological innovations, the emergence of innovative business models, and a host of environmental and sociocultural changes. Further, it offers a glimpse into the overarching themes and contemporary case studies covered in the subsequent chapters of this book. It emphasizes the extensive transformation of the pilgrimage experience in response to the novel challenges posed by our globalized world. The aim of the book is to provide valuable insights into the realms of globalization and virtualization, shifts in religiosity, consumerism, cultural assimilation and the evolving global religious landscape.

Introduction

Pilgrimage could be defined as a journey to the particular places or events of searching and expressing a traveller's values and identities (Liutikas, 2021c). Modern perspectives on identity suggest that it is a product of construction, constantly changing and evolving, influenced by both internal factors within the individual and external factors outside the individual. Internal factors may include personality, motivation, self-concept and implicit religiosity. External factors can encompass the sociopolitical climate, culture, role expectations and religious practices. Since a pilgrimage journey integrates both internal and external constructs, it significantly impacts one's identity. It addresses various effects of the global society on religious tourism and pilgrimage. Do religious tourists need to respond proactively to the challenges of globalization? Is it possible to estimate the transformation of the pilgrimage experience? Griffin and Raj (2015, p. 57) further emphasize the 'societal trends of increasing secularization and decreasing domestic religious travel and wider international consumption trends of globalization and internationalization'.

The book aims to identify both the positive and the negative effects of globalization on religious

*darius.liutikas@gmail.com

DOI: 10.1079/9781800623675.0001

tourism. It highlights the importance of the search for identity and transformative experiences during religious tourism and examines how globalization reshapes the concepts of personal and social identity, as well as the transformative experience of pilgrimage. This book is an essential reassessment of the impact of globalization on religious tourism. Effects of globalization include changes of values, adoption of different new rituals and new pilgrims' values, changes of tradition, acceptance of technologic innovations, development of new business models, and other environmental and sociocultural changes.

Analysis of the values and identities during religious tourism is often based on theoretical models (Raj and Griffin, 2015; Liutikas, 2017, 2021b). This book brings together contributions that address the identity of religious tourists and their transformative experiences in a globalized world. It presents a range of contemporary case studies on the transformations of the pilgrimage experience, in response to new challenges posed by the globalized world, such as artificial intelligence (AI) and virtualization, changes in religiosity, consumerism, cultural assimilation and shifts in the global religious landscape. It is a carefully curated collection of case-based chapters that reflect the contradictory elements of the postmodern globalized society. The primary focus is to provide changing perspectives on religious tourism in the 21st century.

Globalization and Tourism

This book explores the impact of globalization on the experience of religious tourism and pilgrimage. Various political, economic, technological, social, cultural, health, demographic and environmental factors influence religious tourism and pilgrimage in a globalized society. As noted by Timothy (2019), tourism is not only a force for globalization but also a product of it. Sociologist A. Giddens (2002) argued that globalization affects everyday life as significantly as it does global events. Meethan (2001) drew attention to the increasing interconnectivity, growing economic 'depth' and the extension of commodity relations. These interconnections create global mobility for people, whether for tourism or migration (Shaw and Williams, 2004).

The benefits of globalization include increased connectedness, cultural activities, the diffusion of ideas and practices, the adoption of technological improvements, the development of the knowledge economy, and increased revenues (Reisinger, 2009). Acknowledgement of cultural diversity, the spread of democracy and the global market also impact tourism development. However, Giddens (2002) emphasized the challenges of globalization, such as new risks and uncertainties (e.g. global climate change, the global electronic economy and fundamentalism). Globalization can also lead to increased polarization in the world, a negative impact on local communities and environmental degradation (Reisinger, 2009).

In contemporary times, globalization refers to both the compression of the world and the intensification of global consciousness (Robertson, 1992, 2000). The increasing sociocultural density leads to the relativization or challenge of coexisting vastly diverse cultures.

The concept of a shrinking world, as described by Dicken (2007), can be illustrated through innovations in transportation technologies. In the past, travel primarily relied on animals such as horses, donkeys, oxen, camels and elephants. People either rode these animals directly or used carriages and coaches on land, as well as various boats and sailing ships on lakes, rivers and seas during the 16th–19th centuries. The mid-19th century saw a significant transformation in transportation with the advent of steam locomotives and steamships. This was followed by the emergence of motorized cars and buses at the end of the 19th century and the early 20th century.

However, it was after World War II that air transport experienced rapid growth and mass utilization. Propeller aircraft operated in the 1930s and 1940s, while the 1950s witnessed the widespread use of jet passenger aircraft, revolutionizing long-distance air travel. In the 21st century, low-cost airlines have become a defining characteristic of air travel. These developments in transportation technology have played a pivotal role in making travel more accessible and the world more interconnected, effectively reducing the perceived distances between various parts of the globe.

New forms of political, economic, social and cultural organization are emerging, and the production of tourist space involves both physical space and socio-economic context. The growth of mass tourism has led to the standardization of tourist experiences. Well-known elements of this trend include 'McDonaldization' (Ritzer, 1996), which involves the spread of fast-food chains and international cuisine at restaurants, the development of international hotel chains, and the global distribution and service systems such as Booking.com, Expedia, Airbnb and TripAdvisor. This standardization raises the important question of what kind of influence global society has on religious tourism experiences. Can religious tourism experiences be homogenized and quantified? According to Beyer (1994, p. 9), the answer is yes, but global culture will only fully emerge after two or three centuries from now. Friedman (1994) highlighted that social reproduction goes beyond the confines of individual societies. Since social cycles are intrinsically linked to time, they offer a comprehensive framework for understanding cumulative social processes and transformations occurring over time (Friedman, 1994, p. 7). However, Beyer (1994) also emphasized that globalization alters the context within which specific cultures exist, indicating a process of transformation rather than the disappearance of distinct and identifiable identities.

Globalization has created new opportunities for advancements in the tourism sector. It has fundamentally altered the dynamics of competition for competitive advantage, not only among firms and destinations but also within countries and regions. Globalization has led to reduced air-travel expenses and enhanced accessibility to various destinations. Reisinger (2009, p. 11) suggests that a new tourist in the global world 'demands new products, variety, flexibility, and personalization' and 'develops new values and worldviews that stress the importance of family and ecology'.

The distinct feature of contemporary globalization lies in the interconnectedness, multiplexity and hybridization of social life at various levels, including social networking, the organization of work and leisure time, and spatial dimensions.

The positive aspects of globalization on tourism include concerns about global climate change and the safeguarding of natural resources, valuing social relations and diversity, acquiring knowledge and enhancing the overall quality of experiences. Increasingly, tourists are seeking to manifest their identity or express themselves through cultural exploration (Fig. 1.1). They recognize the significance of physical, intellectual, emotional and spiritual well-being. Current travel trends include sustainable, responsible, eco-friendly, green, experience-oriented and transformative tourism. Tourism is moving towards achieving a sense of well-being and mindfulness (Stankov *et al.*, 2020) in a post-coronavirus disease 2019 (COVID-19) world. Raj and Griffin (2020) further highlight that COVID-19 has had notable implications for religious travellers.

Viewing the global as dominant flows of transformation, the chapters in this book take an overview of the crucial interactions of identity formation and the accelerated pace of covariation in relevant tourism activities.

Impact of Globalization on Religion

Globalization significantly impacts the religious and spiritual aspects of people's lives, including religious tourism and pilgrimages. The impact of globalization on religion is multifaceted and complex (Robertson, 2000; Juergensmeyer, 2005; Woodhead *et al.*, 2021). This impact varies across different regions and contexts and includes religious pluralism, the growth of global religious movements, secularization and the revival of religiosity, religious conflicts, or even religiously motivated violence and the exchange of cultural and religious practices, beliefs and traditions.

None the less, it is important to recognize that globalization should be viewed as an evolving phenomenon that results from a combination of economic, political, sociocultural, environmental and technological processes operating at various levels. Globalization is both shaped by existing structures and influences the formation of new structures. Beyer (1994, p. 9) discusses the 'creation of a new global culture with its attendant social structures', the nature of which depends critically on sub-global processes (Jessop, 1999).

Sociologist G. Davie (2007) raised questions about the role of secularization in the modernization process. Her answer is that secularization is extrinsic to modernization because it is possible to be both fully modern and fully religious (Davie,

Fig. 1.1. Religious tourists in Assisi, Italy. Photo taken by D. Liutikas.

2007, p. 247). The argument put forth by Stark and Bainbridge (1985) regarding secularization as a significant concept in modern times, generating two opposing processes – religious revival and religious innovation – remains highly relevant today. Research areas such as the revitalization of religion (Wuthnow, 1992) or the return of religion (Robertson, 2007) have highlighted the deficiencies of secularization theory. The 'un-secularization' of the world is related to the emergence of new religious movements and the revival and renewal of traditional religions (Adams, 1997). Is there actually a rise of traditional religions? Are we living in a post-secular era? Gorski and Altinordu (2008) suggest that in the transition from the 20th century to the 21st century, religion was rediscovered. Bauman (1992, p. x) attributed modernity as disenchantment and despiritualization and postmodernity as re-enchantment.

From an individual's perspective, beliefs have become fragmented, leading scholars to use terms such as 'privatisation of religion' (Beyer, 1994, pp. 86–94) and 'commodification of religion' (Robertson, 2000, p. 65). In the context of globalization and in the era of neoliberalism, religion has transformed into a commodity that exists alongside other goods and services (Gauthier, 2020). Various religious ideas, traditions, groups and movements, along with religious institutions, engage in competition within the religious marketplace. The religious entities that have successfully adapted to these conditions emerge as significant beneficiaries of globalization. Gauthier (2020) claimed that non-institutionalized, entrepreneurial, transnational and media-savvy religious movements, based on emotional, healing-oriented and charismatic qualities, are on the rise.

The diversification of religious practices within globalization leads to a fragmentation of beliefs. The proximity of diverse truths coexisting side by side inevitably results in the relativization of religious claims asserting absolute truths (Robertson, 2000, p. 60). As cultures come into closer contact and societies become more interconnected and exposed to a range of religious ideas and practices, there is a greater likelihood of encountering diverse interpretations and perspectives within religious communities. This diversification of beliefs can contribute to a sense of uncertainty and a perceived threat to traditional religious doctrines and practices. Some individuals or groups may adopt a more rigid and exclusive approach, emphasizing strict adherence to specific interpretations of religious texts or traditions. This phenomenon, known as religious fundamentalism, often seeks to preserve traditional beliefs and resist external influences (Bruce, 2008; Herriot, 2009).

Religious fundamentalism, while driven by a desire to safeguard core religious principles, can also lead to tensions and conflicts, both within religious communities and with the broader society. It may manifest as resistance to cultural changes, social integration or engagement with differing belief systems. Addressing the challenges posed by globalization on religion and the potential for religious fundamentalism requires thoughtful dialogue, interfaith understanding, and respect for diverse perspectives.

Globalization not only has an indirect impact on religion (for instance, through its social effects) but also has direct implications for the way people live their religious lives. Globalization affects various aspects of society, including communication, the dissemination of innovations, the exchange of cultural and religious practices, beliefs, and traditions, and the construction of social values. Certainly, transnational corporations and global lending institutions often embody and promote values that can be in opposition to religious values. These values commonly include consumerism, individualism, materialism, competitiveness, indebtedness and a profit-oriented mentality. Thus, the impact of globalization on the religious or spiritual lives of people is very real.

The exchange of cultural and religious practices, beliefs or traditions is related to interconnection and cultural hybridization (Meethan, 2001). An example of the adoption of diverse cultural and religious traditions within Christianity can be seen in various ecumenical movements and the spread of hymns, such as Taizé, Hillsong or Gospel music. Hartje-Döll (2016) noticed that popular Christian praise and worship music genres move among churches 'in a steady stream'. Other examples illustrate how cultural and religious traditions can transcend geographical boundaries, assimilate into new contexts, and become shared practices related to the spread of cultural feasts (e.g. Christmas, Chinese New Year, Diwali), physical practices (yoga, Tai-Chi), and religious-based culinary traditions (Halal and Kosher food, Indian desserts). In general, it is essential to recognize that globalization is both an external force acting upon individuals and something that individuals are aware of and conscious about (Fig. 1.2).

Technological Aspects of Globalization: AI and ChatGPT

Various aspects of globalization, such as the speed of information transmission, the development of digital technologies and technological innovations, also impact religious tourism. According to Dicken (2007), technology should primarily be understood as a catalyst or facilitator that enables and empowers various developments. It plays a crucial role in enabling the creation of new structures, organizational and geographical arrangements of economic activities, as well as innovative products and processes. However, it is important to note that technology alone does not determine specific outcomes or results. It provides possibilities and potentials, but the actual result depends on human decisions.

As Liutikas (2023) pointed out, the impact of technological innovation on tourism has become a new social and economic reality. Technological innovations, including AI, automation, virtual reality (VR) and augmented reality (AR), have significantly impacted the tourism sector by introducing new services and business models. The COVID-19 pandemic further accelerated the adoption of information and communication technologies in tourism. Smartphones and mobile devices have become essential tools for tourists, enabling them to access relevant information, navigate to destinations and capture their experiences. Social innovations in tourism, such as sharing economy platforms (couch surfing, co-living and co-working) and temporary communities (social networks of people visiting the same places), often integrate with various types of information technology (IT).

Amaro *et al.* (2022) provided an overview of the use of information and communication technologies (ICTs) in religious tourism. The main types of ICTs used for religious tourism

Fig. 1.2. Replica of the Lourdes Grotto in Hue, Vietnam, decorated for the New Year celebration. Photo taken by D. Liutikas.

purposes were identified, including webpages, social media, AR and VR, and mobile devices and applications. The authors conclude that ICTs can have an impact on enhancing tourists' experiences, promoting religious tourism, facilitating accessible tourism, ensuring sustainable tourism, providing safety and advancing religious tourism research (Amaro *et al.*, 2022).

Digital technology and VR proved to be particularly valuable during the COVID-19 pandemic (Liutikas, 2021a). AR provides enriched experiences for pilgrims by offering interactive and digitally enhanced information about their surroundings. Virtual pilgrimages to sacred and valueistic places are possible, allowing individuals to form virtual groups and visit famous pilgrimage sites and events for spiritual or cognitive purposes. Virtual route trips and the replication of rituals, such as virtual repentance, enable travellers to follow in the footsteps of medieval pilgrims. Pilgrim centres invite Internet users to explore their websites and virtually experience the sanctuaries. Virtual pilgrims can communicate and share impressions on virtual chat sites. The combination of physical activity and virtual travel can be achieved through indoor cycling systems, allowing pilgrims to cycle to destinations like Santiago de Compostela or Fátima from the comfort of their homes (Liutikas, 2021a, p. 242). Additionally, new virtual products, such as VR tours of holy sites, have emerged, enhancing the ability of pilgrims to practise religion and find meaning in everyday life.

Virtual visits to ancestral graves in Japan are now possible, offering a unique experience for users of such services. Specialized agencies provide online platforms where visitors can access the gravesites of their family ancestors. Upon clicking on the picture of a grave, the virtual tomb is adorned with digital flowers, and the sacred sutras can be heard. To authenticate their visit, visitors have the option to register and leave messages in the virtual guestbook (Inoue, 2000).

According to authors (MacWilliams, 2004; Helland, 2013), establishing a virtual connection with a physical place can provide individuals with a means to enter a liminal space where they experience an authentic connection to the sacred site. VR can act as a symbolic substitute, enabling the facilitation of religious tourism. Furthermore, social networks and software offer novel avenues for expressing values and identities in the digital realm.

The newest global product, ChatGPT (OpenAI, 2023), was launched in November 2022. It is a natural language-processing tool driven by AI technology. It allows people to have human-like conversations and obtain different answers to various questions, both complex and simple. ChatGPT is more advanced than a usual chatbot and assists you with tasks such as text editing, composing e-mails, essays and code.

However, the use of ChatGPT raises various moral and ethical issues regarding the integration of AI-generated texts in scientific research. For example, when asked to highlight the main aspects of religious tourism and globalization, the free version of ChatGPT-4 generated the text shown in Box 1.1 in 3 seconds.

ChatGPT can answer or lead you on the path to finding answers to pure spiritual or theological questions. However, the most important challenge is that various AI and VR or AR possibilities can replace interpersonal contacts. So, automation, in particular, raises questions about the extent to which religious tourists are willing to embrace technological advancements that replace human interactions.

The Search for Identity and Transformative Experience

Bauman (1997) presented a distinction between the pilgrim and the tourist as metaphors for

Box 1.1. Text produced by ChatGPT-4 on 23 June 2023 when asked to highlight the main aspects of religious tourism and globalization (italics added). From ChatGPT, 2023.

When examining the intersection of globalisation and religious tourism, several main aspects can be considered:

Cultural exchange: Globalisation has facilitated increased cultural exchange through religious tourism. People from various parts of the world visit religious sites, such as pilgrimage destinations, allowing for the sharing of beliefs, practices, and traditions. This exchange can foster understanding, dialogue, and tolerance among diverse religious communities.

Economic impact: Religious tourism has become a significant contributor to local and national economies. Globalisation has made it easier for travellers to reach religious sites, leading to increased visitor numbers and tourism-related businesses. The economic benefits derived from religious tourism can have positive effects on infrastructure development, employment, and local communities.

Religious revival: Globalisation has played a role in the revival and spread of religious practices. Through religious tourism, individuals often can reconnect with their faith, experience spiritual growth, and deepen their religious commitment. This revival can be observed both in traditional pilgrimage sites and the emergence of new religious tourism destinations.

Conflict and controversy: Globalisation and religious tourism can also be accompanied by conflicts and controversies. The influx of tourists to sacred sites can lead to tensions between commercialization and the sanctity of religious spaces. Additionally, cultural clashes may arise when visitors from different religious backgrounds encounter unfamiliar practices or beliefs.

Technological advances: Globalisation has enabled technological advancements that impact religious tourism. The availability of online resources, virtual tours, and travel platforms facilitates planning and information sharing. These technological tools can enhance the religious tourism experience and enable individuals to explore religious sites remotely.

Environmental impact: The growth of religious tourism can have environmental implications. Increased visitor numbers can strain local ecosystems, infrastructure, and resources. Sustainable tourism practices and responsible management of religious sites are crucial to mitigate these environmental impacts.

Overall, the main aspects of the intersection between globalisation and religious tourism encompass cultural exchange, economic impact, religious revival, conflict and controversy, technological advances, and environmental considerations.

modern and postmodern life strategies. The concept of a destination and achieving a purpose is most important for the pilgrim, while they 'wander the land with no destination'. Life as a pilgrimage 'gives form to the formless' and involves 'identity-building' because of the distance between the goal (always in the future) and the present moment (Bauman, 1997, p. 22). In contrast, the tourist, like the vagabond, is on the move, 'everywhere he goes in but nowhere of the place he is in'. However, 'the tourist is a conscious and systematic seeker of experience' (Bauman, 1997, p. 29) until it finally becomes unclear which one of the visiting places is home.

According to Yamin (2008), social and personal identity can be understood as both our own consciousness of ourselves and how others perceive our individuality. While it is acknowledged that people from various backgrounds and historical periods may attribute different values to specific behaviours, it is incorrect to suggest that the acceptability of any behaviour is solely dependent on an individual's perspective. Religion is related to individual and social identity, belonging to a larger group. Tajfel (1981) emphasized that our sense of self is closely tied to our identification with a group and the emotional significance of belonging to that group. Certain affiliations hold particular importance to our self-concept, providing feelings of self-esteem, stability, security or power. Being part of a group allows individuals to experience a sense of belonging, acceptance and solidarity through shared values and life goals.

Many authors relate pilgrimage to the manifestation of values and identity (Morinis, 1992; Norman, 2011, 2020; Liutikas, 2012, 2013, 2014, 2021b, c; Olsen, 2012, 2019; Griffin and Raj, 2015) or transformative experience (Turner and Turner, 1978; Morinis, 1984; Winkelman and Dubisch, 2005; Ozorak, 2006; Warfield and Hetherington, 2019). The concept of valuistic journeys, proposed by Liutikas (2012, 2021b, c), encompasses the expression of valuistic ideals and the confirmation and demonstration of identity throughout the journeys. The driving force behind choosing a pilgrim journey lies in social and personal identity, as well as various value orientations associated with religion, culture, sport, nation or personal experience. This concept comprises three key components: (i) identity; (ii) values; and (iii) their expression or demonstration. Griffin and Raj (2017) state that choosing the religious journey is linked to a vision, for the motives of the travel have remained important for travellers.

Warfield and Hetherington (2019) observed that pilgrimage can involve transformative processes that encompass transitions from illness to well-being, from grief to closure, and from fractured to integrated. Pilgrimage includes possibilities for self-transformation, particularly in relation to the transformation of identity. Pilgrimage is transformational (Winkelman and Dubisch, 2005) because it relates to belief in the impact of holy places, the power of rituals and faith. The idea of transformation was connected to the indulgence (forgiveness of sins), which was one of the main reasons for pilgrimage in medieval Christian Europe.

Identity is a fundamental concept in sociology and holds significant importance in understanding the motivations and behaviour of pilgrims and religious tourists. It is important to note that identity is not a static category but rather a dynamic aspect that can cause travellers to move between diverse types of motivation. Religious tourism and pilgrimage are connected to the demonstration, expression, development or affirmation of one's values and identity. The primary distinction between a recreational journey and a values-rich journey lies in the traveller's motivation and inner disposition (Liutikas, 2012). In a religious pilgrimage, spiritual and religious motives take precedence, while in a secular pilgrimage, cultural, national, community-based and personal motives are more prominent. A values-rich journey involves reaching the destination while simultaneously expressing one's personal or social identity and values, thereby defining the unique characteristics of the self.

Religious behaviour is influenced by factors such as social learning, social identity or social rituals. Socialization plays a vital role in shaping our social identity, as it involves learning societal norms, cultural practices and religious beliefs. Various rituals, feasts and festivals, as well as lectures and sermons, contribute to the process of socialization and re-socialization, providing strong guidelines for our values and behaviours (Fig. 1.3).

Pilgrims engage in diverse types of rites, rituals and behaviours, including internal and

Fig. 1.3. Thousands of pilgrims every year visit Medjugorje, Bosnia and Herzegovina, known for the apparitions of the Virgin Mary, in search of transformative experiences. Photo taken by D. Liutikas.

external actions, external physical actions, the use of attributes, complex actions, and actions to memorialize the journey (Liutikas, 2021b). Rituals play a significant role in pilgrimage, serving as motives for the journey, part of the travel experience and means of socialization. For example, Catholic pilgrimage traditions involve practices like praying the rosary, participating in the Holy Mass, confession, lighting candles and participating in processions. Sacrifice, hardships and restrictions are often seen as acts of purification and devotion. Pilgrims also use external markers of identity, such as clothing, scarves, medallions and symbols. It is worth mentioning that VR and AR have the potential to enhance the pilgrim's experience.

Different travel conceptions can be identified, including those based on religious values and identity, partly religious or non-religious motivation, and instrumental motivation. Understanding the behaviour of religious tourists and their identity manifestation requires considering their religiosity, religious practices, rituals and motivations. Liutikas (2021b) identified 12 types of religious-oriented travellers based on religious to secular motivation and from a low level of religious identity and involvement to a prominent level of religious identity and involvement (Fig. 1.4). This model is based on empirical research conducted in Lithuania, with the aim of identifying recognizable and measurable groups of religious travellers.

Overall, comprehending the interplay between identity and its transformation can offer valuable insights into the nature of pilgrimages and their significance to individuals (Liutikas, 2022).

Overview of the Chapters

In Chapter 2, Elżbieta Bilska-Wodecka provides an analysis of the phenomenon of copying sanctuaries in the Roman Catholic Church. The prevalence of sanctuary replicas and advancements in the IT sector indicate that nowadays, it is the shrine that comes to the pilgrim, rather than the pilgrim embarking on a journey to the shrine. Countless examples of global Catholic pilgrimage sites and their replicas are presented. Replicas can serve as a powerful means to promote the original sanctuary, inspiring a desire to visit it.

In Chapter 3, Dino Bozonelos focuses on geopolitical considerations concerning the future of religious tourism and pilgrimage. The analysis highlights three potential developments: (i) border securitization and its potential impact on global mobility; (ii) changes in information flow due to technological decoupling; and (iii) the resurgence of nationalism, potentially influencing preferences for religious travel within one's country's sphere of influence. The author addresses the phenomenon of

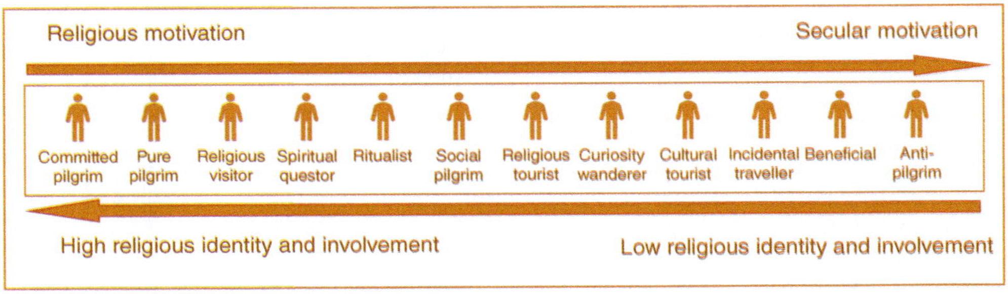

Fig. 1.4. Types of religious-related travellers. From Liutikas, 2021b.

deglobalization, defined as the process of reducing interdependencies among nations. The author suggests that geopolitical considerations will influence religious travel. Rising geopolitical tensions may deter tourists from visiting specific countries or regions.

In Chapter 4, Vitor Ambrosio and Isilda Leitão analyse and categorize the most popular books of the last four decades on the Saint James Way. The authors group published books into seven categories according to their similar characteristics and those who wrote them, including books by: (i) bloggers; (ii) celebrities; (iii) clerics; (iv) guidebook authors; (v) renowned writers; (vi) writers sharing family relationships; and (vii) writers sharing transformative experiences. Typically, all novels refer to a transformative experience of the Camino. Authors point out that even in novels where the Camino is not the main theme, it can still capture readers' attention. Two novels, Paulo Coelho's *The Diary of a Magus* and David Lodge's *Therapy*, are analysed in greater detail.

In Chapter 5, Silvia Aulet explores the transformational potential of visiting sacred sites. Understanding the concept of sacred sites enhances our grasp of their transformative potential. The author presents the potential for transformation in sacred sites or during travel to them from three primary perspectives: (i) personal; (ii) social; and (iii) environmental. The author concludes that religious travel and pilgrimages should serve as transformative and holistic experiences that encourage more conscious modes of travel and living.

In Chapter 6, Stephen Haller discusses pilgrimage as a method of personal and political integration. He examines the perspectives of several philosophers regarding how individuals' responses to crises often lead to a division within themselves, where their personal values do not align with their communal behaviour. The author suggests that modern-day pilgrimage serves as a suitable method for inner transformation. The author concludes that change from within is a personal response to uncertain times and should not be dismissed solely based on its perceived ineffectiveness in consequentialist terms.

In Chapter 7, Mujde Bideci and Caglar Bideci provide a thorough examination of the complexities of immersive IT. Immersive technologies, such as mixed reality (MR), AR and VR, have emerged as tools for enriching the visitor experience by allowing the visitor to engage with pilgrim destinations in innovative ways, bridging the gap between physical and virtual environments. While modern technologies simulate the sights, sounds and even emotions connected to the sacred sites, additional research is required to investigate the precise impact of immersive technologies in creating memorable pilgrimage experiences. However, the immersive technologies in pilgrimage must be considered as a supplement (except for those who may be unable to physically travel to pilgrimage destinations due to health or financial constraints) or support to an actual pilgrimage journey, not a replacement over reality.

In Chapter 8, Eleanor O'Keeffe focuses on virtual pilgrimage in the UK during the COVID-19 pandemic. This chapter synthesizes various social media and media data to provide a macro-level analysis of virtual pilgrimage developments in 2020. The presented evidence indicates the convergence of pandemic-era virtual pilgrimage production with institutional and cultural digital infrastructures. The author challenges the notion that virtual pilgrimage inherently acted as a globalizing or inclusive social force during the pandemic. According to research, virtual pilgrimage demonstrates its entanglement in state–society relations and the complex, at times divisive, politics it contributed to.

In Chapter 9, Nour Farra-Haddad presents the phenomenon of gigantic Christian religious statues and religious monuments in Lebanon. The building of such monuments related to pilgrimage projects, social identity and the tourism development of local communities. However, such monuments as the markers of religious identity can raise sociopolitical divisions and conflicts. The various interests and motivations of the building-project promoters, tourism stakeholders and local religious communities are discussed.

In Chapter 10, Ricardo Nicolas Progano delves into the globalization of sacred sites in regional areas, using Kōyasan, a Shingon Buddhism site situated in Wakayama Prefecture, Japan, as a case study. He noticed that sacred sites in rural areas may choose to enter the global spotlight and cater to international tourism, illustrating the connection between rural regions and globalization. While concerns about the commodification of temples exist, religious institutions have broader objectives beyond visitor growth. This interplay between religion and consumerism forms a crucial aspect of contemporary religious tourism.

In Chapter 11, Polyxeni Moira, Spyridon Parthenis and Dimitrios Mylonopoulos analyse the visitors' experience on Mount Athos, Greece. The authors affirm previous research findings that people often go on pilgrimages with twofold objectives, seeking to satisfy both their spiritual and their material needs. The authors explore the spiritual, cultural, educational, gastronomic, social and environmental experiences associated with visiting Mount Athos.

In Chapter 12, Dimitrios Mylonopoulos, Polyxeni Moira and Spyridon Parthenis present the phenomenon of Passion Plays. Case studies from Oberammergau (Germany), Škofja Loka (Slovenia) and Paros (Greece) are analysed, religious events that serve as attractions for both believers (such as pilgrims and those seeking a spiritual experience) and religious tourists. The authors have observed that throughout the history of the Passion Play, it becomes evident that the re-enactments progressively shed their purely religious character and evolve into more festive events, incorporating secular elements. Those responsible for maintaining this tradition must also consider its tourism-related aspects.

In Chapter 13, Jaffer Idris analyses the Hajj pilgrimage as a tool for seeking identity and transformation. For Muslims, the Hajj pilgrimage is a transformative experience that deepens their connection to their faith. The sense of unity and brotherhood and sisterhood has a profound impact on pilgrims. The most important aspects of a pilgrim's identity after the Hajj is the achievement of the sense of spiritual fulfilment and rejuvenation. The Hajj is also regarded as a transformative experience that extends beyond the spiritual realm, encompassing social and emotional dimensions, and fostering increased compassion, empathy and a heightened sense of social responsibility.

In Chapter 14, Antonietta Ivona and Donatella Privitera explore religious symbols and their commercialization. The authors delve into the transformation of Saint Nicholas into Santa Claus, illustrating the commercialization of a religious symbol that gained widespread recognition for secular festivities. They argue that certain images have ceased to be considered sacred and no longer necessitate religious context for a sacred experience, as they have taken on a secular dimension that sometimes supersedes their original religious significance. While globalization has created new opportunities for cross-cultural exchange and mutual understanding, it has also presented risks to the authenticity and integrity of religious symbols.

In the final chapter, Chapter 15, Alison T. Smith discusses questions of identity and transformation with her students using analysis of non-fiction readings including Ben Montgomery's (2014) *Grandma Gatewood's Walk* and Cheryl Strayed's (2012) *Wild*, the film *The Way* by Emilio Estevez (2010), and the novel *The Unlikely Pilgrimage of Harold Fry* by Rachel Joyce (2013). As a model for the kinds of microjourneys the students would undertake for their final course project, the author presents Clint Smith's (2021) *How the Word is Passed: a Reckoning with the History of Slavery Across America*. Her students read about transformative travel and pilgrimage experiences with the aim of crafting their own microjourney. The students also delve into how travel and pilgrimage can aid individuals in confronting personal and collective traumas, thereby fostering a deeper comprehension of their role in the world and their own identity. While these microjourneys may not strictly qualify as pilgrimages, many of them do share elements commonly associated with pilgrimage.

References

Adams, J.D. (1997) Toward a theological understanding of postmodernism. *Cross Currents* 47(4): 518.

Amaro, S., Barroco, C. and Fonseca, P. (2022) The use of information and communication technologies in religious tourism. In: Olsen, D.H. and Timothy, J.D. (eds) *The Routledge Handbook of Religious and Spiritual Tourism*. Routledge, Abingdon, UK, pp. 372–384.

Bauman, Z. (1992) *Intimations of Postmodernity*. Routledge, London.

Bauman, Z. (1997) From pilgrim to tourist – a short history of identity. In: Hall, S. and du Gay, P. (eds) *Questions of Cultural Identity*. Sage, London, pp. 18–36.

Beyer, P. (1994) *Religion and Globalization*. Sage, London.

Bruce, S. (2008) *Fundamentalism*. Polity Press, Cambridge.

ChatGPT (2023) Main aspects of religious tourism and globalization. Available at: https://chat.openai.com/ (accessed 23 June 2023).

Davie, G. (2007) *The Sociology of Religion*. SAGE Publications, Thousand Oaks, California.

Dicken, P. (2007) *Global Shift: Mapping the Changing Contours of the World Economy*. Guilford Press, New York.

Estevez, E. (dir.) (2010) *The Way* [Film]. Filmax/Elixir Films, USA.

Friedman, J. (1994) *Cultural Identity and Global Process*. Sage, London.

Gauthier, F. (2020) *Religion, Modernity, Globalisation: Nation-State to Market*. Routledge, Abingdon, UK.

Giddens, A. (2002) *Runaway World: How Globalisation is Reshaping our Lives*. Profile Books, London.

Gorski, P.S. and Altinordu, A. (2008) After secularization? *Annual Review of Sociology* 34, 55–85.

Griffin, K. and Raj, R. (2015) The globalization of pilgrimage tourism? Some thoughts from Ireland. In: Raj, R. and Griffin, K. (2015) (eds) *Religious Tourism and Pilgrimage Management: an International Perspective*, 2nd edn. CAB International, Wallingford, UK, pp. 57–78.

Griffin, K.A. and Raj, R. (2017) The importance of religious tourism and pilgrimage: reflecting on definitions, motives and data. *International Journal of Religious Tourism and Pilgrimage* 5(3), article 2. Available at: https://arrow.tudublin.ie/ijrtp/vol5/iss3/2/ (accessed 5 November 2023).

Hartje-Döll, G. (2016) (Hillsong) united through music: praise and worship music and the evangelical 'imagined community'. In: Ingalls, M., Landau, C. and Wagner, T. (eds) *Christian Congregational Music. Performance, Identity and Experience*. Routledge, Abingdon, UK, pp. 137–154.

Helland, C. (2013) Ritual. In: Cambell, A.H. (ed.) *Digital Religion: Understanding Religious Practice in New Media Worlds*. Routledge, Abingdon, UK, pp. 25–40.

Herriot, P. (2009) *Religious Fundamentalism: Global, Local and Personal*. Routledge, Abingdon, UK.

Inoue, N. (2000) From religious conformity to innovation: new ideas of religious journey and holy places. *Social Compass* 47, 21–32.

Jessop, B. (1999) Reflections on globalization and its (il)logic(s). In: Olds, K., Dicken, P., Kelly, P.F., Kong, L. and Yeung, H.W.-C. (eds) *Globalization and the Asia Pacific: Contested Territories*. Routledge, London, pp. 19–38.

Joyce, R. (2013) *The Unlikely Pilgrimage of Harold Fry*. Random House, New York.

Juergensmeyer, M. (2005) Introduction: religious ambivalence to global civil society. In: Juergensmeyer, M. (ed.) *Religion in Global Civil Society*. Oxford University Press, Oxford, pp. 3–10.

Liutikas, D. (2012) Experiences of valuistic journeys: motivation and behaviour. In: Sharpley, R. and Stone, P. (eds) *Contemporary Tourist Experience: Concepts and Consequences*. Routledge, New York, pp. 38–56.

Liutikas, D. (2013) Experiences of pilgrimage in Lithuania: expressions of values and identity at new destinations. *Journal of Tourism Consumption and Practice* 5(2), 43–60. Available at: https://pearl.plymouth.ac.uk/bitstream/handle/10026.1/11719/JTCP-2013-Vol5n2_p43_Liutikas.pdf?sequence=5&isAllowed=y (accessed 20 June 2023).

Liutikas, D. (2014) Lithuanian valuistic journeys: traditional and secular pilgrimage. *Journal of Heritage Tourism* 9(4), 299–316.

Liutikas, D. (2017) The manifestation of values and identity in travelling: the social engagement of pilgrimage. *Tourism Management Perspectives* 24, 217–224.

Liutikas, D. (2021a) Conclusions: pilgrimage during and after the pandemic crisis. In: Liutikas, D. (ed.) *Pilgrims: Values and Identities*. CAB International, Wallingford, UK, pp. 239–250.
Liutikas, D. (2021b) The expression of identities in pilgrim journeys. In: Liutikas, D. (ed.) *Pilgrims: Values and Identities*. CAB International Wallingford, UK, pp. 17–34.
Liutikas, D. (2021c) Introduction: motivation of the pilgrimage journey. In: Liutikas, D. (ed.) *Pilgrims: Values and Identities*. CAB International, Wallingford, UK, pp. 1–16.
Liutikas, D. (2022) Travel motivations of pilgrims, religious tourists, and spirituality seekers. In: Olsen, D.H. and Timothy, J.D. (eds) *The Routledge Handbook of Religious and Spiritual Tourism*. Routledge, Abingdon, UK, pp. 225–242.
Liutikas, D. (2023) Post-COVID-19 tourism: transformations of travelling experience. In: Dube, K., Nhamo, G. and Swart. N.M.P. (eds) *COVID-19, Tourist Destinations and Prospects for Recovery. Volume One: a Global Perspective*. Springer Nature, London, pp. 277–302.
MacWilliams, M. (2004) Virtual pilgrimage to Ireland's Croagh Patrick. In: Dawson, L. and Cowan, D. (eds) *Religion Online: Finding Faith on the Internet*. Routledge, New York, pp. 223–238.
Meethan, K. (2001) *Tourism in Global Society. Place, Culture, Consumption*. Palgrave Macmillan, London.
Montgomery, B. (2014) *Grandma Gatewood's Walk: the Inspiring Story of the Woman Who Saved the Appalachian Trail*. Chicago Review Press, Chicago.
Morinis, A. (1984) *Pilgrimage in the Hindu Tradition: a Case Study of West Bengal*. Oxford University Press, Oxford.
Morinis, A. (1992) Introduction: the territory of the anthropology of pilgrimage. In: Morinis, A. (ed.) *Sacred Journeys: the Anthropology of Pilgrimage*. Greenwood Press, Westport, Connecticut, pp. 1–28.
Norman, A. (2011) *Spiritual Tourism: Travel and Religious Practice in Western Society*. Continuum, London.
Norman, A. (2020) Pilgrimage and tourism. In: Possamai, A. and Blasi, J.A. (eds) *The SAGE Encyclopedia of the Sociology of Religion*. Sage, London.
Olsen, D.H. (2012) Negotiating religious identity at sacred sites: a management perspective. *Journal of Heritage Tourism* 7(4), 359–366.
Olsen, D.H. (2019) Religion, spirituality, and pilgrimage in a globalizing world. In: Timothy, D.J. (ed.) *Handbook of Globalisation and Tourism*. Edward Elgar, Cheltenham, UK, pp. 270–283.
OpenAI (2023) Introducing ChatGPT. Available at: https://openai.com/blog/chatgpt (accessed 6 June 2023).
Ozorak, W.E. (2006) The view from the edge: pilgrimage and transformation. In: Swatos, H.W. Jr (ed.) *On the Road to Being There. Studies in Pilgrimage and Tourism in Late Modernity. Religion and the Social Order.* Vol. 12. Brill, Leiden, the Netherlands, pp. 61–82.
Raj, R. and Griffin, K. (2015) Introduction to sacred or secular journeys. In: Raj, R. and Griffin, K. (eds) *Religious Tourism and Pilgrimage Management: an International Perspective*, 2nd edn. CAB International, Wallingford, UK, pp. 1–15.
Raj, R. and Griffin, K. (2020) Reflecting on the impact of COVID-19 on religious tourism and pilgrimage. *International Journal of Religious Tourism and Pilgrimage* 8, 1–8.
Reisinger, I. (2009) *International Tourism. Cultures and Behaviour*. Elsevier, Amsterdam.
Ritzer, G. (1996) *The McDonaldization of Society*, revised edn. Pine Forge Press, Thousand Oaks, California.
Robertson, R. (1992) *Globalization: Social Theory and Global Culture*. Sage, London.
Robertson, R. (2000) Globalization and the future of traditional religion. In: Stackhouse, L.M. and Paris, J.P. (eds) *God and Globalization. Vol.1. Religion and the Powers of the Common Life*. Trinity Press, Harrisburg, Pennsylvania, pp. 53–68.
Robertson, R. (2007) Global millennialism: a postmortem on secularization. In: Beyer, P. and Beaman, L. (eds) *Religion, Globalization, and Culture*. Brill, Leiden, the Netherlands, pp. 9–34.
Shaw, G. and Williams, M.A. (2004) *Tourism and Tourism Spaces*. SAGE Publications, London.
Smith, C. (2021) *How the Word is Passed: a Reckoning with the History of Slavery Across America*. Little, Brown and Company, New York.
Stankov, U., Filimonau, V. and Vujičič, M.D. (2020) A mindful shift: an opportunity for mindfulness-driven tourism in a post-pandemic world. *Tourism Geographies.* DOI: 10.1080/14616688.2020.1768432.
Stark, R. and Bainbridge, S.W. (1985) *The Future of Religion: Secularization, Revival and Cult Formation*. University of California Press, Oakland, California.
Strayed, C. (2012) *Wild: From Lost to Found on the Pacific Crest Trail*. Vintage, New York.
Tajfel, H. (1981) *Human Groups and Social Categories: Studies in Social Psychology*. Cambridge University Press, Cambridge.
Timothy, J.D. (2019) Introduction to the *Handbook of Globalisation and Tourism*. In: Timothy, J.D. (ed.) *Handbook of Globalisation and Tourism* Edward Elgar Publishing, Cheltenham, UK, pp. 2–11.

Turner, V. and Turner, E. (1978) *Image and Pilgrimage in Christian Culture*. Columbia University Press, New York.

Warfield, A.H. and Hetherington, K. (2019) Introduction. In: Warfield, A.H. and Hetherington, K. (eds) *Pilgrimage as Transformative Process: the Movement from Fractured to Integrated*. Brill, Leiden, the Netherlands, pp. 1–10.

Winkelman, M. and Dubisch, J. (2005) Introduction: the anthropology of pilgrimage. In: Dubisch, J. and Winkelman, M. (eds) *Pilgrimage and Healing*. University of Arizona Press, Tucson, Arizona. pp. ix–xxxvi.

Woodhead, L., Gauthier, F., Cornelio, J. and Martikainen, T. (2021) Introduction: religion in global societies. In: Cornelio, J., Gauthier, F., Martikainen, T. and Woodhead, L. (eds) *Routledge International Handbook of Religion in Global Society*. Routledge, Abingdon, UK, pp. 1–16.

Wuthnow, R. (1992) *Rediscovering the Sacred: Perspectives on Religion in Contemporary Society*. Eerdmans Publishing, Grand Rapids, Michigan.

Yamin, S. (2008) Understanding religious identity and the causes of religious violence, peace prints. *South Asian Journal of Peacebuilding* 1(1), 1–21.

2 Globalization and Pilgrimages in the Roman Catholic Church

Elżbieta Bilska-Wodecka*
Jagiellonian University, Krakow, Poland

Abstract

This chapter explores and evaluates the primary reasons behind the replication of sanctuaries in the Roman Catholic Church within the context of globalization. Among the earliest instances of such replication in history are the recreations of sites in Jerusalem associated with the Passion of the Lord. Subsequently, during the Counter-Reformation, Loreto houses were replicated. In the early 20th century, grottoes resembling Lourdes and sites connected to the Fátima apparitions were established in various locations. Over the last decade there has been dynamic expansion of sanctuaries dedicated to Divine Mercy and St John Paul II. These holy places are not only associated with sacred objects but also feature well-known prayers such as the Chaplet of Divine Mercy, the Stations of the Cross, and the Litany of Loreto, which are familiar in the Catholic world.

Introduction

The opportunity to pray and be in a place considered sacred is a profoundly significant moment in the life of a religious person (referred to as *homo religious* in Latin). The importance of this experience can be greatly emphasized when one has travelled long distances, endured numerous hardships and faced dangers to reach a particular holy place. Such holy sites, also known as sanctuaries, were visited by several million visitors annually in the 1980s and 1990s, with a significant percentage of these visitors coming from abroad. These places are often referred to as 'international pilgrimage sites' or key locations for Christian pilgrimages (Bhardwaj and Rinschede, 1988; Stoddard, 1994). In the modern era, the presence of large numbers of visitors from different countries in each sanctuary is considered a manifestation of globalization (Griffin and Raj, 2015).

Globalization is analysed through various forms of human activity. In the field of religious studies, the examination of globalization focuses on the legal and administrative aspects of the functioning of religious associations and the geographical scope of their influence (Beyer, 1994; Casanova, 2001). However, only a handful of works, such as Notermans (2008), Giacalone and Griffin (2018), Alvis (2021) and Von Klimo (2022), delve into the global aspects of religious devotion, particularly its less-explored connections to the establishment and ongoing evolution of sanctuaries and the concept of religious pilgrimages.

As Griffin and Raj (2015) point out, the presence of visitors at a pilgrimage centre, including both domestic pilgrims and those from

*elzbieta.bilska-wodecka@uj.edu.pl

 Religious Tourism and Globalization. D. Liutikus and R. Raj (eds)
DOI: 10.1079/9781800623675.0002

abroad, is indicative of the current trends of globalization and the internationalization of religious tourism. This is unquestionably one of the key aspects of globalization as a concept that impacts our modern world. However, embarking on a journey to a holy place is not always feasible.

Therefore, the purpose of this chapter is to analyse the processes that lead to a situation where a 'holy place' is made more accessible to pilgrims by constructing regional replicas. This form of 'replication' of holy places eliminates the need for long-distance journeys and allows this type of worship to assume a global form. One of the earliest examples of such replicated pilgrimages involved copying places in Jerusalem associated with the Passion of the Lord (calvaries and Ways of the Cross). Loreto houses were replicated during the Counter-Reformation. Today, Marian sanctuaries such as Lourdes, Fátima, Divine Mercy and St John Paul II's shrines are experiencing dynamic growth.

This chapter will also analyse and describe selected examples of the replication of sanctuaries within the Roman Catholic Church in the context of globalization. The chapter explores the phenomenon of making pilgrimages within the Roman Catholic Church and addresses pilgrimage-related religious devotion as a global phenomenon.

The Place of Organized Religion in the Process of Globalization and the Creation of Global Pilgrimage Centres

Opinions on the place and image of religion in the era of globalization differ among researchers. According to Halík (2003), the Roman Catholic Church was probably the first 'global player' in the history of mankind. It is difficult not to agree with this author's assertion observing the influence of the Church in the realm of political power, culture and social life. Religion can be examined in terms of three unique aspects constituting its empirical image in the context of globalization. According to Durkheim's (1912) definition, religion is the relationship between man and sanctity, defined and interpreted in various ways ('the sacred'), and the divine realm, 'manifesting itself': (i) in the doctrinal dimension (faith); (ii) in religious activity (devotion); and (iii) in the social and organizational realm (Church, religious community).

The analysis of globalization usually focuses on the socio-organizational realm and diffusion of doctrine, manifested physically as a given region inhabited by members of a given religious association (Beyer, 1994; Casanova, 2001; Davie, 2007). One of the most relevant aspects of religious globalization is related to religious devotion – the ritual dimension of religion. Religious devotion includes official obligatory religious practices and those which are optional (May Services, the rosary, Stations of the Cross and pilgrimages to various sanctuaries) (Nadolski, 1991).

The features of globalization observable at a pilgrimage centre may be analysed in terms of two different aspects. The first aspect is dynamic and refers to the demographic characteristics of visitors. It describes the national or ethnic structure of the pilgrims, with reference to foreign visitors, in relation to the total number of visitors. This aspect of religious globalization was demonstrated in a study by Griffin and Raj (2015).

The demographic aspect of globalization at a pilgrimage centre can be influenced by a range of factors. These include: (i) essential religious factors (such as the current popularity and rank of a specific form of religious devotion within the Catholic Church); (ii) economic factors (including economic crises and the income levels of individuals desiring to embark on pilgrimages to specific locations); (iii) social factors (such as the influence of elites in shaping pilgrimage trends); and (iv) political factors (including armed conflicts and social unrest within the country where a particular place of worship is situated).

The second aspect of globalization is static and refers to the object of worship venerated at a pilgrimage centre. This aspect is represented through architecture, liturgy, transnational prayer or formation of groups, and the media. A visible sign of the globalization of a given religious devotion may be the establishment of sanctuaries that are similar in terms of the architecture of the chosen place, or the presence of a relic or image as in the original place. Copies of an object of devotion or even places considered sacred (e.g. the Holy Family's house in Loreto, the figure of Our Lady of Fátima, places associated with the Passion of Jesus Christ in Jerusalem) often, but not always, give rise to new sanctuaries, thus relocating the pilgrimage movement to the new sanctuary (e.g. calvary).

When analysing the historical influence of sanctuaries, it becomes evident that only certain types of religious devotion have attained international prominence. From a religious standpoint, both doctrinal and specific devotion-related factors play a significant role. First, the message that provides religious hope to contemporary worshippers is crucial. Second, the recognition of key supernatural events by the Pope, endorsing the authenticity of apparitions in line with the teachings of the Church, has been instrumental in the designation of holy days, formulation of public prayers, endorsing the public veneration of an image or statue, and granting indulgences for prayers or pilgrimages to specific sanctuaries. Alongside the key religious aspect, the social impact of miraculous events associated with apparitions or holy places, such as victories in battle or healings, has also been noteworthy.

The aftermath of apparitions often consists of the generation of vital messages about worshippers' way of life and their relationship with God. People interested in spreading the message often form different groups. Spiritual communication is then established through the functioning of various associations. Prayer or the formation of groups of a living spirituality is propagated at these sanctuaries, such as 'Faustinum' – the Association of the Apostles of Divine Mercy (Faustinum. PL, 2022).

A very important role in this situation is played by the flow of information in the process of 'the shrine coming to the pilgrim'. It can take many different forms, such as printed periodicals, magazines in different languages or websites. First, the very availability of the Internet and then, the global coronavirus disease 2019 (COVID-19) pandemic forced us to look for new ways to provide pilgrims with the possibility of a 'spiritual pilgrimage' and prayer at a holy place. As a result, the sanctuaries of Vilnius, Krakow, Lourdes and Fátima offer online broadcasts 24 hours a day, and the sanctuary in Loreto provides a broadcast of the Holy Mass and other prayer meetings

Examples of Pilgrimage Sites with Global Features

A pilgrimage site can be local, regional, national or international in nature. Owing to the number of visitors, until the end of the 20th century, the following sites were most often indicated as international pilgrimage centres: (i) the Basilica of the Holy Sepulchre and the Way of the Cross in Jerusalem; (ii) the burial place of Saints Peter and Paul and the venerated image of *Salus Populi Romani* in Rome; (iii) Marian shrines in Lourdes, Fátima, Loreto, Czestochowa and Guadalupe; and (iv) the burial place of Saint James the Great in Santiago de Compostela (Bhardwaj and Rinschede, 1988; Stoddard, 1994). In the first decade of the 21st century, this group has come to include two sanctuaries of the Divine Mercy (Krakow and Vilnius) and the Sanctuary of St John Paul II (Krakow).

The general devotion to the person of Jesus Christ

Catholic spirituality is most affected by pilgrimages to sacred centres related to the life of Jesus (Nativity, Childhood, Transfiguration, Passion and Death, Resurrection) as well as the attributes of God (Body and Blood of Christ, the Sacred Heart of Jesus, and Divine Mercy) in the context of globalization. In this group of sanctuaries, only two have achieved global significance. The first is Jerusalem, known for the devotion to Our Lord's Passion, and it holds the distinction of being the oldest sanctuary in this context. The second is a relatively new sanctuary, jointly represented by Krakow and Vilnius, both of which are recognized for the devotion to the Divine Mercy. The holy sites located in the cities of Jerusalem, Krakow and Vilnius have attained worldwide prominence.

The specific devotion to the Passion of Jesus Christ

The most important destination for Christians has been Jerusalem – regardless of historical period. Jerusalem, the city where Jesus Christ lived, died and was resurrected, has held immense significance for Christian pilgrims throughout the last two millennia of history. In the early days of Christianity, visitors were driven to visit Jerusalem mostly by the desire to witness the places mentioned in the Bible and to explore new geographic territories. However, it was not until AD 326 when St Helena's efforts led to the discovery of the relics of the Holy Cross that these journeys began to take on a more distinctly religious character.

Over the centuries, political turmoil, including the capture of Jerusalem by Arab armies in the year 637, led to various restrictions on pilgrimages. However, in 1333, the pastoral services provided by the Franciscans in the Basilica of the Holy Sepulchre in Jerusalem significantly improved conditions for pilgrims (Mruk, 2001). This continuous presence and care by the Franciscans not only facilitated pilgrimages but also played a crucial role in promoting and solidifying the devotion to Our Lord's Passion across regions inhabited by Catholics. Through their efforts, the veneration of the places associated with Jesus' Passion gained popularity and became an integral part of Christian and later Catholic religious practices.

Since the Middle Ages, numerous graces and indulgences granted by the Holy See have been associated with Jerusalem, and those granted for visiting the Lord's Sepulchre and places associated with Jesus's Way of the Cross (Mruk, 2001). The result of such efforts by the Pope has been an increase in the number of pilgrims arriving, starting at the beginning of the 14th century. This also coincided with dynamic changes in Christian spirituality, including a stronger focus on reliving the events of Jesus's Passion. In order to make it possible to obtain Jerusalem indulgences for a broader group of people, substitute Jerusalem buildings were built (e.g. copies of Our Lord's Sepulchre). In German cities, they were built in the second half of the 15th century and took the form of a prayer route marked with chapels, the so-called seven falls of Christ. In 1486 the Franciscans in Varallo Sesia (Piedmont region of Italy) began the construction of the first 'sacred mountain', which later contributed to the establishment of a calvary. During the period of Counter-Reformation, the first calvaries were built (e.g. Kalwaria Zebrzydowska in 1602) in Poland and elsewhere (Fig. 2.1). The distribution of calvaries in Europe was determined by various historical, political and sociocultural factors. Calvaries were built and still are being built in countries where the Roman Catholic faith is dominant. Most European calvaries are located in Austria, Hungary, Germany and Poland. Some of them are independent local pilgrimage centres dedicated to

Fig. 2.1. Christ's Sepulchre built according to the model brought from Jerusalem (on the left) in Kalwaria Zebrzydowska (Poland). Photo taken by E. Bilska-Wodecka.

the devotion to Our Lord's Passion (Bilska-Wodecka, 2003).

It was the emergence of places making possible 'the reflection on the painful Passion of the Saviour', such as ways of prayer, Italian calvaries and the possibility of obtaining Jerusalem indulgences outside Jerusalem that had a significant impact on the establishment of the Stations of the Cross service. During this prayer, Christ's way from Pilate's house to the Calvary of Jerusalem is considered, stopping to pray at 14 places (stations) marked with a cross and often with visual representations. This service received Church approval in the form of indulgences like those obtained by pilgrims visiting the places of Our Lord's Passion in Jerusalem, granted in the *breve* announced by Pope Paul V on 5 November 1606. Initially, the Way of the Cross could only be established in Franciscan churches, but starting from 1741, by the decision of Pope Benedict XIV, it could be established in every parish church. On 17 July 1931, Pope Pius XI announced that anyone who prays the Stations of the Cross can obtain a plenary indulgence. In the liturgy of the Church, the Stations of the Cross service is solemnly celebrated in Catholic churches every Friday during Lent, and especially solemnly on Good Friday.

The devotion to the Divine Mercy

An awareness of the Divine Mercy has been present in the Roman Catholic Church since the beginning of its existence. However, for centuries, the teaching of the Church has paid more attention to God as a just judge. The revelations experienced by Saint Sister Faustina (Faustyna is the Polish spelling) Kowalska (1905–1938) and the message that Jesus directed to people through her, reminded people of, and made it possible for them to get to know, the attribute of the merciful God. The development of the devotion to the Divine Mercy in the forms proposed by Sister Faustina, despite some initial difficulties, became known and popular not only in Poland, but also throughout the Catholic Church. Apart from Saint Faustina, Blessed priest Michał Sopoćko (1888–1975), Polish emigrants during World War II and Cardinal Karol Wojtyła (later Pope John Paul II) have contributed to its development. The essence of the devotion to the Divine Mercy is the attitude of trust towards God. The second basic element of this devotion is the performance of acts of mercy towards our neighbours through our deeds, words and prayer. According to Sister Faustina Kowalska, new forms of the devotion to the Divine Mercy include the following:

- The image 'Jesus, I Trust In You', shown in a vision on 22 February 1931 (Saint Maria Faustina Kowalska, 2005, pp. 47–48). The first picture was painted according to Sister Faustina's instructions by Stanisław Kazimirowski in 1934 in Vilnius, first exhibited in 1935. The Krakow one from 1943 was painted by Adolf Hyła and placed in the chapel of the convent of the Sisters.
- The Feast of Divine Mercy on the first Sunday after Easter, indicated in a vision on 22 February 1931 (Saint Maria Faustina Kowalska, 2005, p. 699). On 5 May 2000, the Dicastery for Divine Worship and the Discipline of the Sacraments issued a decree establishing the Feast of Divine Mercy, which is binding throughout the Catholic Church (Dekret, 2005).
- The Chaplet of Divine Mercy, which Christ gives to the faithful as the source of Divine Mercy for the world, is celebrated every day exactly at 3:00 p.m. It was given to Sister Faustina in Vilnius on 13–14 September 1935 (Saint Maria Faustina Kowalska, 2005, p. 687).
- The Hour of Mercy, at which Jesus promises to grant all the graces that the faithful will ask for through God's Mercy for sinners. In October 1937, in the city of Krakow, under circumstances not described in detail by Sister Faustina, the Lord Jesus commanded to worship the hour of his death (Saint Maria Faustina Kowalska, 2005, p. 1320).
- Spreading the devotion to the Divine Mercy (Saint Maria Faustina Kowalska, 2005, pp. 1074–1075).

The most important events affecting the formation of the devotion to the Divine Mercy in Kraków and the pilgrimage movement in general include: (i) the *Notification* of the Holy See prohibiting the propagation of the devotion to the Divine Mercy (1959); (ii) the initiation of the information process regarding the heroic virtues of Sister Faustina (1965); (iii) listing of the monastery chapel by Cardinal Karol Wojtyła in the index of diocesan sanctuaries (1968); (iv) the appeal against the *Notification* (1978); and (v) beatification (1993) and then the canonization (2000) of Sister Faustina. During his pilgrimage in 2002, Pope John Paul II consecrated the Basilica and performed the act of entrusting all humanity to the Divine Mercy. On 11 April 2015, Pope Francis declared an extraordinary Jubilee of the Divine Mercy (8 December 2015–20 November 2016).

Today the most important sanctuaries related to the devotion to the Divine Mercy are Krakow (Poland) and Vilnius (Lithuania) (Fig. 2.2). The sanctuary in Krakow has already been visited by three popes: (i) John Paul II twice (1997 and 2002); (ii) Benedict XVI (2006); and (iii) Francis during World Youth Day celebrations in 2016.

It is estimated that between 2.0 and 2.5 million pilgrims come to the sanctuary in Krakow each year, while 300,000–500,000 visit the sanctuary in Vilnius (Lithuania). By the end of 2022 there were 29 sanctuaries of the Divine Mercy in Poland, and 74 outside Poland, with most being in Europe (22), North America (17), South America (14), followed by Asia (13), Africa (six) and two in Australia (Socha, 2001; Sanktuaria Bożego Miłosierdzia, 2023).

The Marian Devotion

The official approval of the Marian Devotion took place at the Council of Ephesus in AD 431, when Mary was granted the title of *Mother of God*. Since then, the development of the Marian Devotion has been observed across Europe and the world. In the Western Church, the devotion to the various miraculous images of the Virgin Mary, intensifying from the end of the 15th century, reached its apex after the Council of Trent (1545–1563), where, among other things, the rules of the veneration of images were established. The intercessory function of the Mother of God, both in matters of the Church and nations or countries, began to play a primary role at that time. In the propagation of the Marian Devotion, especially during the last two centuries, an important role of Marian apparitions has been observed.

The Pontifical Sanctuary of the Holy House of Loreto

Loreto is a well-known pilgrimage site in Catholic Europe. Currently, the sanctuary is visited by

Fig. 2.2. The copy of the first image 'Jesus, I Trust In You' in the House of Saint Faustina (Vilnius, Lithuania). Photo taken by E. Bilska-Wodecka.

about 3.5 million pilgrims a year (Rinschede, 1995; Soljan, 2012). Since 1294 the object of worship here has been the Loreto house (in Italian *Santa Casa*). According to tradition, this is the house where the Holy Family lived in Nazareth. It was subsequently miraculously carried by the Angels to Loreto. Since then, this special place has become the destination of numerous pilgrimages. In 1387, Pope Urban VI granted it the privilege of indulgence for the first time. In the Jubilee Year of 1450, the Loreto sanctuary was visited by Pope Nicholas V. Owing to the development of the pilgrimage movement in Europe, in 1469 the construction of a basilica began around the House. In the eastern part of the House there is an altar and a niche with a statue of Our Lady of Loreto (Prejs, 2003).

The rank of the Loreto sanctuary was raised by the *breve* of Pope Leo X of 1520, in which a pilgrimage to Loreto was considered equivalent to a pilgrimage to either Jerusalem or Rome. Pilgrimages to Loreto were made not only by ordinary people, but also by elites of the Catholic world, for example by popes (Pius II, Gregory XVI and John Paul II), heads of state (including the Emperor of Austria Ferdinand II), philosophers, poets, writers, saints and the blessed including Francis Xavier, Charles Borromeo, Teresa of the Child Jesus and Mother Teresa of Calcutta, as well as by various travellers and scholars.

The Litany to the Mother of God, composed in France in the 12th century, has been recited in the Loreto sanctuary since at least 1531, which is why it is known in the Church as the Litany of Loreto. Its text was approved in 1578 by Pope Gregory XIII (Santi, 1910). In 1587, Pope Sixtus V granted the 'Litany of Loreto' 200 days of indulgence. Since the end of the 18th century, it has been the main part of the service held throughout the Roman Catholic Church in May in honour of the Virgin Mary and is known as the May Service. In 2019, Pope Francis added the memorial of Our Lady of Loreto on 10 December to the Roman Calendar, which means it will be celebrated throughout the Roman Catholic Church (Vatican News, 2019).

Further development of the devotion of Our Lady of Loreto was facilitated by the victory coalition of Catholic states in the Battle of Lepanto against Turkey in 1571, and the introduction in the mid-16th century of the Counter-Reformation throughout the Church (Chélini and Branthomme, 1982). After 1620, in the Kingdom of Bohemia, ruled by the Habsburg dynasty especially by Emperor Ferdinand II (1578–1637), copies of the Loreto House appeared (Matsche, 1978). Building physical copies of holy places was in line with the policy of the Catholic League, which believed that in this way it would strengthen the position of the Catholic Church. The construction of Loreto chapels was one of the forms of the recatholicization of Bohemia (Matsche, 1978; Bukovský, 2000). During this period, 51 chapels were built in Bohemia and Moravia (Bukovský, 2000), 50 in Bavaria (Pötzl, 1979) and in Tirol (Grass, 1979). Indeed, even despite the Counter-Reformation policy, churches and chapels dedicated to Our Lady of Loreto were constructed in various European cities (Fig. 2.3). For example, King Władysław IV of Poland and Queen Cecylia Renata built such places of worship in Warsaw between 1640 and 1642. Additionally, a church in honour of Our Lady of Loreto was established at the Capuchins' premises in Krakow between 1712 and 1719.

Sanctuary of Our Lady of Lourdes

The origin of the sanctuary is related to apparitions of Our Lady to Bernadette Soubirous (1844–1879). In 1858, a total of 18 apparitions were recorded in the Massabielle Grotto over 5 months (from 11 February to 16 July). The Virgin Mary asked for prayers for sinners, especially by saying the rosary. She also expressed the desire to make the grotto a place of special worship. The culminating moment took place on 25 March, when the 'Beautiful Lady' introduced herself as the 'Immaculate Conception'. For Church authorities, these words were an important factor in favour of the authenticity of the said apparitions on 18 January 1862. Bernadette Soubirous entered the convent of the Sisters of Charity of Nevers in 1866. She was beatified in 1925 and canonized in 1933.

In 1866, the Chapel of the Grotto (lower church) was consecrated at the site of the apparitions. Near the basilicas, one can find the most visited holy site: the Massabielle Grotto. In 1864, a statue of Our Lady of the Immaculate Conception, the work of Józef Fabisch, made according to Bernadette's instructions, was placed there. The coronation of the statue took place in 1876.

(b) (a)

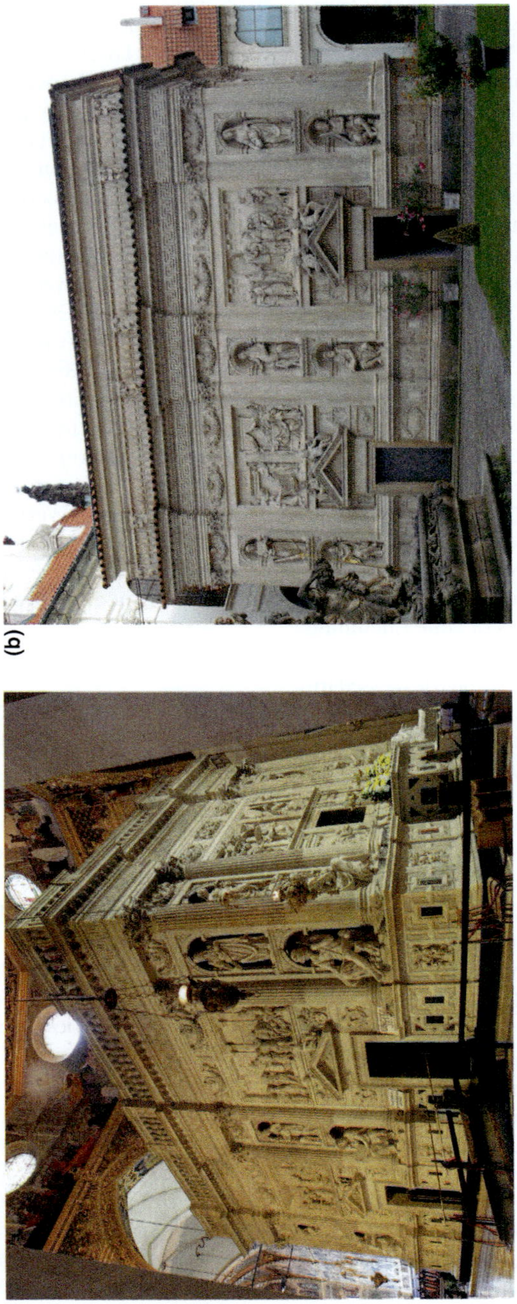

Fig. 2.3. The Holy House of Loreto in Italy (a) and the replica of the Holy House of Loreto in Prague (Czechia) (b). Photos taken by E. Bilska-Wodecka.

In 1891, Pope Leo XII instituted the Feast of the Apparition of Our Lady in Lourdes, which St Pius X in 1907 extended to the whole Church and celebrated on 11 February in the liturgical calendar. An event reminding Catholics of the message coming from Lourdes is World Day of the Sick – celebrated annually on 11 February – established by Pope John Paul II on 13 May 1992.

Since 1873, sick people have been a special group of pilgrims (currently about 70,000 people a year). The arrivals of the sick result mainly from a deep faith in the healing properties of the water from the miraculous spring flowing in the Massabielle Grotto. Healings are investigated by the special Office of Medical Certificates operating in Lourdes since 1882. By 1995, over 6000 healings had been reported (Fig. 2.4).

Pilgrims appeared almost immediately after the apparitions (in 1872 – 60,000 pilgrims; in 1908 – in excess of 1 million). Currently, more than 7 million pilgrims and tourists are recorded here every year. The first foreign groups from Europe appeared in Lourdes in 1873 (Belgium) and in 1875 (Germany, Poland, Italy), and from outside Europe in 1873 (Canada) and 1874 (the USA). Today, foreigners make up 50–60% of Lourdes pilgrims (Rinschede, 1985; Giuriati and Lazi, 1994; Soljan, 2012).

The celebration of the 50th anniversary of the apparitions served as a special impulse to build monuments in honour of the Mother of God on the model of the one found in Lourdes. Copies of the Lourdes grotto are built next to some churches, monasteries and sanctuaries in Catholic countries.

Sanctuary of Our Lady of the Rosary of Fátima

On 28 June 1914 World War I began, and on 23 February 1917 the February Revolution broke out in Russia. Both events gradually changed the political landscape of Europe. In such a turbulent sociopolitical situation, from 13 May to 13 October 1917, always on the 13th of the month (except August), apparitions took place in fields called Cova da Iria ('Peace Valley'). Their target audience consisted of three children – Lucia dos Santos (1907–2005) and two of her younger cousins, Francisco Marto (1908–1919) and Jacinta Marto (1910–1920). Francisco and Jacinta were beatified on 13 May 2000 and canonized on 13 May 2017. Lucia devoted herself

Fig. 2.4. Pilgrims gathered at the cave of Massabielle in Lourdes (France). Lithograph available in the public domain from the Wellcome Collection, London.

to religious life starting in the year 1925. She died on 13 February 2005, at the Discalced Carmelites convent in Coimbra, Portugal. With the consent of Pope Benedict XVI in 2008, 3 years after her death, her canonization process began.

The said apparitions are associated with the so-called 'Three Secrets of Fátima' and the service of the first five Saturdays of the month. The Virgin Mary recommended the daily saying of the rosary for the intention of ending the war and for peace in the world. In June, Our Lady asked for the establishment of a service to Her Immaculate Heart. The first secret concerned visions of heaven, purgatory and hell. The second spoke about the need to entrust the world, and especially Russia, to the Immaculate Heart of Mary, which was to save humanity from another terrible war and the emergence of communism across the world. In the last apparition, in October, all those gathered (70,000 people) witnessed an unusual atmospheric phenomenon known as the 'dance of the Sun'.

The first chapel was erected at the site of the apparitions in 1919 and a basilica was erected in the years 1928–1953. A statue of the Virgin Mary made of cedar wood, according to the seers' instructions, was placed in it. In 1930, the apparitions were officially recognized as real, and thus the devotion to Our Lady of Fátima was officially allowed. On 13 May 1946, the legate of Pope Pius XII solemnly crowned Our Lady of Fátima *Queen of Peace*. Pope John XXIII instituted the Feast of Our Lady of Fátima (13 May).

In the period from May to October, every evening in the Chapel of the Apparitions and in many churches around the world there is a rosary service with a procession (with candles) with the statue of Our Lady to the Basilica and back.

After the end of the apparitions, pilgrims continued to come to this place. In 1921, the local bishop allowed the celebration of the Holy Mass for pilgrims. A particularly rapid development of the pilgrimage movement took place after 1965, when attendance began to increase from about 1 million pilgrims to 5–6 million today. It was then that the international character of Fátima was crystallized (in 1965, there were pilgrims from 22 countries, and today there are pilgrims from over 100 countries). In 1987, there were 3.6 million pilgrims, of whom about 10% of the people listed in organized groups came from abroad. At the same time, in Lourdes, as many as 60% of pilgrims arriving on organized pilgrimages came from abroad (Rinschede, 1988; Giuriati and Lazi, 1994; Sołjan, 2012). Every year on 13 August, the great 'National Pilgrimage of Portuguese Emigrants' takes place (Von Klimo, 2022).

The Devotion to Our Lady of Fátima, which is constantly expanding throughout the world, together with an almost universal reception of the key messages from Fátima, were sufficient arguments for starting the peregrination of the statue in 1942, initially in Portugal, and from 1947 also abroad. To date, the statue has visited over 60 countries on all the continents. The Sanctuary of Fátima also has several 'copies' in other countries (e.g. Poland, the USA).

Devotion to the saints

The devotion to saints in the Roman Catholic Church has a long tradition and takes various forms. More than 7000 saints are recorded in the Roman Martyrology. In this group, only a few are commonly recognized throughout the entire Roman Catholic Church such as Saint Peter the Apostle in Rome, Saint James the Great in Santiago de Compostela, Saint Pio of Pietrelcina in San Giovanni Rotondo, Thérèse of the Child Jesus and the Holy Face in Lisieux, Saint Rita in Cascia, and Saint Faustina in Kraków. And an even smaller group of saints have their own local sanctuaries.

Pope John Paul II (1920–2005; pontificate 1978–2005) is undoubtedly one of the most recognized saints today, and in line with the theory of globalization, his pontificate coincided with a time of the rapid development of the media. St John Paul II used these technological achievements many times during his papacy. The great popularity of Pope John Paul II, his long-term pontificate, a total of 104 trips abroad during which he visited over 130 countries, including Poland eight times, became hallmarks of his service to the Church.

More than 6 months after the Pope's death, the Bishop of Krakow Cardinal Stanislaw Dziwisz signed the Decree establishing the John Paul II Centre in Krakow on 2 January 2006. In 2006, during a pilgrimage to Poland, Pope Benedict XVI (1927–2022) consecrated the foundation stone for the future church dedicated to John Paul II. After his quick beatification on 1 May 2011, the Sanctuary of St John Paul II was established on 11 June 2011. On the same day, the relics of the pope's blood were placed in the chapel, and this date is now considered the official beginning of the devotion to St John Paul II. Among the memorabilia in the sanctuary there are also the tombstone of John Paul II from the Vatican Grottoes of the Basilica of St Peter, a papal pastoral cross, mitre and chasuble. The canonization of John Paul II on 27 April 2014, significantly affected the growth of the devotion to his person, which already has a global dimension at the present time.

Since the Pope's beatification, a total of 25 parishes and churches dedicated to him have already been founded in Poland. His first-class relics are already found in over 100 churches and chapels in Poland. In Poland there are now ten St John Paul II sanctuaries, two are located in the USA (Washington and Clifton), with one each in Italy, the Philippines and Australia. He is also the patron saint of schools, the international airport in Krakow, certain hospitals and some streets. His monuments are being erected (about 800 in Poland) and trails are marked on an ongoing basis. In Wikipedia one can find the entry 'Pope John Paul II in popular culture' (Wikipedia, 2023). These and many other aspects of John Paul II's presence in the public realm were defined by Przybylska and Sołjan (2015) as the process of 'John-Paul-the-Second-isation' (Fig. 2.5).

Fig. 2.5. Monument to St John Paul II on the square in front of the Sanctuary in Marija Bistrica (Croatia). Photo taken by E. Bilska-Wodecka.

The record-breaking pace of the action of Church authorities was possible mainly thanks to the widespread recognition of the sanctity of John Paul II. The exceptionally quick start of the beatification process by Pope Benedict XVI was also due thanks to the attitude of the faithful, who already during his lifetime had surrounded John Paul II with reverence and a kind of real devotion. Cardinal Stanisław Dziwisz, the longtime secretary and associate of John Paul II, became the main guardian of his legacy after the death of the pope.

Summary

Pilgrimages possess a long tradition in the Roman Catholic Church as an optional devotional practice. Hence, various pilgrimage sites have become very popular. Some destinations have served as pilgrimage sites for Catholics worldwide, maintaining their status as international pilgrimage centres for many years. Only a few have replicas built in various countries, becoming places of local prayer and worship for the local community; they are known globally, and celebrated with a profound connection to the associated holy place. Such a multifaceted (object of worship, devotions/services, indulgences) presence of the main pilgrimage centre in a different place in the form of a copy can certainly be regarded as a sign of the globalization of the pilgrimage centre. On the one hand, a copy of a centre of worship makes it possible for pilgrims to be more frequently spiritually closer to the sanctuary without destroying the desire to make a pilgrimage 'to the original'. On the other hand, a replica is also a great way to popularize the original, arousing a desire to visit it, or contribute to its maintenance, or even increase the number of pilgrims in the main (original) sanctuary.

The process of the globalization of pilgrimage centres, understood as the presence of a copy of a sanctuary in the architectural sense and its message, is extremely complex, and in some cases spread out over time. In a historical sense, it may be said that this is not a feature of just modern times. Analysing the development of some religious devotions, it has been shown that it had been occurring for a long time, but owing to the slower flow of information, it lasted longer. The emergence of new forms of information transfer, from book printing, public press, through radio, television and now the Internet, has undoubtedly accelerated the process of globalization in the religious realm, with a particular emphasis on new content.

Besides the speed of information transfer, the primary factors contributing to the dissemination of a religious devotion to various countries inhabited by Catholics include a message that aligns with the teachings of the Church, validated through formal approval of apparitions, first by the local bishop and subsequently by the Pope. The involvement of higher Church hierarchs (local bishops and successive popes) and secular authorities has undoubtedly had a positive impact on the development of specific religious devotions. It is also important for the faithful to be aware of the importance of the message. Their desires and fidelity to a given religious devotion diminish the scepticism of Church authorities. Often it is the memory of ordinary people, contrary to the opinion of the clergy, that sustains, preserves and develops a new religious devotion. This was the case, for example, with the devotion to the Divine Mercy in the years 1959–1978.

Over time, awareness of the link between the replica and the original has, in some cases, become obscured. In the case of certain rituals and services, this leads to a gradual forgetting of the connection to the place of origin, and they are treated as conventional prayers, akin to practices like the Litanies of Loreto and the Way of the Cross.

References

Alvis, R.E. (2021) The tenacity of popular devotions in the age of Vatican II: learning from the Divine Mercy. *Religions* 12(1): 65. http://dx.doi.org/10.3390/rel12010065

Beyer, P. (1994) *Religion and Globalization*. Sage Publications, London.

Bhardwaj, S.M. and Rinschede, G. (1988) Pilgrimage – a world-wide phenomenon. *Geographia Religionum* 4, 11–19.

Bilska-Wodecka, E. (2003) *Kalwarie Europejskie: Analiza Struktury, Typów i Genezy*. Instytut Geografii i Gospodarki Przestrzennej UJ, Kraków, Poland.
Bukowský, J. (2000) *Loretánskié Kaple v Čechách a na Moravě*. Nakladatelstí Libri, Praha, Czechia.
Casanova, J. (2001) Religion, the new millennium, and globalization. *Sociology of Religion* 62(4), 415–441.
Chélini, J. and Branthomme, H. (1982) *Les Chemins de Dieu: Histoire des Pèlerinages Chrétiens des Origines à nos Jours*. Hachette, Paris.
Davie, G. (2007) Globalization and the study of religion. In: *The Sociology of Religion*. SAGE Publications, Thousand Oaks, California, pp. 202–223.
Dekret (2005) Dekret w sprawie Niedzieli Miłosierdzia Bożego 5.05.2000. Available at: https://opoka.org.pl/biblioteka/W/WR/kongregacje/kkultu/dekret_05052000.html (accessed 2 September 2022).
Durkheim, E. (1912) *Les Formes Elémentaires de la Vie Religieuse: le Système Totémique en Australie*. Lib. Félix Alcan, Paris.
Faustinum. PL (2022) Official website of the Association of the Apostles of Divine Mercy "Faustinum". Available at: https://www.faustinum.pl/en/en-2/ (accessed 2 September 2022).
Giacalone, F. and Griffin, K.A. (eds) (2018) *Local Identities and Transnational Cults within Europe*. CAB International, Wallingford, UK.
Giuriati, P. and Lazi, G. (1994) Pilgrims to Fátima as compared to Lourdes and Medjugorje. *Geographia Religionum* 8, 57–79.
Grass, N. (1979) Loreto im Bergland Tirol. *Jahrbuch für Volkskunde* 2, 161–186.
Griffin, K.A. and Raj, R. (2015) The globalization of pilgrimage tourism? Some thoughts from Ireland. In: Griffin, K.A. and Raj, R. (eds) *Religious Tourism and Pilgrimage Management: an International Perspective*, 2nd edn. CAB International, Wallingford, UK, pp. 57–78.
Halík, T. (2003) Religia i globalizacja. *Znak* 4(575), 12–21.
Matsche, F. (1978) Gegenreformatorische Architekturpolitik: Casa Santa-Kopien und Habsburger Loreto-Kult nach 1620. *Jahrbuch für Volkskunde* 1, 81–118.
Mruk, W. (2001) *Pielgrzymowanie do Ziemi Świętej w drugiej połowie XIV wieku*. Towarzystwo Wydawnicze Historia Iagiellonica, Kraków, Poland.
Nadolski, B. (1991) *Liturgika. Liturgia i czas*, Vols 1–4. Pallottinum, Poznań, Poland.
Notermans, C. (2008) Local and global icons of Mary: an ethnographic study of a powerful symbol. *Anthropos* 103(2), 471–481.
Pötzl, W. (1979) Loreto in Bayern. *Jahrbuch für Volkskunde* 2, 187–218.
Prejs, R. (2003) Loreto. In: *Encyklopedia Katolicka*. Wyd. TNKUL, Lublin, Poland, t. 10, kol. pp. 1382–1384.
Przybylska, L. and Sołjan, I. (2015) 'John-Paul-the-Second-isation' of cultural landscape in Poland. *Geographia Polonica* 88(4), 669–686.
Rinschede, G. (1985) Das pilgerzentrum Lourdes. *Geographia Religionum* 1, 195–256.
Rinschede, G. (1988) The Pilgrimage Center of Fátima/Portugal. *Geographia Religionum* 4, 65–98.
Rinschede, G. (1995) The Pilgrimage Centre of Loreto, Italy: a geographical study. In: Dubey, D.P. (ed.) *Pilgrimage Studies: Sacred Places, Sacred Traditions*. The Society of Pilgrimage Studies, Allahabad, India, pp. 157–178.
Saint Maria Faustina Kowalska (2005) *Diary of Saint Maria Faustina Kowalska. Divine Mercy in My Soul.* Marian Press, Stockbridge, Massachusetts.
Sanktuaria Bożego Miłosierdzia (2023) Sanktuaria Bożego Miłosierdzia. Available at: https://www.faustyna.pl/zmbm/blog/category/sanktuaria/sanktuaria-bozego-milosierdzia/ (accessed 2 September 2022).
Santi, Angelo de (1910) Litany of Loreto. In: Herbermann, C. (ed.) *Catholic Encyclopedia*. Vol. 9. Robert Appleton Company, New York. Available at: http://www.newadvent.org/cathen/09287a.htm (accessed 19 April 2023).
Socha, P. (2001) Development of the worship of Divine Mercy in Poland and abroad. *Peregrinus Cracoviensis* 11, 111–142.
Sołjan, I. (2012) *Sanktuaria i Ich Rola w Organizacji Przestrzeni Miast na Przykładzie Największych Europejskich Ośrodków Katolickich*. IGiGP UJ, Kraków, Polska.
Stoddard, R.H. (1994) Major pilgrimage places of the world. *Geographia Religionum* 8, 17–36.
Vatican News (2019) Pope adds feast of Our Lady of Loreto to Roman Calendar. Vatican News, 25 March. Available at: https://www.vaticannews.va/en/church/news/2019-10/feast-of-our-lady-of-loreto-added-to-general-calendar.html (accessed 2 September 2022).
Von Klimo, A. (2022) The Cult of Our Lady of Fátima—modern Catholic devotion in an age of nationalism, colonialism, and migration. *Religions* 13(11): 1028. http://dx.doi.org/10.3390/rel13111028
Wikipedia (2023) Pope John Paul II in popular culture. Available at: https://en.wikipedia.org/wiki/Pope_John_Paul_II_in_popular_culture (accessed 2 September 2022).

3 Religious Tourism after Globalization: Geopolitics, Deglobalization and its Considerations

Dino Bozonelos*
California State University, San Marcos, California

Abstract
This chapter discusses how the invasion of Ukraine has signalled a potential end to the current era of globalization. What is potentially replacing it is a new era of geopolitics, where classical geopolitical ideas regarding state power and territorial control have resurfaced. While geopolitical considerations have always existed in religious tourism and pilgrimage, such as ensuring the security of sacred sites, the fundamental understandings regarding neo-liberalism have fractured. The intensification of geopolitical competition, the rise of geostrategic tourism, and the reimagining of tourism have brought several geopolitical considerations into focus regarding the future of religious tourism and pilgrimage. Three potential developments are explored in this chapter: (i) the securitization of borders, which may lead to decreased global mobility; (ii) a decrease in the flow of information, where technological decoupling could alter patterns of religious tourism; and (iii) renewed nationalism, where religious tourists may preference travel within their country's sphere of influence.

Introduction

Religious tourism and pilgrimage have truly become an important example of what globalization aspires to. If globalization is defined as an increasing interdependence among countries, then the interactions and encounters that accompany religious tourism are central to the success of this process. The field of global studies divides the study of globalization into several dimensions, including economic, political and cultural. First, religious travel provides a positive economic impact through the development of infrastructure (Timothy, 2022). Second, it can provide a space for political rapprochement, where non-violence conflict resolution norms are incorporated as practices in faith-based tourism, specifically pilgrimages (McIntosh, 2017). The Director-General of the World Trade Organization highlighted that religious tourism could foster inclusivity and build cultural understanding (Griffin and Raj, 2017). Third, religious tourism and pilgrimage is a form of cultural tourism, which some countries have incorporated as part of their foreign policy. This religious soft-power approach, that favours attraction over hard power, has become more common in a globalized world (Mandaville and Hamid, 2018; Bozonelos, 2023).

However, the invasion of Ukraine has signalled a potential end of the current era of globalization. Russian President Vladimir Putin has signalled an end to the current liberal international order (LIO), a rules-based system supported

*dbozonelos@csusm.edu

DOI: 10.1079/9781800623675.0003

by American hegemony. Instead, President Putin posits that the world has entered a multipolar world where US unilateralism will no longer be tolerated (Putin, 2022). This sentiment is echoed by Chinese President and General Secretary Xi Jinping, who has made it clear that the People's Republic of China has 'taken a clear-cut stance against hegemonism' (Xi, 2022). In response, the USA has labelled China as the main 'geopolitical competitor' to the USA and sees Russia's war in Ukraine as a direct assault on the international system (Biden, 2022; US Department of Defense, 2022).

What is replacing it is potentially a new era of geopolitics, where classical geopolitical ideas regarding state power and territorial control have returned. While geopolitical considerations have always existed, such as the security of sacred sites from potential terrorist attacks, the fundamental understandings regarding neoliberalism have fractured. Intensifying geopolitical competition, geostrategic tourism, soft power and the reimagining of tourism, particularly pilgrimage, are aspects of geopolitics that will impact global religious tourism (Cordoba Azcarate *et al.*, 2021). This chapter will explore these geopolitical considerations regarding the future of religious tourism and pilgrimage. Three potential developments will be explored: (i) the *securitization of borders*, which may lead to decreased global mobility; (ii) a *decrease in the flow of information*, where technological decoupling could alter patterns of religious tourism; and (iii) *renewed nationalism*, where religious tourists may preference travel within their country's sphere of influence.

Globalization and Tourism

Concepts of free trade, the free flow of capital and free travel are fundamental aspects that have characterized global tourism since the collapse of the Soviet Union in 1990 with the declaration of independence by Lithuania and Latvia. Indeed, Telfer (2015) notes that an expectation has developed that tourists and tour companies should be free to operate as they see fit across international borders. Any state intervention is seen as onerous and should be either minimized or eliminated altogether. In addition, if there is state involvement, it is to the benefit of tourism, as all states have invested in the infrastructure to facilitate the economic development that comes with global tourism (C.M. Hall, 2017). This understanding of globalization is often referred to as *neoliberalism* and represents the political-economic approach that was taken in quite a few economic sectors. Indeed, Hall *et al.* (2019) argue that the United Nations World Tourism Organization (UNWTO) has adopted a neoliberal frame for the governance structures of global tourism. This is despite the potential ecological challenges and sustainable development issues that arise from such an adoption.

Globalization has been accepted as a fait accompli for the tourism industry. Tourism would seem to benefit tremendously from the economic interdependence and integration that are features in economic globalization. Dwyer and Čavlek (2019) note the growth of the 'leisure society' where people, particularly tourists, have adopted a more international or global outlook regarding their recreational activities. Wearing *et al.* (2019) writes that neoliberalism, the theoretical cousin of globalization, has promoted a consumer culture, where certain natural phenomena and cultural practices have become commodified. Tourist consumption is the most important aspect, and all other factors, including employee job security, job safety and benefits, such as health care and retirement, are seen as less important. The needs of the tourist come above the needs of the tourist employee. Neoliberalism is where market competition is proposed as the correct political-economic model for countries to adopt if they choose to modernize and develop. From an international political economy perspective, the global push for free trade and liberalization has transformed the global tourism industry (Giampiccoli, 2007). More importantly, it has impacted the management of religious tourist destinations regarding the development of transnational tourism corporations (Bianchi, 2002).

As late as 2019, Timothy (2019a) posits that the 'deterioration of state sovereignty, nationalism, and the protective barriers associated with international borders constitute another modern phenomenon that is now a conventional process' (p. 4). This process of denationalization has been aided through the emergence of supranational organizations, such as the European

Union (EU) or the Association of Southeast Asian Nations (ASEAN). The organizations are specifically organized to promote neoliberal concepts, such as free trade areas, customs unions, or in the case of the EU, a single market where people, goods and services can move freely within the EU irrespective of national borders (Timothy, 2019b). Hall *et al.* (2019) discuss how financial deregulation and improvements in transportation and information technologies have led to a dramatic rise in contemporary mobilities. Coupled with liberalization of goods and services through the reduction of trade barriers, there should be no 'place' where a tourist should not be able to visit. Williams (2009, 2019) offers the notion of enfolded mobility, where one type of travel is wrapped up in another kind of travel. This is a consequence of a world with fewer restrictions.

Globalization and Religious Tourism

Religion itself is seen as playing a developmental role in globalization. Esposito *et al.* (2008) write that 'world religions' provided a foundation for global travel before the current era of globalization. Followers of a faith would often travel in the wake of a conquest, such as the Catholic missionaries who followed the conquistadors in Mesoamerica. Others travelled for strictly religious reasons, such as pilgrimage to a holy site (Hall, 2006). Still others travelled as a community, fleeing persecution. Similarly, Juergensmeyer (2003) demarcates three patterns of religious globalization, the first includes diasporic religious patterns, the second transnational religious patterns, and the third the pattern of religious tourism.

Given the globalizing forces of religion and tourism, the combination of the two, religious tourism, should have an interactive, or synergistic effect. Interestingly, religious tourism and globalization has been understudied (Griffin and Raj, 2015). This may be due to the conceptual distinction between religious tourism and pilgrimage as some scholars consider religious tourism only to be a niche within cultural tourism studies, to be studied primarily as a touristic phenomenon.

Olsen (2019) offers a thorough analysis of globalization and religious tourism. The author identifies three effects, including: (i) the migration of religious groups to create diasporic religious communities; (ii) the democratization of access and visitation to sacred sites; and (iii) the commodification and commercialization of religion, including the selling of spiritual souvenirs and the spread of popular religion. In the first effect, diasporic communities use pilgrimage as a method of connection to their homelands and ancestral origins. In addition, these same communities recreate buildings in their new surroundings, to recreate similar religious and sacred spaces. These new locations can become religious destinations as well, particularly if the diasporic community attributes to the space spiritual value. A good example is the study by Orthodox Christian pilgrimages in North America. Parishes, founded by migrant communities and now attended by their native-born children and converts, routinely organize visits to monasteries. Founded by monks from the world-renowned Mount Athos, monasteries such as St Anthony's in Arizona, have become religious destination sites, not just for the communities in North America, but also for pilgrims from Orthodox countries, such as Greece and Russia (Klimova, 2011, 2016).

The second effect in Olsen's analysis involves the discussion of how the dramatic improvements in communication, transportation and lodging have led to a democratization of religious travel. By this the author explains that religious tourism and pilgrimage is more available to more people than in previous eras. Traditional pilgrimages were often treacherous where the pilgrims would face physical, environmental and personal challenges. A good example is the Camino de Santiago where pain and suffering, and even death, was an accepted part of the experience (Barlar, 2022). Pilgrimages have now shifted, where advances have changed the entire experience. Pilgrims on the Camino can now check in digitally, using quick response (QR) technology where QR stamps substitute for physical stamps that were collected over time in a physical passport (Lois-González and Somoza-Medina, 2022). The change is even more dramatic among religious tourism, where the motivations differ.

The final effect is the commodification of religion and spirituality, where 'cultural deregulation' is driving aspects of the global economy. Olsen (2019) cites co-optation of religious practices and sacred sites by newer spiritual movements, in addition to the efforts by more traditional religious groups to commercialize aspects of their rituals and practices. The latter could lead to a sense of inauthenticity, which can then leave religious travellers without a sense of meaning. Kitiarsa (2010) describes how 'commercial religious culture, religious consumption, marketized faith, and other forms of religious commodification are proliferating' and this is intimately tied to modernization and globalization (p. 564). Consumerism, which is a major feature of neoliberalism, is a central feature in religious commodification. As globalization deepened, so did this market-religion convergence. Indeed, Kitiarsa (2010, p. 573) posits that 'religious commodification is a joint invention by religious institutions, market forces, and demand from consumer-cum-religious pleasure and identity seekers,' which suggests commercialization is a two-way street.

Deglobalization?

Deglobalization is defined by Witt (2019) as 'the process of weakening interdependencies among nations' (p. 1054). Witt measures deglobalization, either through trade, defined as world average levels of imports of goods and services as a percentage of gross domestic product (GDP), weighted by GDP, or through foreign direct investment (FDI), defined as the world average levels of inward FDI flows as a percentage of GDP, weighted by GDP. Irwin (2022) prefers to call it 'slowbalization' where economic integration is clearly in retreat. However, Irwin suggests that the global economy is at an inflection point, and that there is still time for the global economy to recover. Lagarde (2023), the president of the EU Central Bank suggests that the global economy is fragmenting more than deglobalizing. A result of this fragmentation will be 'system competition' where multipolarity will lead to competing blocs in the short run. Overall, she is concerned that fragmentation will eventually give way to bloc coalescence around the two major economies, the USA and China. Arezki (2023) suggests that instead of deglobalization, the world will re-globalize, where the author posits that countries will develop multilayered relationships, rather than the tight, interconnected global-level relationships that characterize the current state of globalization.

Still, evidence exists that the world is indeed in a deglobalization phase. García-Herrero and Tan (2020) posit that deglobalization is a megatrend that has existed before, primarily at the end of the Age of Empires in the lead-up to the World War II. In that era, increased protectionism stemming from the entry of the major powers into conflict led to a reversal of the trends that had been witnessed for a century. In the current era, they write that 'de-globalization of trade is happening in terms of value and volume of gross trade and in terms of the importance of global value chains' (p. 15). This is like Witt's (2019) analysis and reinforces the emphasis that deglobalization is a process and not a condition. Link (2018) refers to James's 'pendulum theory' of globalization, where economic liberalization is then followed by protectionism. In this view, there have been several periods of global openness, each one circumscribed by global compression, where nationalists see interconnectedness as a condition to avoid at all costs. The author notes the 'distinctive political and institutional architectures' is what allows for globalization to exist. Thus, the *politics of globalization* are replaced by the *politics of deglobalization*.

For Witt (2019), politics is the key driver of deglobalization. USA-China tensions, populism, nationalism in Europe and the apparent risk of a global recession suggest that the world is entering the rest of a decade characterized by rising economic complexity and geopolitical divisions. Still, deglobalization does not require that every single country become less interdependent, or that we reach a hypothetical end point of a completely deglobalized world economy, where autarky is the norm. As discussed by Witt, deglobalization is a process, not an outcome. Thus, what matters is the trend over time. Rising levels of interdependence indicate globalization, whereas declining levels signify deglobalization.

Most of the recent discourse on deglobalization comes from international business. However, the concept of deglobalization has a longer discourse in global studies, where scholars, such as Bello (2004), understand deglobalization as a negative reaction to the economic consequences of global capitalism. Contrasting with Witt, Bello suggests that it is economics, and not politics that drives deglobalization. Neoliberalism, as an applied ideology, produces social and ecological distortions that negatively affect a society once having entered the global economy. If the push is to increase economic growth through efficient production, then destabilization will be the result. The process of globalization involves certain actions such as moving production to less developed countries, taking away resources from communities, and cuts in government spending that leads to less social investment. In Bello's understanding of deglobalization, the market becomes 'disembodied' from society, where the needs of the market no longer align with the needs of society.

Again, for Witt, it is politics, and not economics, which drives deglobalization. Witt (2019) discusses hegemony as only belonging to the international relations (IR) theory of realism. However, Witt fails to recognize that neoliberal institutions, an accepted IR variant of liberalism, is in and of itself a hegemonic approach. Neoliberalism itself is a political-economic ideology championed by the USA and her allies. The LIO that both China and Russia have opposed is rooted in an interdependent global economic system, sustained by US political and military might. For example, Witt *et al.* (2023) dig deeper and postulate that geopolitical competition between the USA and China is leading to decoupling, which they define as: 'the process of weakening interdependence between two nations or blocs of nations' (p. 101399). In this instance, the authors are primarily focused on the decoupling between the two largest economic partners, China and the USA. Decoupling between these two great powers could negatively impact all global industries.

The Geopolitics of Tourism

Geopolitics is defined by Caldara and Iacoviello (2022) as the 'study of how geography affects politics and the relations among states' (p. 1197). Given this, geopolitics can be formed by numerous variables. These include traditional variables, such as relationships with other countries, geographic location, natural and man-made resources, socio-economic and demographic factors, and human capital levels. Geopolitics varies from 'politics' through its emphasis on geography, and the role of geography in explaining IR. The study of geopolitics before the Cold War was considered the era of classical geopolitics. According to Ó Tuathail (1992) this involved the analysis of 'invasions, battles and the deployment of military force' (p. 191). During the Cold War, the term geopolitics fell out of usage in IR, mostly due to Fascist appropriation of the term to explain their expansionist visions in the 1930s. In this aftermath, much of the discourse on geopolitics was in political geography, though there was quite a bit of pushback there as well. Geographers often contested the non-discursive approach of geography in politics, where geographical realities were treated as immutable (Ó Tuathail, 1992). Geopolitics never quite disappeared from the global stage and the Cold War was anything but geopolitical. In IR, the term was replaced with other concepts, such as containment, deterrence, compellence, the 'ABC' or 'Anything but Communism' approach, and détente (Jentleson, 2014).

Geopolitics, however, has made a comeback, mostly after the 2008 Global Financial Crisis, where the global political economy was brought to its knees and the global interconnectedness that had been triumphed now exposed vulnerabilities. As a term, geopolitics has become much more prevalent in international business (IB), with Bozonelos and Tsagdis (2023) showing a fourfold increase in IB articles within the top three business journals referencing either geopolitics or geopolitical since 2010. Another good example includes Munoz's (2013) *Handbook on the Geopolitics of Business*. This edited volume records how the Global Financial Crisis fundamentally changed business practices and how practitioners operate. The study of geopolitics and its corollary, geopolitical risk, has become a cottage industry where numerous scholars explore how a wide range of geopolitical events affect IB practices (Bremmer and Keat, 2010; Rice and Zegart, 2018).

Concurrently, there has also been an increase in the use of the term geopolitics in tourism as well. As much of tourism research is conducted by management scholars and practitioners, this a logical extension. For example, D. Hall (2017) stresses the geographical context of tourism sites and how geopolitics has always been at the heart of this activity. D. Hall (2017) also conceptualizes tourism as a form of soft power, where heritage is an intrinsic part of a country's foreign policy. In addition, the interconnection of tourism activities leaves the tourism subjected to political and economic instability. This exposure through tourism has led C.M. Hall (2017) and other tourism scholars to incorporate geopolitical themes in their analyses. In addition, other scholars have developed *critical* geopolitics, which views tourism as the exercise of power through physical spaces (Ó Tuathail, 1994).

A good example is the previous conversation on geopolitical risks. Recently, geopolitical risk, as a more general concept, has been applied to tourism, whereas previous articles were more concerned about the effect of terrorism on tourism, which is a specific kind of geopolitical risk (see Enders *et al.*, 1992; Thompson, 2011; Araña and León, 2008). Balli *et al.* (2019) explore the impact of geopolitical risk on tourism demand, concluding that the effect is heterogenous. Some countries are resistant to geopolitical events, whereas other countries are more affected. Demiralay and Kilincarslan (2019) provide evidence that geopolitical risks impact the stock performance of leisure and tourism companies. Their analysis suggests that European and North American stocks are more susceptible than stocks in the Asian markets. In addition, Demir *et al.* (2019) find that geopolitical risk negatively impacts inbound tourism, particularly for countries that are much more dependent on tourism. Finally, Gozgor *et al.* (2022) systematically study the impact of geopolitical risk and capital investment in tourism, defined as tourism supply. They conclude that geopolitical risks are harmful, and that tourism policy makers should be cognizant of the potential negative effects.

Mostafanezhad *et al.* (2021) provided what will be become a seminal work on the geopolitics of tourism. The authors discuss various frameworks from which to analyse the geopolitics of tourism. These include: (i) feminist and everyday geopolitics; (ii) popular geopolitics; (iii) mobility geopolitics; (iv) environmental geopolitics; and (v) security geopolitics. Each potential framework has been drawn from the existing geopolitics literature and each provide a way forward in tourism research. However, religious tourism does not fall neatly into one of these frameworks. At some level, elements of all five frameworks manifest themselves at sacred sites. More so than other forms of tourism, religious tourism, and specifically pilgrimage can be more phenomenologically based. Lived experiences and how such experiences are consumed and interpreted (Pernecky and Jamal, 2010) are the basis for what gives religious destinations their value. This value is in turn harnessed by various entities to power projects and promote national interests.

Religious Tourism after Globalization

As mentioned previously, one of the most salient trends in this deglobalization discourse is that of decoupling, or the 'process of weakening interdependence between two nations or blocs of nations' (Witt *et al.*, 2023, p. 1). While much of the focus has been on the technology sector, and rightfully so, this process of decoupling will affect global tourism. Chinese outbound tourism became the largest tourist segment globally before the coronavirus disease 2019 (COVID-19) pandemic, reaching that status in 2012 (*The Economist*, 2019). Indeed, Chinese outbound tourists totalled 170 million in 2019, which was larger than any other national outgroup (i.e. group of people that leave the country). Recent research suggests that the number of outbound Chinese tourists is forecasted to rebound to two-thirds of pre-pandemic levels in 2023 (Chen *et al.*, 2023). Still as D. Hall (2017) notes 'tourism is subject to the outcomes of geopolitical activity' (p. 5). Thus, even if Chinese outbound tourism rebounds to pre-pandemic levels, there may be a 'chilling' effect due to rising geopolitical tensions. It is plausible that some Chinese tourists may be less willing to travel to countries that have been critical of Chinese policy. Patriotic tourism, where tourists visit places sympathetic to their country or country's foreign policy may become more commonplace.

A more critical concern involves the potential technological delinking that may accompany decoupling. This would also negatively impact global tourism as ease of access to information is one of the major drivers of individualized travel (Hall *et al.*, 2019). The significant challenges of technological decoupling have led several world leaders to shift from referencing the term **decoupling**, and instead use **de-risking** (von der Leyen, 2023). For example, technological decoupling is seen as an either/or situation, where there is an 'undoing of cross-border trade in high-tech goods and services' (Cerdeiro *et al.*, 2021). De-risking is meant to be a more moderate approach, or akin to a more surgical form of decoupling. Traditionally, de-risking occurs when 'large international banks had begun exiting relationships with countries and market segments they perceived to be high in risk or else not worth the rising cost of compliance' (Woodsome *et al.*, 2015, p. vii). However, for European Commission and EU leaders, de-risking is seen as less impactful as they are nervous that a binary approach to decoupling with China will bring more harm than benefit (Cave, 2023). Still, Cave mentioned that to Chinese officials, de-risking does not sound that different from decoupling.

If deglobalization is indeed a potential future direction in IB and in tourism, then what are the potential implications for religious tourism? Three future developments are outlined here. The first is how a decrease in global mobility will mean less religious travel, if not necessarily in total volume. Indeed, total volume may increase whereas the who, what, where, when and how elements may become circumscribed due to geopolitical considerations. The second effect is the decrease in the flow of information, which is a direct result of technological decoupling between China and the USA and its larger effects on the global communications and information sharing. As these two countries evolve distinct spheres of influence, the communication and transportation networks between the two spheres will weaken, leading to a scenario where Chinese tourists will travel within their sphere and Western tourists will travel within the US sphere. The third and final effect will be religious travel and pilgrimage in an age of renewed nationalism. In this unfolding era, tourists from certain countries will be viewed with suspicion, and their motives will be questioned. In general, religious tourism scholars never query the motives of a religious traveller regarding harm or subterfuge. While researchers will quibble on the exact motives of religious tourism, whether it is rooted more in spiritual desires, or in more non-religious tourist expectations, negative motivations are not considered. This could shift as a country might question the intentions of a religious traveller. Table 3.1 provides a summary of these potential future developments.

The securitization of borders: a decrease in global mobility?

Global mobility simply refers to how people travel. As Hall *et al.* (2019) state, 'mobility is a quintessential feature of tourism' (p. 55). One of the most troubling aspects of an age of geopolitics could be a potential decrease in the number of people travelling for tourist reasons, including those for religious reasons and motivations. Simply stated, mobility may be reduced over geopolitical considerations. C.M. Hall (2017) refers to

Table 3.1. Potential future developments for religious tourism after globalization.

Development	Effect	Outcome
Securitization of borders	Decreased global mobility	Countries could block the travel of people from certain countries, simply out of geopolitical concerns.
Decrease in flow of information	Technological decoupling could alter patterns of religious tourism	Division of the world into two different 5G systems, with different sets of protocols and standards, could affect access to information regarding sacred sites.
Renewed nationalism	Pushback by national populists against tourists of a certain faith in their country *and* Onset of a second Cold War between the USA and China	Religious tourists may: (i) prefer to travel to countries with friendlier dispositions; *or* (ii) prefer to travel within their country's sphere of influence.

this as re-territorialization, where movement across borders becomes less liberalized and more constrained. This would be the opposite of Timothy's (2019a) assertion about the decline in state sovereignty and would entail restrictions on travel, including tourism. Re-territorialization is at some level, an expression of state power, and could include limitations on either overall inbound or outbound travel. Concurrently, it could involve restraints on specific types of travellers, or visitors from specific countries and/or regions.

The notion that geopolitical considerations affect tourist mobility had been approached by scholars before the 2022 Russian invasion of Ukraine. Bianchi *et al.* (2020) note the inherent tension in an 'ideology of perpetual movement' and the 'fixity' of borders. The neoliberal dream is dualistic. There is a cosmopolitan, mobile world for elites, which includes paying tourists, contrasted with a world of barriers, exploitation and security controls for everyone else. Castles (2011) notes that liberalization has never truly been complete. The underlying assumption is that there should be near complete freedom of capital, goods and ideas. However, that has never been the case for labour. Market forces demand freedom of movement, yet political forces demand control. This 'securitization' of the border and of borderlands, which has existed to prevent irregular migration, took on new significance with the COVID-19 pandemic. Biosecurity and the prevention of the spread of the coronavirus took importance over the 'right to travel'.

The concern is that the geopolitical considerations could impact religious travel. Glimpses of this have already been seen with access to contested sites, such as the Babri Masjid/Ram Janmabhoomi in India, and the Western Wall, the Church of the Holy Sepulchre (Tomb), and the Temple Mount/Haram al-Sharif in Jerusalem (Jobani and Perez, 2020). It is conceivable that countries could block the travel of people from certain countries, simply out of geopolitical concerns. This most recently took place with restrictions placed on Russian tourists into the EU, with some countries, notably Estonia, Latvia, Lithuania, Finland, Poland and the Czech Republic, imposing an outright ban (York, 2023). Pilgrimages to historically significant Orthodox pilgrimage sites in these countries such as the Alexander Nevsky Cathedral in Tallin, Estonia, have been disrupted. Further conflict, or the hint of a conflict, could lead to further travel restrictions.

Decrease in flow of information, technological decoupling and religious tourism

The technological decoupling between China and the USA could have several effects, primarily regarding global communications and information sharing. Witt *et al.* (2023) suggest that the war in Ukraine has 'widened the rift between China and the United States' (p. 101399) and that the development of a Sino-Russian bloc would seek to advance a distinct sphere of influence. As these two spheres of influence progress, the communication and transportation networks between the two spheres will commensurately weaken. Mbaiwa *et al.* (2019) identify technology as a major driver of global tourism. The evolving interconnectedness of the world has diminished the barriers that constrained tourism establishment and has fundamentally shifted destination management. It makes sense then, that a technological decoupling will either alter this trajectory in tourism, the 'slowbalization' referred to earlier in this chapter, or become a central feature in the process of deglobalization.

García-Herrero and Tan (2020) point to how technological decoupling will lead to technology protectionism, where complete bans on hardware and software related to information will progress over time. Some of these actions will be part of future export control regimes on high-end technology products. Whereas other restrictive actions have already begun. For example, under the Trump administration the USA introduced 'entity lists', which forbid American companies from doing business with selected Chinese counterparts. Another major action has been the banning of Huawei, a Chinese telecommunications firm, from providing 5G platforms in the USA. The UK, Denmark, Sweden, Estonia, Latvia and Lithuania have also banned Huawei from their 5G network build, with Germany and Portugal edging closer to doing so. These moves will divide the world into two different 5G systems, with different sets of protocols and standards (Gross and Jopson, 2023). As communication and information sharing is a cornerstone of an expanded global tourism market, such a bifurcation will naturally divide tourism. This could affect how a potential religious tourist would access information regarding a sacred site.

Another potential area of impact will include virtual or cyber pilgrimages. The concept of a cyber pilgrimage has existed for some time, with Hill-Smith (2011) defining it as the 'practice of undertaking pilgrimage on the Internet' (p. 236). The use of virtual pilgrimages exploded during the COVID-19 pandemic where pilgrimages were livestreamed to the faithful around the world. Bailey (2022) notes that a 15-hour live online event at Lourdes France attracted *80 million* viewers. While many of these rituals, devotionals and practices have returned in person, these online features are likely to persist. These novelties have the effect of 'modernizing' religious travel and will prove to be attractive to several tourists and pilgrims who may lack mobility. The potential separation of global society into different spheres will funnel end users of online religious consumption, such as cyber pilgrims, into two distinct 5G worlds.

Religious travel and pilgrimage in an age of renewed nationalism

There are two correlated developments that have occurred in a post-globalization era. The first development is the wave of national populism that has mostly manifested in democracies. Populism is understood as politics that represents the general will of the people, rather than the interests of the elites. In this definition, populists dichotomize society into two groupings, 'the pure people' versus 'the corrupt elite' (Mudde, 2017, p. 29). National populism is simply a combination of populism with precepts from nationalism, where populists believe that the dishonest elites serve the globalist interests at the expense of the nation. In national populism, the country is under attack. Populism has, for the lack of a better description, become 'popular'. National populism and nationalist-populist parties have surged in established democracies, particularly since the Global Financial Crisis (Judis, 2016). The election of Donald Trump in the USA, Brexit in 2016, and the election to parliament of a sweeping number of far-right parties in Europe are evidence of this wave.

National populism pushes back against the de-territorialization of the nation state. Mazzoleni (2023) discusses the fundamental importance of territorial borders to national populists. For these politicians, the defence of the country starts with the defence of the border. Who comes in and out of the country is an existential question. Sovereignty is encompassed at the border and these territorial lines play a crucial role for populists in defining the 'people'. In this discourse, it is likely that as geopolitical conflicts become more acute, national-populist leaders adopt restrictive entry policies towards those who are identified as being antithetical to the national identity of the country. There is a potentially additive effect as religiosity often overlaps with nationalism. A more sobering example is the restriction by the Ukrainian government of Russian Orthodox clergy and monastics to visit the Kyiv-Pechersk Lavra, which for the Russian Orthodox is considered the cradle of their faith (Eckel, 2023). In this setting, a likely outcome is that religious tourists may preference travel to countries that have a friendlier disposition, though this may not be the case for pilgrims who tend to be more committed to travel to a particular sacred site.

The second development is what Witt *et al.* (2023) suggest will be a second Cold War, which they define as: 'a fierce contest for economic dominance, the commanding heights of science and technology, and control of the institutional infrastructure of the world' (p. 9).

This reasoning leads to a scenario where Chinese and Russian tourists will travel within their sphere of influence, whereas tourists from Western and US-allied countries will travel within the US sphere. This is not to suggest that travel will dissipate entirely. Even during the Cold War, religious travel took place between the Soviet Union and non-communist countries (Bechmann Pedersen and Noack, 2019). However, tourism by default is entwined in competing globalizing and nationalizing forces. This sentiment was captured in a 2022 survey of Chinese travellers by Morning Consult. Moskowitz, a senior analyst with the surveying firm notes that, 'Nearly two-thirds of Chinese adults (65%) say China's fraying relationship with the United States is a reason to avoid US travel, while 43% cite growing enmity between the two nations as a "major reason" to stay away' (Moskowitz, 2022).

Thus, a second Cold War could significantly affect religious travel. As noted above, a religious tourist, more interested in acquiring cultural knowledge, would be more affected by such developments. A pilgrim may be less deterred.

Summary

Religious tourism and pilgrimage are an important example of globalization. The increase of those who travel for religious reasons represents the ideal conditions of the processes of economic, political and cultural globalization. However, the invasion of Ukraine has signalled a potential end of the current era of globalization. The rules-based system that is supported by the USA and Western allies are being challenged by Russia and China. Their vision of a multipolar world, where various great powers compete for influence and check each other through a balance-of-power system, will affect religious travel. At the very least, we are witnessing the formation of two major blocs, one dominated by the USA and the other dominated by a Sino-Russian alliance. Both blocs have started competing with each to convert non-aligned countries to their cause. Each bloc is seeking to shore up its positions throughout several regions, most notably in the Indo-Pacific.

In sum, this new geopolitical era will fundamentally alter religious travel. For most of the past 30 years, the major concern among religious tourism scholars has been the motivations of the religious traveller. Many surveys question faith-based tourists on their intention, whether they are a religious tourist, a pilgrim, or something in between. In a geopolitical world, the motivations and intentions of the state may matter more. Your citizenship, national identity and religious background may impact your ability to cross a border for religious tourism. A ‘right to tourism’ will be questioned in this new era. Finally, geopolitical considerations will impact religious travel. As geopolitical tensions rise, so will tourist motivations not to travel to certain countries or regions. Religious tourists will travel to where they can get through, where they have more information, and where they feel welcomed. While these considerations have always lurked underneath all forms of travel, they have the possibility of becoming more acute in this new geopolitical reality.

References

Araña, J.E. and León, C.J. (2008) The impact of terrorism on tourism demand. *Annals of Tourism Research* 35(2), 299–315. DOI: 10.1016/j.annals.2007.08.003.

Arezki, R. (2023) Not your grandfather’s cold war. *Aljazeera.com*, 23 May. Available at: https://www.aljazeera.com/opinions/2023/5/23/not-your-grandfathers-cold-war (accessed 29 May 2023).

Bailey, A.E. (2022) Pilgrimage during Covid-19: impacts, adaptations, and recovery. *International Journal of Religious Tourism and Pilgrimage* 10(1), 189–202. DOI: 10.21427/hjtc-4m56.

Balli, F., Uddin, G.S. and Shahzad, S.J.H. (2019) Geopolitical risk and tourism demand in emerging economies. *Tourism Economics* 25(6), 997–1005. DOI: 10.1177/1354816619831824.

Barlar, S.H. (2022) Experiences along the Camino de Santiago. In: Olsen, D.H. and Timothy, D.J. (eds) *The Routledge Handbook of Religious and Spiritual Tourism*. Routledge, London, pp. 254–265.

Bechmann Pedersen, S. and Noack, C. (2019) Crossing the Iron Curtain: an introduction. In: Bechmann Pedersen, S. and Noack, C. (eds) *Tourism and Travel during the Cold War: Negotiating Tourist Experiences Across the Iron Curtain*. Routledge, New York, pp. 1–20.

Bello, W. (2004) *Deglobalization: Ideas for a New World Economy*. Zed Books, London.

Bianchi, R.V. (2002) Towards a new political economy of global tourism. In: Sharpley, R. and Telfer, D. (eds) *Tourism and Development: Concepts and Issues*. Channel View Publications, Clevedon, UK, pp. 265–299.

Bianchi, R.V., Stephenson, M.L. and Hannam, K. (2020) The contradictory politics of the right to travel: mobilities, borders and tourism. *Mobilities* 15(2), 290–306. DOI: 10.1080/17450101.2020.1723251.

Biden, J.R. (2022) *National Security Strategy*. The White House, Washington, October 2022. Available at: https://www.whitehouse.gov/wp-content/uploads/2022/10/Biden-Harris-Administrations-National-Security-Strategy-10.2022.pdf (accessed 29 May 2023).

Bozonelos, D. (2023) Soft power and sacred sites: the geopolitics of prayer. In: Bozonelos, D. and Moira, P. (eds) *The Politics of Religious Tourism*. CAB International, Wallingford, UK, pp. 42–56.

Bozonelos, D. and Tsagdis, D. (2023) From fragmented geopolitics to geopolitical resilience in international business. *Academy of International Business (AIB) Insights* 23(2), 1–7. DOI: 10.46697/001c.73803.

Bremmer, I. and Keat, P. (2010) *The Fat Tail: the Power of Political Knowledge in an Uncertain World*. Oxford University Press, Oxford.

Caldara, D. and Iacoviello, M. (2022) Measuring geopolitical risk. *American Economic Review* 112(4), 1194–1225. DOI: 10.1257/aer.20191823.

Castles, S. (2011) Migration, crisis, and the global labour market. *Globalizations* 8(3), 311–324. DOI: 10.1080/14747731.2011.576847.

Cave, D. (2023) How 'decoupling' from China became 'de-risking'. *New York Times*. Available at: https://www.nytimes.com/2023/05/20/world/decoupling-china-de-risking.html (accessed 21 May 2023).

Cerdeiro, D.A., Eugster, J., Mano, R.C., Muir, D. and Peiris, S.J. (2021) Sizing up the effects of technological decoupling. *IMF [International Monetary Fund] Working Paper* WP/21/69.

Chen, G., Chen, Z., Saxon, S. and Yu, J. (2023) Outlook for China tourism 2023: Light at the end of the tunnel. *McKinsey & Company*, 9 May. Available at: https://www.mckinsey.com/industries/travel-logistics-and-infrastructure/our-insights/outlook-for-china-tourism-2023-light-at-the-end-of-the-tunnel#/ (accessed 22 November 2023).

Cordoba Azcarate, M., Mostafanezhad, M. and Norum, R. (2021) Introduction: tourism geopolitics. In: Mostafanezhad, M., Cordoba Azcarate, M. and Norum, R. (eds) *Tourism Geopolitics: Assemblages of Infrastructure, Affect, and Imagination*. University of Arizona Press, Tucson, Arizona, pp. 29–32.

Demir, E., Gozgor, G. and Paramati, S.R. (2019) Do geopolitical risks matter for inbound tourism? *Eurasian Business Review* 9(2), 183–191.

Demiralay, S. and Kilincarslan, E. (2019) The impact of geopolitical risks on travel and leisure stocks. *Tourism Management* 75, 460–476. DOI: 10.1016/j.tourman.2019.06.013.

Dwyer, L. and Čavlek, N. (2019) Economic globalisation and tourism. In: Timothy, D.J. (ed.) *Handbook of Globalisation and Tourism*. Edward Elger, Cheltenham, UK, pp. 12–26.

Eckel, M. (2023) Holy eviction: What's going on with the standoff at Kyiv's famous Monastery of The Caves? *Radio Free Europe/Radio Liberty*. Available at: https://www.rferl.org/a/ukraine-kyiv-pechersk-lavra-standoff-monastery-russia-monks/32343746.html (accessed 23 May 2023).

Enders, W., Sandler, T. and Parise, G.F. (1992) An econometric analysis of the impact of terrorism on tourism. *Kyklos* 45(4), 531–554.

Esposito, J.L., Fasching, D.J. and Lewis, T. (2008) *Religion and Globalization: World Religions in Historical Perspective*. Oxford University Press, New York.

García-Herrero, A. and Tan, J. (2020) Deglobalisation in the context of United States-China decoupling. *Policy Contribution 2020/21*. Bruegel, Brussels.

Giampiccoli, A. (2007) Hegemony, globalisation and tourism policies in developing countries. In: Novelli, P.M. (ed.) *Tourism and Politics: Global Framework and Local Realities*. Elsevier, Oxford, pp. 175–191.

Gozgor, G., Lau, M.C.K., Zeng, Y., Yan, C. and Lin, Z. (2022) The impact of geopolitical risks on tourism supply in developing economies: the moderating role of social globalization. *Journal of Travel Research* 61(4), 872–886. DOI: 10.1177/00472875211004760.

Griffin, K. and Raj, R. (2015) The globalization of pilgrimage tourism? Some thoughts from Ireland. In: Raj, R. and Griffin, K. (eds) *Religious Tourism and Pilgrimage Management: an International Perspective*, 2nd edn. CAB International, Wallingford, UK, pp. 57–78.

Griffin, K. and Raj, R. (2017) The importance of religious tourism and pilgrimage: reflecting on definitions, motives and data. *International Journal of Religious Tourism and Pilgrimage* 5(3), ii–ix.

Gross, A. and Jopson, B. (2023) Portugal paves way to a Huawei ban on country's 5G network. *Financial Times*. Available at: https://www.ft.com/content/4facc4e8-5bcc-4dc3-9a3c-6327927f0b6d (accessed on 26 May 2023).

Hall, C.M. (2006) Travel and journeying on the sea of faith: perspectives from religious humanism. In: Timothy, D.J. and Olsen, D.H. (eds) *Tourism, Religion and Spiritual Journeys*. Routledge, London, pp. 64–77.

Hall, C.M. (2017) Tourism and geopolitics: the political imaginary of territory, tourism, and space. In: Hall, D. (ed.) *Tourism and Geopolitics: Issues and Concepts from Central and Eastern Europe*. CAB International, Wallingford, UK, pp. 15–24.

Hall, C.M., Amore, A. and Arvanitis, P. (2019) The globalizing force of human mobilities. In: Timothy, D.J. (ed.) *Handbook of Globalisation and Tourism*. Edward Elger, Cheltenham, UK, pp. 270–283.

Hall, D. (2017) Bringing geopolitics to tourism. In: Hall, D. (ed.) *Tourism and Geopolitics: Issues and Concepts from Central and Eastern Europe*. CAB International, Wallingford, UK, pp. 3–14.

Hill-Smith, C. (2011) Cyberpilgrimage: the (virtual) reality of online pilgrimage experience. *Religion Compass* 5(6), 236–246. DOI: 10.1111/j.1749-8171.2011.00277.x.

Irwin, D.A. (2022) Globalization is in retreat for the first time since the Second World War. Peterson Institute for International Economics, 28 October. Available at: https://www.piie.com/research/piie-charts/globalization-retreat-first-time-second-world-war (accessed 28 May 2023).

Jentleson, B.W. (2014) *American Foreign Policy: the Dynamics of Choice in the 21st Century*. W.W. Norton & Company, New York.

Jobani, Y. and Perez, N. (2020) *Governing the Sacred: Political Toleration in Five Contested Sacred Sites*. Oxford University Press, Oxford.

Judis, J.B. (2016) *The Populist Explosion: How the Great Recession Transformed American and European Politics*. Columbia Global Reports, New York.

Juergensmeyer, M. (2003) *Global Religions: an Introduction*. Oxford University Press, New York.

Kitiarsa, P. (2010) Toward a sociology of religious commodification. In: Turner, B.S. (ed.) *The New Blackwell Companion to the Sociology of Religion*. Blackwell Publishing, Oxford.

Klimova, J. (2011) Pilgrimages of Russian Orthodox Christians to the Greek Orthodox Monastery in Arizona. *Tourism* 59(3), 305–318.

Klimova, J. (2016) 'As if the road there is covered with honey': inquiries into the seductiveness of a Greek Orthodox Monastery in Arizona for Russian Orthodox parish pilgrims. In: Di Giovine, M.A. and Picard, D. (eds) *The Seduction of Pilgrimage: Sacred Journeys Afar and Astray in the Western Religious Tradition*. Routledge, New York, pp. 99–119.

Lagarde, C. (2023) Central banks in a fragmenting world. European Central Bank, 17 April. Available at: https://www.ecb.europa.eu/press/key/date/2023/html/ecb.sp230417~9f8d34fbd6.en.html (accessed 22 April 2023).

Link, S. (2018) How might 21st-century de-globalization unfold? Some historical reflections. *New Global Studies* 12(3), 343–365. DOI: 10.1515/ngs-2018-0024.

Lois-González, R.C. and Somoza-Medina, X. (2022) The necessary digital update of the Camino de Santiago. In: Calabrò, F., Spina, L.D. and Piñeira Mantiñán, M.J. (eds) *New Metropolitan Perspectives: Post COVID Dynamics: Green and Digital Transition, Between Metropolitan and Return to Villages Perspectives*. Springer, Cham, Switzerland, pp. 268–277.

Mandaville, P. and Hamid, S. (2018) *Islam as Statecraft: How Governments Use Religion in Foreign Policy*. Berkley Center for Religion, Peace and World Affairs, Washington, DC.

Mazzoleni, O. (2023) Introduction: national populism and the politicisation of borders in a changing Europe. In: Mazzoleni, O., Biancalana, C., Pilotti, A., Bernhard, L., Yerly, G. and Lauener, L. (eds) *National Populism and Borders: the Politicisation of Cross-border Mobilisations in Europe*. Edward Elgar, Cheltenham, UK, pp. 1–15.

Mbaiwa, J.E., Mogomitsi, P.K., Mbaiwa, T. and Siphambe, G.B. (2019) Globalisation, tourism, and ecosystems management. In: Timothy, D.J. (ed.) *Handbook of Globalisation and Tourism*. Edward Elger, Cheltenham, UK, pp. 270–283.

McIntosh, I.S. (2017) Pilgrimages and peace-building on the global stage. In: McIntosh, I.S. and Harman, L.D. (eds) *The Many Voices of Pilgrimage and Reconciliation*. CAB International, Wallingford, UK, pp. 13–18.

Moskowitz, S. (2022) Souring Sino-American relations may drive chinese tourists away. Morning Consult Pro, 20 September. Available at: https://pro.morningconsult.com/analysis/souring-sino-american-relations-may-drive-chinese-tourists-away (accessed 23 May 2023).

Mostafanezhad, M., Cordoba Azcarate, M. and Norum, R. (2021) *Tourism Geopolitics: Assemblages of Infrastructure, Affect, and Imagination*. University of Arizona Press, Tucson, Arizona.

Mudde, C. (2017) Populism: an ideational approach. In: Rovira Kaltwasser, C., Taggart, P. and Ostiguy, P. (eds) *The Oxford Handbook of Populism*. Oxford University Press, New York, pp. 27–47.

Munoz, J.M.S. (2013) *Handbook on the Geopolitics of Business*. Edward Elgar, Cheltenham, UK.

Olsen, D.H. (2019) Religion, spirituality, and pilgrimage in a globalising world. In: Timothy, D.J. (ed.) *Handbook of Globalisation and Tourism*. Edward Elger, Cheltenham, UK, pp. 270–283.

Ó Tuathail, G. (1992) Geopolitics and discourse: practical geopolitical reasoning in American foreign policy. *Political Geography* 11(2), 190–204. DOI: 10.1016/0962-6298(92)90048-X.

Ó Tuathail, G. (1994) The critical reading/writing of geopolitics: re-reading/writing Wittfogel, Bowman and Lacoste. *Progress in Human Geography* 18, 313–332.

Pernecky, T. and Jamal, T. (2010) (Hermeneutic) phenomenology in tourism studies. *Annals of Tourism Research* 37(4), 1055–1075.

Putin, V. (2022) Address to participants of 10th St Petersburg International Legal Forum. President of Russia, 30 June. Available at: http://en.kremlin.ru/events/president/news/68785 (accessed 23 May 2023).

Rice, C. and Zegart, A. (2018) *Political Risk: How Businesses and Organizations Can Anticipate Global Insecurity*. Twelve Books, New York.

Telfer, D.J. (2015) The evolution of development theory and tourism. In: Sharpley, R. and Tefler, D.J. (eds) *Tourism and Development: Concepts and Issues*. Channel View Publications, Bristol, UK, pp. 31–73.

The Economist (2019) Holidaying for the motherland. *The Economist* 430(9131), 61–62.

Thompson, A. (2011) Terrorism and tourism in developed versus developing countries. *Tourism Economics* 17(3), 693–700. DOI: 10.5367/te.2011.0064.

Timothy, D.J. (2019a) Introduction to the *Handbook of Globalisation and Tourism*. In: Timothy, D.J. (ed.) *Handbook of Globalisation and Tourism*. Edward Elger, Cheltenham, UK, pp. 2–11.

Timothy, D.J. (2019b) Supranationalism and tourism: free trade, customs unions, and single markets in an era of geopolitical change. In: Timothy, D.J. (ed.) *Handbook of Globalisation and Tourism*. Edward Elger, Cheltenham, UK, pp. 100–113.

Timothy, D.J. (2022) Sociopolitical and economic implications of religious and spiritual tourism. In: Olsen, D.H. and Timothy, D.J. (eds) *The Routledge Handbook of Religious and Spiritual Tourism*. Routledge, London, pp. 301–314.

US Department of Defense (2022) National Defense Strategy. U.S. Department of Defense. Available at: https://www.defense.gov/News/News-Stories/Article/Article/3202438/dod-releases-national-defense-strategy-missile-defense-nuclear-posture-reviews/ (accessed 23 May 2023).

von der Leyen, U. (2023) Speech by President von der Leyen on EU-China relations to the Mercator Institute for China Studies and the European Policy Centre. European Commission, 30 March. Available at: https://ec.europa.eu/commission/presscorner/detail/en/speech_23_2063 (accessed 23 May 2023).

Wearing, S., McDonald, M., Taylor, G. and Ronen, T. (2019) Neoliberalism and tourism. In: Timothy, D.J. (ed.) *Handbook of Tourism and Globalisation*. Edward Elgar, Cheltenham, UK, pp. 27–43.

Williams, A.M. (2009) Migration, tourism, and globalization. In: Timothy, D.J. (ed.) *Handbook of Globalisation and Tourism*. Edward Elger, Cheltenham, UK, pp. 2–11.

Williams, A.M. (2019) International migration, uneven development, and polarization: an introduction. *European Urban and Regional Studies* 16(3), 309–322.

Witt, M.A. (2019) De-globalization: theories, predictions, and opportunities for international business research. *Journal of International Business Studies* 50, 1053–1077. DOI: 10.1057/s41267-019-00219-7.

Witt, M.A., Lewin, A.Y., Ping Li, P. and Gaur, A. (2023) Decoupling in international business: evidence, drivers, impact, and implications for IB research. *Journal of World Business* 58: 101399. DOI: 10.1016/j.jwb.2022.101399.

Woodsome, J., Ramachandran, V., Lowery, C. and Myers, J. (2015) Policy Responses to De-risking: Progress Report on the CGD Working Group's 2015 Recommendations. Center for Global Development, Washington, DC.

Xi, J. (2022) Hold high the great banner of socialism with Chinese characteristics and strive in unity to build a modern socialist country in all respects. Report to the 20th National Congress of the Communist Party of China, Xinhua News Agency, 16 October. Available at: http://my.china-embassy.gov.cn/eng/zgxw/202210/t20221026_10792358.htm (accessed 16 October 2022).

York, J. (2023) Russian tourists seek new destinations as Europe shuts its doors over Ukraine war. France 24, 23 May. Available at https://www.france24.com/en/europe/20230523-russian-tourists-seek-new-destinations-as-europe-shuts-its-doors-over-ukraine-war (accessed 23 May 2023).

4 Novels Promoting the Camino de Santiago

Vitor Ambrosio[1,2]* and Isilda Leitão[2,3]
[1]*Escola Superior de Hotelaria e Turismo do Estoril, Estoril, Portugal;* [2]*CiTUR – Centre for Tourism Research, Development and Innovation, Estoril, Portugal;* [3]*Instituto de Estudos de Literatura e Tradição (IELT), Universidade NOVA de Lisboa, Portugal*

Abstract
This chapter identifies and categorizes novels with common characteristics that have revolved around the Camino de Santiago, or Saint James Way, an ancient pilgrimage route to the city of Santiago de Compostela in Galicia, Spain. The Camino de Santiago was, and continues to be, a symbolic religious attraction. More recently, it has also become a tourist attraction, whose fascination did not escape intellectuals of different nationalities and faiths. The Camino can be seen as a physical, spiritual and religious challenge. This reality has been transferred to travel literature by many writers. This chapter points out specific characteristics of some of the most sold and popular novels of the last four decades on the Saint James Way. Likewise, the novels in which the Camino is not the main theme can also trigger the intention to undertake the challenge. Findings suggest that novels either referring directly or indirectly to the Camino are in general a transformative experience for the readers and they can also become a strong influence for those who hear the call to set out on the Camino.

Introduction

Since the Middle Ages, the Camino de Santiago has been, and continues to be, a symbolic religious attraction. More recently, it has also become a tourist attraction, captivating intellectuals of different nationalities and faiths. The Camino can be viewed as a physical, spiritual and religious challenge, sometimes combining two or even all three of these aspects. Many writers in travel literature have captured this reality.

According to Mattoso (2000), pilgrimage is fundamentally at odds with scientific rationality, which sought to discredit it. There is a profound impulse within pilgrimage that compels individuals to journey through space and time. As Branthomme (1982) notes, the pilgrim travels not only physically but also morally. The journey, with its disruptions of daily life, introduces unforeseen events, risks and sacrifices, yet it broadens the pilgrim's perspective beyond the familiar horizon, providing clarity about oneself.

Ostrowski (2000) highlights the complexities associated with pure pilgrimages solely focused on religious acts. Many specialists believe that pilgrims require physical, mental and spiritual well-being, as curiosity often drives them to explore new places, meet new people and even seek entertainment, incorporating tourist practices into their journey.

Undoubtedly, these observations offer a rich foundation for literary creativity. Literature can awaken the latent desire to embark on a pilgrimage. In this chapter, we explore how literature, or

*vitor.ambrosio@eshte.pt

 Religious Tourism and Globalization. D. Liutikus and R. Raj (eds)
DOI: 10.1079/9781800623675.0004

related forms of it, inspire people to undertake such adventures.

In this chapter we identify and categorize novels with similar characteristics that have been among the most popular and best-selling works over the past four decades centred on the Saint James Way. Furthermore, we demonstrate that even novels where the Camino is not the central theme can capture readers' attention with references to the Way. To fulfil this purpose, we provide brief analyses of two novels: (i) *The Diary of a Magus* by Paulo Coelho (1992), which focuses on the Camino; and (ii) *Therapy* by David Lodge (2011), in which the Way serves as a supplement to the narrative. The findings suggest that novels directly or indirectly related to the Saint James Way are transformative experiences for readers and can be a powerful influence for those who feel the call to embark on the Camino.

Reflections on the History of Saint James Way

Going back in time, historians state that alongside the conversion of the Scandinavian people around the end of the 12th century, Europe had turned completely Christian. According to Ambrósio (2016), in the 13th century, both in the East and in the West, the tombs of certain saints became important centres of pilgrimage. Irrespective of the great competition between the sites home to relics, there were two great sanctuaries in the medieval West that exceeded all others in fame and reputation: (i) the tombs of Saint Peter and Saint Paul in Rome (Italy); and (ii) Saint James, in Compostela (Spain). Those pilgrimages were equivalent to going to Jerusalem, particularly following the conquest of that city by Saladin (1187) and the defeat of the third crusade (1189).

Taking into consideration the long distances pilgrims had to cope with, in the 13th century, an international regulatory framework was set out both in terms of Church law (the canonical), and civil law (whether the Roman or consuetudinary). This served to protect the pilgrim from arbitrary imprisonment as well as economic exploitation and to the extent such individuals were exempt from toll fees and other taxes (Rapp, 1982).

With greater demand came the publication of good travel guides with exacting descriptions considering the most minute of details, one of the best examples being *Iter pro peregrinis ad Compostellam* written in the mid-12th century by the monk Aymeric Picaud in the *Códice Calixtino* (García Blanco, 2016), documenting his own pilgrimage to Compostela.

Although the first pilgrims walked devotedly, with the increased number of accommodation facilities where food and wine could be purchased and various forms of entertainment enjoyed in exchange for money, many did not stand up to the temptations and contributed towards the notoriety of the false pilgrim. It is worth emphasizing that only rich pilgrims could read (the guides) and could afford the total costs of a pilgrimage without depending on charity if they wished to maintain a certain well-being. According to Rapp (1982), Sebald Rieter of Nuremberg spent more than 1300 florins on his pilgrimage to Jerusalem and 100 florins at the time represented the annual earnings of a very well-paid university professor!

If, prior to the end of the Middle Ages, the practice of pilgrimage had already drawn criticism, in the 14th century such criticism abounded in copious and ever more severe terms. Learned academics, whose star was rising with the spread of universities, denounced that which was wrong with pilgrimages. Among the humanists were Geoffrey Chaucer, author of *The Canterbury Tales* (Chaucer, 1987), written between 1387 and 1400, and Erasmus the author of *In Praise of Folly* (Erasmo, 1973), written in 1509, both of whom provided a sarcastic perspective. Also, some theologians, such as Thomas de Kempis in *Imitation of Christ* (Kempis, 1989), written between 1418 and 1427, observed that rare were the faithful who found sanctity at the end of the travels.

Within the scope of Protestantism, the pilgrimage to the sanctuaries no longer made sense. In the territories dominated by the new faith, many sanctuaries were abolished although some were to re-emerge in the 19th century. Meanwhile, in Catholic countries, the Council of Trent and the Christian Reformation attributed legitimacy to the pilgrimage.

Despite the reacceptance of pilgrimage, supported by Baroque pastoral, criticism of such practices was still to be heard. In the late 17th century, and throughout the following century, two forces stood in opposition towards each other (particularly as regards the less correct practices): (i) the spirit of the epoch (Enlightenment); and (ii) a better educated clergy.

A reflection of these offensives are some descriptions to be found in travel writings. Chateaubriand noted on his *Itinerary from Paris to Jerusalem* (Chateaubriand, 1884) in 1806 that Latin pilgrims in the Holy Sepulcher were rare. Furthermore, the English traveller Townsend had emphasized in *A Journey Through Spain in the Years 1786 and 1787* (Townsend, 1791), that the enthusiasm for pilgrimage had fallen away greatly. In fact, at the beginning of the 19th century, the pilgrimage, spurned by the educated elites and persecuted by the civil and religious authorities, was effectively moribund (Pierrard, 1982).

Nevertheless, during these same years, within the scope of Romanticism, a change of spirit began to break through. In Germany, Achim von Arnim, who worked with Clemens Brentano, gathered the different forms of popular poetic expression in *Des Knaben Wunderhorn* [The Boy's Magic Horn] (Arnim and Brentano, 1845), written in 1805, associating pilgrimages with a hidden soul. In the same period in France, Chateaubriand in his *Genius of Christianity* (Chateaubriand, 1829), written in 1802, far from sharing the displeasure of intellectuals in popular acts of devotion, rehabilitated pilgrimages affirming that when they were undertaken within a spirit of innocence, they were more effective against unhappiness than all the desperate theories of dry-hearted thinkers.

Within an ambience again favourable to pilgrimages, there was a resurgence of interest in the Camino but now the experience is central, with the objective almost peripheral. Paraphrasing Vidal (1996), many of those walking the 'former route of the stars' are agnostic and, among the faithful, many do not believe that the body lying in the cathedral is that of Saint James.

Also, nowadays, the Saint James Way has become more democratic, as it is not only the rich who follow the pilgrimage route and the information is accessible to almost everyone (literacy is a reality among those who walk the Camino). Bearing that in mind, this chapter analyses how novels can be crucial to trigger the motivation or to keep a link with an already performed pilgrimage.

Travel Literature

Literary tourism, seen as a form of tourism that promotes knowledge of places which have relevant tangible and intangible heritage, was developed mainly since the 18th century in countries such as France or the UK (Herbert, 2001; Deprest, 2004; Watson, 2006; Robinson, 2007).

According to Robinson (2007), literature and travel accounts from previous centuries continue to play an important role in shaping contemporary tourists' attitudes towards the world. For this author the more elaborate the trip or the more remote or exotic the destination, the greater the reader's feeling of adventure. He also believes that the consumption of travel literature can act as an antidote against mass tourism, which is often portrayed as stereotypical, unadventurous and unimaginative. In this sense, travellers and tourists can shape geographies, histories, cultures and societies and each one may experience it in different ways. For these reasons the author considers that the emerging research agenda in tourism studies should give more attention to the relationships between travel writing and the production/consumption of tourist spaces, and between travel narratives and tourists' motivations, experiences and their performances.

Literature (poetry or prose) and travel literature were, at the beginning, conceived as a form of quality narratives performed by authors with different points of view (e.g. chroniclers, navigators, scientists, politicians, merchants, ecclesiastics, artists, aristocrats, writers and journalists). However, nowadays researchers such as Herbert (2001), Watson (2006) and Robinson (2007) emphasize the relevance of these works as sources that contributed, and continue to contribute, to the increased tourism flows.

Even if it is still not possible to establish with precision the extension of the role of literature and travel writings in the 'creation of spaces' and 'patterns of tourism' (Robinson, 2007), a new interest seems to facilitate a continuum between the different areas of study, mainly regarding the last decades of the 20th century and the beginning of the 21st century (Watson, 2006). In fact, nowadays, literature and art are recognized as having the capacity to interconnect fictional places with real ones, enhancing tourist experiences (Deprest, 2004). Literature and art become less a subsidiary resource of tourism destinations and more a form of promotion and development of those places (Leitão, 2018).

According to Pine and Gilmore (2011), for tourist experiences to be memorable, they must be associated with a certain theme, with the stimulation of the senses, with a certain space and they must be positive, resulting in a set of learning and memories that become remarkable for the individual involved in them. For Watson (2006), the literary tourist experience is deeper when the place of composition coincides with the fictional setting.

We also believe that the literary experience (i.e. the reading of authors and works that make someone grow spiritually) can trigger the pilgrimage, and this may be independent of the pilgrimage being a religious or cultural performance (or combination of both), and that walking along the pilgrimage ways increases and transfigures the feelings attained while reading those works.

Literature on the Camino de Santiago

At this point, we intend to draw attention to the travel accounts written about Santiago de Compostela, from the Middle Ages to the contemporary period, which may have influenced pilgrims/travellers to undertake the Camino.

Among the most important books, which served as a basis for pilgrimages, whether from a religious perspective or from a more spiritual and/or more or less secular pilgrimage, we must mention the *Códice Calixtino* (García Blanco, 2016). This is a work of the 12th century, where Santiago (Saint James) appears as the protector saint in the fight against Islam (the infidels) in the Iberian Peninsula, which ends in Portugal in 1249 and in Spain in 1492. The *Códice Calixtino* is considered one of the first travel books of Western European culture, where the Santiago Way is reported. In the aforementioned Book V (*Códice Calixtino*; García Blanco, 2016), written by Aymeric Picaud and known as the 'Medieval Pilgrim's Guide', in addition to reports of miracles and liturgical texts we find advice given to pilgrims. Also there are descriptions of the habits and customs of the territories, their landscapes and people, in which the religious part combines with descriptions of art and local cultures, ending the book (as the Way) in the cathedral. This cathedral has become symbolic and the focus for poetry, prose, architecture and painting in the 19th and 20th centuries, during which in more secular times religious pilgrimage declined in popularity yet pilgrimage for cultural and/or spiritual reasons was favoured.

Unamuno (2002), in his poem 'Caminito de Santiago' [Little Way of Santiago] in Cancionero, written in 1928, is a good example of expressing a religious mysticism, where doubt is mixed with the need to believe. Also, many other Iberian writers (marked by European secularism) travelled to Santiago, on a cultural pilgrimage, probably searching for their inner Grail.

Another tireless pilgrim to Santiago worth mentioning is Basilio Losada the author of *La Peregrina* [The Pilgrim] (Losada, 1999). In a radio interview in 2005 he was asked what the Camino de Santiago is, to which the writer replied:

> I did it twice, the first in 1947, from Roncesvalles, it took me 40 days; the second from Paris, in 1948, it took me 3 months. In a certain way it is to recover the measure of man, to be hungry, to be thirsty, to feel fatigue, to sleep in a forest on the ground, to meet people, and also the opportunity for someone to find himself.
> (words of Basilio Losada cited in Leitão, 2008, pp. 139–140)

When asked if deep down he was religious, Losada answered: 'the one who does not doubt does not believe, I doubt and I look for God, but I don't find Him, although I keep looking for Him' (Leitão, 2008, p. 140).

Up until the third quarter of the 20th century, works on the Camino de Santiago were authored by renowned writers, but from that time onwards many others ventured into the field of writing. According to the websites that list the books with the greatest impact (more sales), and which probably have been read by a larger number of people, we present a selection of 28 books published in the last four decades (*Follow the Camino*, nd; *Love at First Adventure*, nd), although the list could be much longer. Even without a direct relationship, these are the books which have possibly influenced more people to go on the Camino or made them relive the experiences they had when they walked the pilgrimage way.

For a better understanding and explanation, we decided to group the books into seven categories according to their similar characteristics and those who wrote them, including books by: (i) bloggers; (ii) celebrities; (iii) clerics; (iv) writers sharing family relationships; (v) guidebook authors; (vi) writers sharing transformative experiences; and (vii) renowned writers. Seven tables are presented here in this chapter, one for each of these categories. In the tables, the name of the author, the year of publication of the work, the title of the book and a synopsis of each work are provided.

There are a huge number of blogs about the Camino. Many of those bloggers succeeded in having a very large number of followers (most of them sharing the same kind of emotions). For that reason, it is easy to understand that successful bloggers have gone on to turn their blogs into books, some having the support of well-known publishers. Two examples are provided in Table 4.1.

Table 4.2 presents three novels written by celebrities. According to The New Dork Review of Books (2010), the reason why publishers see these novels as good investments is simple, they sell well; people are obsessed with celebrity, and many of these novels are thinly veiled memoirs. Probably the best example is the book written by Hape Kerkeling, it has sold over 3 million copies and been translated into 11 different languages (Amazon, nd).

From the description of the three books shown in Table 4.3, dedicated to novels written by clerics, it is easy to understand that the emphasis, as expected, is the spirituality of the Camino. Walking along the Saint James Way allows one to admire God's work. It is also an opportunity for everyone to reflect on how to be part of the pastoral care in the church.

To walk the Camino accompanied by our loved ones raises the curiosity and the interest of many who dream to do the same. Table 4.4 includes books about couples sharing the difficulties encountered on the Camino, allowing them, at the same time, to better value each other. It also includes one book depicting father and daughter experiences on the Saint James Way. Beth Jusino's work, listed here in Table 4.4, could also be included in the renowned writer's table as she has become an award-winning writer.

Regardless of being guidebooks, the way they are written, in terms of descriptions of the Way's resources, is within the context of literary language (Table 4.5). Perhaps for this reason, especially Brierley's work, has had successive (updated) editions, with a concern to maintain the book's usefulness in terms of information, but also making it lighter and handier. According to Waterstones (nd) the 18th edition (2022) has 288 pages and weighs 282 g. Some travel guidebooks resemble brochures more than books (Fig. 4.1).

Table 4.1. Bloggers who have turned their blogs about the Camino into books. Based on information from Follow the Camino (nd) and Love at First Adventure (nd).

Author	Year	Book title	Synopsis
Leslee, A.	2019	*The Way of Love: On the Camino de Santiago*	A memoir inspired by the author's blog posts, this Camino book fills the niche for a growing population of women pilgrims, aged over 60 years, walking solo. In this work, she shares her vulnerabilities and confronts her fears, revealing the life-changing lessons she learnt along the way while sharing details of the Camino path itself.
Powell, C.	2018	*Ordinary Magic: Promises I Kept to My Mother Through Life, Illness, and a Very Long Walk on the Camino de Santiago*	In this loving memoir, fierce and funny Powell describes life's most difficult voyage: his mother's cancer diagnosis. Accompanied by his mother and his cousin they undertook the Camino. Blog/journal entries by each family member augment an amusing travelogue.

Table 4.2. Novels written about the Camino by celebrities. Based on information from Follow the Camino (nd) and Love at First Adventure (nd).

Author	Year	Book title	Synopsis
Bennett, B.	2014	*The Way, My Way*	When Bill Bennett, an Australian-based film director, decided to embark on this 800 km pilgrimage he had no clue about why he was doing it. In the middle of his journey, the pain of his injured knee and all the experiences that occurred on the Camino, allowed him to find his own way.
Kerkeling, H.	2009	*I'm Off Then: Losing and Finding Myself on the Camino de Santiago*	In 2001, the German comedian Hape Kerkeling decided to quit his sedentary life to embark on an 800 km journey to Santiago de Compostela. Struggling with his physical limitations, he honestly and humorously recounts his physical and spiritual journey towards forgiveness along the French Camino.
MacLaine, S.	2001	*The Camino: a Journey of the Spirit*	When the American film actress Shirley MacLaine walked the Camino, she was in her 60s. There is some disagreement as to whether this book is a little too 'over the top' or has just the right touch of spirituality, but there is no doubt it inspired many women to commence the journey solo.

Table 4.3. Books written about the Camino by clerics. Based on information from Follow the Camino (nd) and Love at First Adventure (nd).

Author	Year	Book title	Synopsis
Codd, K.	2008	*To the Field of Stars: a Pilgrim's Journey to Santiago de Compostela*	This is a book about the fascinating story of a Catholic priest's personal journey on foot to Santiago. Day by day, Codd examines the physical, emotional and spiritual ups and downs of pilgrimage in a candid chronicle. Accordingly, he invites the reader to walk and share the pilgrimage experience with him.
Frazer, R.	2019	*Travels With a Stick: a Pilgrim's Journey to Santiago de Compostela*	Frazer, a minister influential in the revival of pilgrimage in Scotland, seeks to get at the heart of pilgrimage through his thoughtful descriptions, compassion and humour. He reminds readers that the modern-day pilgrimage to Santiago can be a nourishing spiritual journey for those who are honest and open to transformation.
Pivonka, D.	2009	*Hiking the Camino*	Father Dave invites readers to walk with him on the ancient spiritual pilgrimage. When exhausted, sore and perhaps a little lost, he reminds us that God is ever present and eager to care for us, whether we acknowledge it or not. The author also provides practical life lessons mixed with humorous insights.

Table 4.6 shows books written about the Camino by authors sharing transformative experiences. This category has the largest number of entries, something that is understandable because the Camino is sought after, above all, by people struggling with some kind of existential crisis, often seeking the answer for their doubts. Bradley Chermside, a multi-award-winning singer and entertainer, could also be considered in the books written by celebrities' category and

Table 4.4. Books written about the Camino by authors sharing family relationships. Based on information from Follow the Camino (nd) and Love at First Adventure (nd).

Author	Year	Book title	Synopsis
Jusino, B.	2018	*Walking to the End of the World: a Thousand Miles on the Camino de Santiago*	Beth and Eric Jusino walked the 1000-mile long-distance on the Camino. In this book, she narrates how an average couple manages to face the challenging physical trek. Her entertaining description reminds us that we can all complete the journey if we put our hearts, minds and souls to it.
Murtagh, P and Murtagh, N.	2011	*Buen Camino: a Father and Daughter Journey from Croagh Patrick to Santiago de Compostela*	This book is written jointly by a father and daughter. They started their adventure in Ireland by walking up Croagh Patrick on St James Day, 2010. When Peter found two scallop shells near the sea, he and his daughter attached them to their backpack and officially proclaimed themselves to be pilgrims.
Scott, R.	2019	*A Furnace Full of God: a Holy Year on the Camino de Santiago*	Captured by the spirit of the pilgrimage trail, Rebekah and Patrick left their stressful jobs as journalists and moved to Spain. Rebekah hilariously tells their story that will especially appeal to those wishing to return to a simpler life through pilgrimage.
Spittle, R.	2018	*Camino Sunrise – Walking With My Shadows*	Anyone who has struggled with both debilitating anxiety and life-long self-doubt can empathize with Spittle's Camino story. The author tackles the trek and gives an honest portrayal of the joys and trials of pilgrim life. Walking with his wife Sue, he consequently reminds us of the simple human kindness that exists in the world.
Strange, K.	2019	*It's Your Camino: One Couple's 500-mile Pilgrimage Across Spain*	Kenneth Strange and his wife embarked on a 31 days' journey along the Camino. This American couple share a profound love for Spanish history, culture and gastronomy. The book accurately and entertainingly shows what a pilgrim might expect when walking the Camino.

C.W. Lockhart, as a book-prize winner, could be included in the renowned writer's table.

As stated before, until the 1980s almost only renowned writers wrote books inspired by the Camino de Santiago, but since then others have ventured into the novels' literary world, many of them achieving considerable success. However, there are still renowned authors who feel inspired by the Camino such are the ones referred to in Table 4.7. Of particular note is Paulo Coelho, whose book is considered by many the best-selling novel about the Saint James Way.

Paulo Coelho's *The Diary of a Magus/The Pilgrimage* and David Lodge's *Therapy*

In the previous section, our concern was to consider literature as a motivator to walk the Camino de Santiago. Even though there is a very wide range of novels that are inspired, in different ways, by the Saint James Way, it is also true that there are novels where the Way is not the main theme. Probably whoever hears the call to set out on the Camino will turn to specific books

Table 4.5. Guidebooks about the Camino. Based on information from Follow the Camino (nd) and Love at First Adventure (nd).

Author	Year	Book title	Synopsis
Brierley, J.	2003	*A Pilgrim's Guide to the Camino de Santiago*	There's no doubt, John Brierley is one of the most well-known Camino guidebook authors. This practical work encloses all the information required by modern-day pilgrims wanting to walk the ancient pilgrimage to Santiago de Compostela. This popular guide is regularly updated.
Connor, J.	2019	*Savoring the Camino de Santiago: It's the Pilgrimage, Not the Hike*	Julie Connor does not focus on the physical and spiritual walk but offers useful side information for planning the journey. The author details the many pleasures of the Camino. Emphatically, she reminds pilgrims that there is no one way to experience the Camino, and that it is accessible to all who wish to savour its charms.

Fig. 4.1. Various travel guidebooks to Santiago de Compostela. Photo courtesy of D. Liutikas.

about the Way, but we believe that novels, where the Camino is addressed as a complement to the narrative, can also have the effect that triggers the intention to undertake the Camino de Santiago.

At this point we will analyse (briefly) two novels that illustrate this: (i) one written by Paulo Coelho, *The Diary of a Magus* (first called *The Pilgrimage*), which focuses on the Camino; and (ii) one written by David Lodge, *Therapy*, where the Way is a supplement to the narrative.

The Diary of a Magus

The Diary of a Magus/*The Pilgrimage* is possibly the most publicized and widely read Camino-based novel (Fig. 4.2). After its first edition, in 1987, according to Paulo Coelho and the Christina Oiticica Foundation (nd) the novel was translated into 40 languages, having become a bestseller (it was this book that resulted in the author becoming internationally recognized). According to the same source, to date, Paulo Coelho has published 30 books, his work has been published in over 170 countries, has been translated into 88 languages and sold over 320 million copies.

The Diary of a Magus reflects the 3-month pilgrimage, along the French Way, to Santiago de Compostela that the writer Paulo Coelho made in 1986. At that time the Way was almost forgotten, far from the prominence and number of pilgrims that it currently has.

This book depicts a true adventure, facing the (un)known, the self-knowledge, and the search for ancestral wisdom. The struggle between good and evil, Catholicism and irrationalism, translated into dreams or magic are present throughout the work.

Starting in Saint-Jean-Pied-de-Port, between reality and imagination, the author walks about 800 km to Santiago de Compostela. We say 'reality' because there are clear allusions in the novel to places that appear in that religious itinerary such as Ponferrada, O Cebreiro or Villafranca del Bierzo, and also references to historical facts that occurred in Spain, namely the Battle of Clavijo (9th century) where the Apostle James appears in dreams to Ramiro I, ensuring their presence in battle and consequent victory. According to the legend, the following day, the armies of Ramiro I, accompanied by the presence of the Apostle mounted on a white steed, defeated the enemy (the Muslims). We say 'imagination' because the reason that takes him, after being initiated in

Table 4.6. Books written about the Camino by authors sharing transformative experiences. Based on information from Follow the Camino (nd) and Love at First Adventure (nd).

Author	Year	Book title	Synopsis
Chermside, B.	2019	*The Only Way is West: a Once in a Lifetime, 500 Mile Adventure Walking Spain's Camino de Santiago*	Bradley Chermside walks the Camino to ponder some profound life changes. Hiking 900 km, he discovers the utopia of his fantasies, befriends a Hungarian who speaks English in song titles, and has his *raison d'être* revealed to him by a barefoot Mayan mystic.
Christmas, J.	2007	*What the Psychic Told the Pilgrim: a Midlife Misadventure on Spain's Camino de Santiago*	A Canadian divorced mother of teenage children joins a group of 14 women to celebrate her 50th birthday. She describes what it is like to walk the Camino, and what it meant to her. It recounts her battles with loneliness, hallucinations, as well as picturesque villages and even the fair-haired man.
Devaney, P.	2017	*Two Million Steps: Band-aids, Cocktails, and Finding Peace along Spain's Camino de Santiago*	Plagued with a propensity for negativity and a sense of dissatisfaction, Devaney heads out on the Camino to find purpose and peace along the path. Experiencing physical pain and working through memories and emotions, he recounts the moments that led to healing.
Eanes, R.	2019	*The Walk of a Lifetime: 500 Miles on the Camino de Santiago*	Realizing his old dream to walk the Camino, Eanes set off during a sabbatical to re-orient his hurried life at 61 years old. He lets his story slowly and mindfully unfold on each page, bringing the reader along with him on a life-changing journey that begins on the Camino and continues within each pilgrims' soul.
Koontz, K.	2013	*A Million Steps*	Kurt Koontz thought he was prepared (like many a pilgrim) for his 500-mile trek to Santiago. This diary-travelogue is a wonderful reminder that although you can pack all the right gear, you can never predict the internal and physical transformation that you will experience on the Camino.
Lockhart, C.	2018	*Blanket of Stars: Thru-Hiking the Camino de Santiago*	Looking to face a mid-life crisis, an unlikely pilgrim sets off on the adventure of a lifetime. Carrying far too much emotional baggage, Lockhart manages to shed emotional weight, share laughter and kindness, and honestly examine her past traumas.
Mullen, R.	2010	*Call of the Camino: Myths, Legends and Pilgrim Stories on the Way to Santiago de Compostela*	This book is about the experiences of an ordinary man who walks the Camino. The history of the Way is recounted, as well as its myths, legends and miracles. It also includes stories from the pilgrims, underlying that this pilgrimage will play a defining role in their lives.
Soper, K.	2013	*Steps Out of Time, One Woman's Journey on the Camino*	Descriptive and narrative, you will feel like you are on the Camino path. Many people – especially women – will relate to the pressures and uncertainties that come with natural life changes. Hence, if you've heard the call of the Camino, Soper's alluring writing style and story will draw you in and inspire you.

Continued

Table 4.6. Continued.

Author	Year	Book title	Synopsis
Tranchant, C.	2012	*Boots to Bliss*	Tranchant is no stranger to the Camino, hiking alone as a 64-year-old woman for 100 days through France and Spain. This book depicts the author's personal growth, determination and self-empowerment and reminds the reader to take life at one's own pace and to cherish personal inner strength.
Walker, J.	2019	*It's About Time: a Call to the Camino de Santiago*	Having completed the pilgrimage multiple times, the author highlights the spirit of the Camino by sharing the stories of multiple pilgrims. This book is a great resource full of practical advice for those that have heard the call and who are joyously open to the change pilgrimage can bring.

Table 4.7. Books written about the Camino by renowned authors. Based on information from Follow the Camino (nd) and Love at First Adventure (nd).

Author	Year	Book title	Synopsis
Coelho, P.	1992 (first published in 1987)	*The Diary of a Magus*	The story depicts the author's experiences on the Camino de Santiago. This trip was the catharsis that made him stop dreaming about being a writer to finally becoming one. Undoubtedly the most famous work about the Saint James Way, this enchanting parable highlights the extraordinary in the ordinary.
Moore, T.	2006	*Travels with My Donkey: One Man and His Ass on a Pilgrimage to Santiago*	One of the most entertaining books about the Santiago Way. It recounts the story of the British travel writer Tim Moore, who walked the Camino with his travel companion a donkey (Shinto). He was looking to release his load, but he just added a 200 kg package of obstinacy, phobias and more-or-less stoic forbearance.

Brazil as 'Master and Knight of the Order of RAM .... R for rigour, A for adoration, and M for mercy; R for regnum, A for agnus, and M for mundi' (Coelho, 1992, p. 2), is the search for his initiate's sword, normally considered a heroic symbol related to power and conquest, in the 'Strange Road of Santiago' (Coelho, 1992, p. 4).

As in the stories of Don Juan and his disciple Carlos Castañeda (in works from the writer Castañeda, 1968), during the Way, between Petrus, the master, and his disciple, the author, the surreal runs parallel with the real and often mixes with it; the religion and 'witchcraft' often intersect. But it is on the Camino that solitude is as important as companionship.

It is a path of physical pain, typical of those who walk the Camino and who overcome this great obstacle: 'One final push was all that I needed ... In Ponferrada, we will change the bandages' (Coelho, 1992, pp. 185–186). But above all the Way is spiritual learning, self-knowledge, distinctive of travel: 'When you travel, you experience, in a very practical way, the act of rebirth' (Coelho, 1992, p. 32). To Petrus:

> The true path to wisdom can be identified by three things ... First, it must involve agape (spirit) second, it must have practical application in your life. Otherwise, wisdom becomes a useless thing and deteriorates, like a sword that is never used. ... And finally, it has to be a path that can be followed by anyone. Like the road we are walking now, the Road to Santiago.
>
> (Coelho, 1992, p. 25)

Fig. 4.2. The cover of the Paulo Coelho's *The Diary of a Magus* (1992). Photo courtesy of D. Liutikas.

The secret of this path, of this learning of wisdom seems to be the following:

> the great secret of the journey ... a secret that I'm going to reveal to you now ... It is a secret that has to be experienced to be understood. ... You can learn only through teaching. We have been together here on the Road to Santiago, but while you are learning the practices, I learned the meaning of them. In teaching you, I truly

> learned. By taking on the role of guide, I was able to find my own true path. ... If you succeed in finding your sword, you will have to teach the Road to someone else.
>
> (Coelho, 1992, pp. 191–192)

Teaching and learning simultaneously is one of the paths pointed out by Petrus. The narrator further states that 'I have walked so many miles to discover things I already knew, things that all of us know but are so hard to accept' (Coelho, 1992, p. 220). After all:

> the search for happiness is a personal search and not a model we can pass on to others. Before finding my sword, I had to discover its secret and the secret was so simple: it was to know what to do with it. With it and with the happiness that it would represent to me.
>
> (Coelho, 1992, p. 220)

Therapy

According to Mullan (2012), David Lodge was brought up a Catholic and has described himself as an 'agnostic Catholic'. Many of his characters are Catholic and their Catholicism, particularly the relationship between Catholicism and sexuality, is a major theme. When reading Lodge's novels in chronological order we would see a once orthodox Roman Catholic becoming 'less and less so as time went on'. According to Penguin Random House (nd), his books have been translated into 25 languages and he has sold over 2.5 million books.

Therapy was first published in 1995 and it depicts the life change of Laurence Passmore (Tubby). To all appearances, 58-year-old Laurence Passmore's life is going rather well. He is a successful television writer with two grown children, he has an exclusive house, a superb car, and he enjoys a vigorous sex life with his wife, Sally. What money cannot buy, and his many therapists cannot deliver, is contentment. It's not the trouble behind the scenes of his TV show that's bugging him or even the persistent pain in his knee; it's this deeper, nameless unease. Is Passmore living a spiritual crisis or just one of the midlife varieties? (Goodreads, nd). After his marriage fails, he has a breakdown. He reminisces about his adolescence and in an attempt to atone for bad behaviour towards his first girlfriend (Maureen) at a Catholic youth club, he seeks her out.

As previously mentioned, this novel is not about the Camino de Santiago, although it is used as the scenery for the epilogue. The first time the Camino is mentioned is p. 274 (total pages 321), when Passmore, after a prolonged search, finally finds Maureen's home, but not her. Bede (Maureen's husband) informs Tubby that she went on a pilgrimage and shows him a 'brightly coloured picture postcard of Saint-Jean-Pied-de-Port in the French Pyrenees. On the other side was a brief message from Maureen to Bede' (Lodge, 2011, p. 274). Almost all Chapter Four (from p. 285 to p. 317) presents a narrative developed in the context of the Camino de Santiago. In this, Lodge presents the path considering historical aspects, landscapes, resources of the Way and above all the various types of pilgrims, using a somewhat sarcastic language, characteristic of Passmore (and in some way the author's own).

Based on Kierkegaard's philosophy (for whom Tubby develops an obsession throughout the novel) Lodge introduces us to what he considers to be the three types of pilgrims:

> The aesthetic type was mainly concerned with having a good time, enjoying the picturesque and cultural pleasures of the Camino. The ethical type saw the pilgrimage as essentially a test of stamina and self-discipline. He (or she) had a strict notion of what was correct pilgrim behaviour (no staying in hotels, for instance) and was very competitive with others on the road. The true pilgrim was the religious pilgrim, religious in the Kierkegaardian sense. To Kierkegaard, Christianity was "absurd": if it were entirely rational, there would be no merit in believing it. The whole point was that you chose to believe without rational compulsion – you made a leap into the void and in the process chose yourself. Walking a thousand miles to the shrine of Santiago without knowing whether there was anybody actually buried there was such a leap. The aesthetic pilgrim didn't pretend to be a true pilgrim. The ethical pilgrim was always worrying whether he was a true pilgrim. The true pilgrim just did it.
>
> (Lodge, 2011, pp. 304–305)

Regardless of his humorous view of the pilgrimage, the author, through the character Maureen, shows a positive opinion of its effects, 'a kind of kink in time, when the ordinary rules of life don't apply'; however, he also has the perception of reality, 'when I go home, I'll be married with Bede again' (Lodge, 2011, p. 316).

In the last paragraph, the character's life change that took place while on the Camino allows everyone to reach happiness. Passmore rents a house in the neighbourhood of Bede and Maureen, where:

> every now and then Maureen and I have a siesta. I don't ask her how she squares it with her conscience – I've got more sense. My own conscience is quite clear. The three of us are the best of friends. We're going off together for a little autumn break. To Copenhagen. It was my idea. You could call it a pilgrimage.
>
> (Lodge, 2011, p. 321)

Summary

The pilgrimage route to Santiago de Compostela has evolved over the centuries in tandem with social, economic and technological developments. From the 9th century to the 15th century, the Way gained significance, receiving accreditation from religious and civil authorities. During the Middle Ages, it was esteemed by popes as one of the most important Christian pilgrimages, offering a plenary indulgence to those who undertook it. Kings and queens, including Alfonso II of Asturias 'The Chaste' (760–842), Alfonso III of Asturias 'The Great' (852–910), Queen and Saint Elizabeth of Portugal (1282–1325), Isabella I of Castile (1451–1504) and Ferdinand II of Aragon (1452–1516) walked the Way, often accompanied by chroniclers. It is worth noting that only distinguished writers were appointed to this role, resulting in chronicles of high literary quality.

From the 16th century onwards, societal changes led to the gradual separation of civil power from religious authority, making it almost inconceivable for a reigning king or president to embark on the Camino today. The chronicles penned by chroniclers were increasingly supplanted by travel literature, particularly from the 17th century onwards. In the 18th century, many of these works reached a wider readership due to the growing number of literate individuals and translations of esteemed works.

In the 19th and 20th centuries, the allure of the Camino waned, with very few literary works paying attention to it. According to CaminoWays.com (nd), it wasn't until the 1980s that the Camino experienced a resurgence, thanks to the visionary Spanish parish priest Don Elías. In 1984, he initiated a mission to reclaim, clean and mark the trails along the Saint James Way with yellow arrows, starting in Roncesvalles in the Pyrenees.

In 1987, the Council of Europe designated the Way as the 'First European Cultural Route'. In 1989, Pope John Paul II visited the Cathedral of Santiago during the World Youth Days, which brought together half a million young people from around the world. In 1993, the United Nations Educational, Scientific and Cultural Organization (UNESCO) classified the Camino de Santiago as a World Heritage Site. Works like *The Diary of a Magus* by Paulo Coelho, published in 1987, along with others by different authors, also contributed to increased media coverage of the Way.

Today, various forms of writing, including movie scripts and blogs, play vital roles in promoting the Camino. According to Zhang *et al.* (2021), around 40% of Chinese pilgrims first learned about the Camino from the movie *The Way*, directed by Emilio Estevez and starring Martin Sheen (Estevez, 2010). Blogs enhance the experience as a rite of passage and a site of communitas, shaping representations of the pilgrims themselves, contributing to diverse pilgrimage practices (Polus and Carr, 2021).

In the near future, probably well-written movie scripts and blogs might also be considered literature, as according to its definition the name literature is applied to imaginative works of poetry and prose distinguished by the intentions of their authors and the perceived aesthetic excellence of their execution (Britannica, nd). Presently, novels remain the primary vehicles for promoting the Camino, allowing readers to dream of embarking on it for spiritual transformation, or enabling those who have walked the Way to relive their transformative experiences.

Acknowledgement

This work is financed by National Funds provided by FCT – Foundation for Science and Technology through project ref. UIDB/04470/2020 CiTUR.

References

Amazon (nd) *I'm Off Then: Losing and Finding Myself on the Camino de Santiago*. Available at: https://www.amazon.com/Im-Off-Then-Finding-Santiago/dp/1416553878 (accessed 31 July 2023).

Ambrósio, V. (2016) Peregrinações Cristãs e turismo religioso antes e depois da reforma. In: Franco, J., Ruiz, J., Abreu, J. and Cieszynska, B. (eds) *Concílio de Trento, Inovar na Tradição, História, Teologia, Projeção*. Universidad de Alcalá, Alcalá de Henares, Spain, pp. 343–353.

Arnim, A. and Brentano, C. (1845) *Des Knaben Wunderhorn*. Egbert Bauer, Charlottenburg, Germany.

Bennett, B. (2014) *The Way, My Way*. Createspace Independent Publishing Platform, Scotts Valley, California.

Branthomme, H. (1982) Introduction. In: Chélini, J. and Branthomme, H. (eds) *Les Chemins de Dieu: Histoire des Pèlerinages Chrétiens, des Origines à nos Jours*. Hachette, Paris, pp. 17–20.

Brierley, J. (2003) *A Pilgrim's Guide to the Camino de Santiago*. Findhorn Press, Findhorn, UK.

Britannica (nd) Literature. Available at: https://www.britannica.com/art/literature (accessed 5 August 2023).

CaminoWays.com (nd) The story of the yellow arrows on the camino. Available at: https://caminoways.com/the-story-of-the-yellow-arrows-along-the-camino (accessed 6 August 2023).

Castañeda, C. (1968) *The Teachings of Don Juan: a Yaqui Way of Knowledge*. University of California Press, Berkeley, California.

Chateaubriand (1829) *Génie du Christianisme*. Lefèvre, Paris.

Chateaubriand (1884) *Itinéraire de Paris a Jérusalem*. Alfred Mame et Fils, Tours, France.

Chaucer, G. (1987) *The Canterbury Tales*. Longman, Harlow, UK.

Chermside, B. (2019) *The Only Way is West: a Once in a Lifetime, 500 Mile Adventure Walking Spain's Camino de Santiago*. Independently published.

Christmas, J. (2007) *What the Psychic Told the Pilgrim: a Midlife Misadventure on Spain's Camino de Santiago*. Greystone Books, Vancouver, Canada.

Codd, K. (2008) *To the Field of Stars: a Pilgrim's Journey to Santiago de Compostela*. Eerdmans, Grand Rapids, Michigan.

Coelho, P. (1992) *The Diary of a Magus*, 1st edn in English. HarperCollins, New York.

Connor, J. (2019) *Savoring the Camino de Santiago: It's the Pilgrimage, Not the Hike*. Bayou City Press, Houston, Texas.

Deprest, F. (2004) *Inquérito Sobre o Turismo de Massa: a Ecologia Face ao Território*. Edições Piaget Lda, Lordosa, Portugal.

Devaney, P. (2017) *Two Million Steps: Band-aids, Cocktails, and Finding Peace along Spain's Camino de Santiago*. CreateSpace, Scotts Valley, California.

Eanes, R. (2019) *The Walk of a Lifetime: 500 Miles on the Camino de Santiago*. The Walker Press, London.

Erasmo (1973) *Elogio da Loucura*. Europa América, Lisboa.

Estevez, E. (dir.) (2010) *The Way*. Film. Distributed by Ocean Avenue Entertainment. Filmax/Elixir Films, USA.

Follow the Camino (nd) Top 10 Camino books. Available at: https://followthecamino.com/en/blog/top-10-camino-books/ (accessed 26 July 2023).

Frazer, R. (2019) *Travels With a Stick: a Pilgrim's Journey to Santiago de Compostela*. Birlinn, Edinburgh, UK.

García Blanco, M.J. (ed.) (2016) *Códice Calixtino, Libro V, Siglo XII: Guía del Peregrino Medieval*. Traducción y Notas de A. Moralejo, C. Torres y J. Feo. [Translation and notes by Moralejo, A., Torres, C. and Feo, J.]. Alvarellos edition, Santiago de Compostela, Spain. (in Spanish)

Goodreads (nd) *Therapy*. Available at: https://www.goodreads.com/book/show/69928.Therapy (accessed 27 July 2023).

Herbert, D. (2001) Literary places, tourism, and the heritage experience. *Annals of Tourism Research* 28(2), 312–333.

Jusino, B. (2018) *Walking to the End of the World: a Thousand Miles on the Camino de Santiago*. Mountaineers Books, Seattle, Washington State.

Kempis, T. (1989) *The Imitation of Christ*. Ave Maria Press, Notre Dame, Indiana.

Kerkeling, H. (2009) *I'm Off Then: Losing and Finding Myself on the Camino de Santiago*. Free Press, Florence, Massachusetts.

Koontz, K. (2013) *A Million Steps*. Independently published.

Leitão, I. (2008) Pelos Caminhos de Santiago. Breve viagem com alguns escritores do Século XX. Turismo Cultural e Religioso. In: Vilaça, A. and Pereira, V. (eds) *Congresso Internacional de Turismo Cultural e Religioso, Oportunidades e Desafios para o Século XXI*. TurelTRC, Braga, Portugal, pp. 135–143.

Leitão, I. (2018) Literary sites and literary tourism in some European countries and in Portugal: some reflections. In: Marinov, V., Vodenska, M., Assenova, M. and Dogramadjieva, E. (eds) *Traditions and Innovations in Contemporary Tourism*. Cambridge Scholar Publishing, Newcastle upon Tyne, UK, pp. 304–320.

Leslee, A. (2019) *The Way of Love: On the Camino de Santiago*. Independently published.

Lockhart, C. (2018) *Blanket of Stars: Thru-Hiking the Camino de Santiago*. GF Books, Inc., Hawthorne, California.

Lodge, D. (2011) *Therapy*. Vintage, London.

Losada, B. (1999) *La Peregrina*. Grijalbo, Barcelona, Spain.

Love at First Adventure (nd) 22 must-read Camino de Santiago books to inspire you. Available at: https://loveatfirstadventure.com/camino-de-santiago-books/ (accessed 20 July 2023).

MacLaine, S. (2001) *The Camino: a Journey of the Spirit*. Enterprise Books, Chicago, Illinois.

Mattoso, J. (2000) Peregrinar. In: Silva, H. (ed.) *Caminho do Tejo*. Selecções Reader's Digest e Centro Nacional de Cultura, Lisbon, pp. 4–5.

Moore, T. (2006) *Travels with My Donkey: One Man and His Ass on a Pilgrimage to Santiago*. St. Martin's Griffin, New York.

Mullan, J. (2012) *Small World* by David Lodge. *The Guardian*, 27 January. Available at: https://www.theguardian.com/books/2012/jan/27/small-world-david-lodge-book-club (accessed 2 August 2023).

Mullen, R. (2010) *Call of the Camino: Myths, Legends and Pilgrim Stories on the Way to Santiago de Compostela*. Findhorn Press, Forres, UK.

Murtagh, P. and Murtagh, N. (2011) *Buen Camino: a Father and Daughter Journey from Croagh Patrick to Santiago de Compostela*. Gill & Macmillan, Dublin.

Ostrowski, M. (2000) Pilgrimages or Religious Tourism? *Peregrinus Cracoviensis* 10, 53–61.

Paulo Coelho and Christina Oiticica Foundation (nd) Paulo Coelho, *The Pilgrimage*. Available at: https://paulocoelhofoundation.com/books/paulo-coelho-the-pilgrimage/ (accessed 1 August 2023).

Penguin Random House (nd) David Lodge. Available at: https://prhinternationalsales.com/author/?authorid=17923 (accessed 2 August 2023).

Pierrard, P. (1982) La Renaissance des Pèlerinages au XIX [e] siècle (1814–1914). In: Chélini, J. and Branthomme, H. (eds) *Les Chemins de Dieu: Histoire des Pèlerinages Chrétiens, des Origines à nos Jours*. Hachette, Paris, pp. 295–343.

Pine, B. and Gilmore, J. (2011) *The Experience Economy*. Harvard Business Press, Boston, Massachusetts.

Pivonka, D. (2009) *Hiking the Camino*. Franciscan Media, Cincinnati, Ohio.

Polus, R. and Carr, N. (2021) The role of communication technologies in restructuring pilgrimage journeys. *International Journal of Religious Tourism and Pilgrimage* 9(5), 37–48. https://doi.org/10.21427/MG4M-E412

Powell, C. (2018) *Ordinary Magic: Promises I Kept to My Mother Through Life, Illness, and a Very Long Walk on the Camino de Santiago*. Mascot Books, Herndon, Virginia.

Rapp, F. (1982) Mutations et difficultés du Pèlerinage à la fin du Moyen Age (XIV [e]–XV [e] siècle). In: Chélini, J. and Branthomme, H. (eds) *Les Chemins de Dieu: Histoire des Pèlerinages Chrétiens, des Origines à nos Jours*. Hachette, Paris, pp. 209–334.

Robinson, M. (2007) Narrativas de estar noutro sítio: turismo e literatura turística. In: Lew, A., Hall, M. and Williams, A. (eds) *Compêndio de Turismo*. Instituto Piaget, Lisbon, pp. 341–354.

Scott, R. (2019) *A Furnace Full of God: a Holy Year on the Camino de Santiago*. Peaceable Publishing, Haverhill, Massachusetts.

Soper, K. (2013) *Steps Out of Time, One Woman's Journey on the Camino*. Stellaire Press, Ann Arbor, Michigan.

Spittle, R. (2018) *Camino Sunrise – Walking With My Shadows*. Independently published.

Strange, K. (2019) *It's Your Camino: One Couple's 500-mile Pilgrimage Across Spain*. Independently published.

The New Dork Review of Books (2010) Renaissance or ridiculous: When celebs write fiction. Available at: https://www.thenewdorkreviewofbooks.com/2010/05/renaissance-or-ridiculous-when-celebs.html (accessed 3 August 2023).

Townsend, J. (1791) *A Journey Through Spain in the Years 1786 and 1787*. Available at: https://www.bibliotecavirtualdeandalucia.es/catalogo/es/catalogo_imagenes/grupo.do?path=1002939 (accessed 2 August 2023).

Tranchant, C. (2012) *Boots to Bliss*. Independently published.

Unamuno, M. (2002) *Cancionero. Obras Completas V*. Biblioteca Castro, Madrid.

Vidal, J. (1996) Le Pèlerinage dans la Tradition Chrétienne. *Concilium* 266, 53–67.
Walker, J. (2019) *It's About Time: a Call to the Camino de Santiago.* Redemptionist Publications, Chawton, UK.
Waterstones (nd) *A Pilgrim's Guide to the Camino de Santiago (Camino Frances): St. Jean Pied de Port – Santiago de Compostela*. Available at: https://www.waterstones.com/book/a-pilgrims-guide-to-the-camino-de-santiago-camino-frances/john-brierley/9781912216154 (accessed 3 August 2023).
Watson, N. (2006) *The Literary Tourist: Readers and Places in Romantic and Victorian Britain*. Palgrave Macmillan, London.
Zhang, K., Labajo, V., Ramos, I. and González, A. (2021) A guest at home: the experience of Chinese pilgrims on the Camino de Santiago. *Sustainability* 13: 10658. https://doi.org/10.3390/su131910658

5 The Transformational Potential of Visiting Sacred Sites

Silvia Aulet*
University of Girona, Girona, Spain

Abstract

Visiting sacred sites offers a unique opportunity for individuals to experience profound transformations, influenced by a range of factors including religion, spirituality, culture and the natural environment. This chapter investigates the potential for such transformations, encompassing personal evolution, spiritual growth and overall well-being, drawing insights from previous studies (Timothy and Olsen, 2006; Haq and Wong, 2009; Sharpley, 2009; Raj and Griffin, 2015). Furthermore, the chapter explores how religious tourism can promote sustainable travel practices by highlighting the intrinsic connection between nature and sacred sites, as demonstrated by Collins-Kreiner and Kliot (2017). The impact of three transformations on the visitor's experience is investigated: (i) personal transformation; (ii) social transformation; and (iii) ecological transformation. The research methodology used in-depth on-site observations and interviews conducted at various sacred sites across Europe, with a particular focus on those located in Spain.

Introduction

Religious tourism and pilgrimages, meaning travelling to sacred sites, has been widely explored. Several authors have theorized on the concepts of religious tourism and pilgrimages (Collins-Kreiner, 2010; Raj and Griffin, 2015; Iliev, 2020; Olsen and Timothy, 2022). Most scholars today do not differentiate between pilgrims and tourists, considering pilgrimage a form of tourism because it shares the same features in relation to transport, services and infrastructure (Jackowski and Smith, 1992; Rinschede, 1992; Jackson and Hudman, 1995; Vukonic, 1996). Other scholars argue that there is a clear distinction between pilgrimage and religious tourism, particularly concerning the ultimate motivation and intensity of experiences (Cohen, 1992; Ostrowski, 2000; Santos, 2002; Pardellas, 2005). This discussion could be the subject of an entire chapter, but this section will focus on what they have in common, mainly that both have a sacred site as the main destination.

This chapter aims to explore and reflect on the potential for transformation that visiting sacred sites can have on people. Exploring the concept of sacred sites, different values can be identified, connected to religion, spirituality, culture and nature (Bremer, 2006; Shackley, 2006). When pilgrims and tourists visit these sites, they may be aware of these values, and visiting them might have an impact on visitors that can change them in several ways.

*silvia.aulet@udg.edu

DOI: 10.1079/9781800623675.0005

This chapter is structured into three sections after this introduction. First, the concept of sacred sites is defined to better understand the transformational potential that they can have. Secondly, the potential areas of transformation are presented and analysed using different examples. Thirdly, the chapter concludes with some final reflections. The methodology is based on previous research work developed by the author. The different examples presented have been analysed using qualitative methods, in-depth interviews and on-site observation.

Travelling to Sacred Sites

The concept of the sacred has been studied across various disciplines, including sociology, anthropology and the comparative history of religions (Ries, 1989; Aulet and Sureda, 2019). Sacred sites are spaces transfigured by a *hierophany* (Eliade, 1965); a profane space becomes sacred due to the manifestation of divinity. Sacred sites can be understood as a place of worship and veneration because divinity has made itself present. Mircea Eliade's work in the field of comparative history of religions demonstrates the existence of a spiritual unity of humanity based on the idea of the manifestation of the sacred: it can be described because it manifests itself in space and time as a reality that belongs to a different order than the natural one. The sacred never presents itself in a pure state; it does so through symbols. The manifestation of the sacred through symbols allows us to speak of a universal concept that is represented and symbolized in diverse ways according to the religious tradition in which it is inscribed (Eliade, 1965; Otto, 1965; Severino Croatto, 2002). Despite this, there are a series of symbolisms that are shared, such as the concept of the axis mundi, the symbolism of the mountain or the sacredness of water as a purifying element (Guénon, 1995; Burckhardt, 2000; Olsen, 2019).

One of the distinctive elements of the sacred is that it can be clearly differentiated from the profane: sacred space and profane space; sacred time and profane time. Sacred and profane represent two different ways of being in the world. Sacred spaces are delimited in such a way that the crossing of the profane-sacred threshold is noticeable. The idea of separation is essential, sacred is everything that is separated from the rest because it belongs to the divine realm, to the absolutely different (Eliade, 1965; Aulet and Sureda, 2019). This element is key to understanding their transformative potential.

People's lives take place in profane spaces. The reflection or distinction between the concepts of space and place led the anthropologist Marc Augé to speak also of non-spaces. Thus, people's lives are spent between non-spaces (transit spaces that lack identity and leave no mark on people, such as an airport or a motorway service area), profane spaces (everyday, ordinary) and sacred spaces (extraordinary) (Augé, 1993; Bremer, 2006).

The concept of everydayness is also key to defining tourism. Tourism is a journey that takes place outside one's usual place of residence. In the case of a journey to a sacred space, it implies a journey to an extraordinary place, clearly separated from the profane. Pilgrimage scholars consider the journey to a sacred site as a ritual, passing through the three phases identified by van Gennep (2019 (first published in French in 1909)): (i) separation; (ii) liminality; and (iii) incorporation (Schnell and Pali, 2013; Greenia, 2018).

In the same direction, sociologists and anthropologists have applied the pilgrimage model of study to tourism, understanding that the traveller follows a ritual behaviour in which he or she leaves home (separation ritual) to undertake a journey of discovery (liminality or initiatory death phase) to return to society with some benefits (transformation) (Graburn, 1989; Cohen, 1992; MacCannell, 1999).

In the separation phase, there is the call to visit a sacred site, the reason for starting a pilgrimage. This call can respond to various motives, among which there can be religious, spiritual or cultural reasons. To begin the journey, it is necessary to prepare not only physically (luggage or training for a long journey on foot) but also spiritually. Remember, for example, that many pilgrims in the Middle Ages made a will before leaving for the Holy Land. This phase of separation also implies moving away from everyday space, leaving one's comfort zone (Turner and Turner, 1989).

Sacred spaces have to be far from profane spaces (especially metaphorically speaking) because this journey, this separation, is necessary to prepare oneself for the encounter with divinity. Thus, one enters the phase of liminality, which means neither being in one place (physical or mental) nor in another, it is being on the threshold between something that has gone and something that is yet to come. It is the moment when one completely leaves the comfort zone to enter the sacred space and connect with divinity and transcendence. It is in this experience that the potential for transformation lies, everything that has been lived and experienced takes on a new meaning (Devereux and Carnegie, 2006; Beckstead, 2021).

However, the ritual is not over, as there is a final phase of incorporation. Once the transformation has taken place, it is necessary to reincorporate into society to return home. To return home transformed by this experience. This part is one of the least explored or studied, as there are few studies that tell us how religious tourists or pilgrims transformed by this experience have come to change things in their previous life.

Areas of Transformation

The transformation can take place in sacred sites or during the period of travelling to them. Transformation can be considered from three main perspectives: (i) the personal one; (ii) the social one; and (iii) the environmental one (Aulet, 2020b).

1. Personal transformation

The personal aspect encompasses everything that relates to the traveller and their experiences. Visiting sacred sites has the potential for significant personal development, and Gandhi's (1964) quote 'If you want to change the world, start with yourself' further reinforces this idea. Both religious tourism and pilgrimages can be driven by the individual quests of pilgrims. Numerous scholars have investigated the correlation between pilgrimage and spirituality, including Timothy and Olsen (2006), Haq and Wong (2009), Sharpley (2009) and Raj and Griffin (2015), among others.

The spiritual quest relates not only to religious tourism and pilgrimage, but it can also be extended to other typologies referring to people who seek out places and activities that can contribute to fulfilling their spiritual needs (Haq and Wong, 2009). Nevertheless, it is in the field of religious tourism and pilgrimages that has been most studied, being one of the main motivations for undertaking a transformative journey (Aulet, 2020b; Álvarez Orozco and Aulet, 2021). In fact, the separation phase starts with a call. When a pilgrim decides to undertake the journey there is a kind of inner call that pushes him or her; sometimes it is hard to explain or define. In some interviews with pilgrims on the Camino de Santiago, they mentioned that 'they felt it was the time to do it' (Aulet, 2020b).

According to certain philosophers, spirituality entails a quest for the meaning of life, a pursuit that all humans are inherently compelled to undertake since the essence of humanity is bound up with the spirit (Willson *et al.*, 2013). Spiritual consciousness is a pervasive and innate feature of human nature. The pursuit of spirituality has emerged as a prominent aspect of life in the late 20th century. It is a result of collective uncertainty and characterizes contemporary Western society as a spiritual revolution (Sharpley and Jepson, 2011).

According to Willson *et al.* (2013), spirituality influences the upper tiers of Maslow's hierarchy of needs by enabling personal growth and development, which is synonymous with self-fulfilment (Sánchez Sánchez and Fernández Herreira, 1996). Pigem (2015) suggests that self-fulfilment should serve as a guiding principle for individual and collective living: 'An economy in tune with life should focus not on the growth of monetary goods but on the growth of people and life: it should be an economy of self-realisation' (Pigem, 2015, p. 122).

It is widely acknowledged that the pursuit of spirituality today is strongly associated with the search for well-being, which is a multidimensional concept comprising various facets of human life (Bimonte and Faralla, 2012), not solely related to religious convictions. Spirituality can be defined as the quest for connectedness or connection with oneself, other human beings, the Earth and nature, and often with a higher power or transcendence (Schulz, 2005; Willson *et al.*, 2013; Raj and Griffin, 2015). For instance, the World Health Organization (WHO) defines health as a comprehensive state of physical, mental and social well-being, not merely the lack of illness or disease. In 2013, spiritual health was acknowledged as the fourth dimension among the physical, emotional and social dimensions. Despite tourism often being associated with disconnection, religious tourism enables individuals to find a moment of connection thereby promoting the quest for spirituality, inner peace and well-being.

On the other hand, managers of sacred sites need to be aware of this and should try to offer proper facilities to enrich this spiritual experience. For example, it is important to have: (i) silent places; (ii) the offer to participate in some rituals; and (iii) spaces for prayers. The management of sacred sites has been studied a great deal, and the main conclusion is that sometimes the balance between devotees and visitors is hard to find (Shackley, 2006; Raj and Griffin, 2015; Aulet *et al.*, 2019).

One of the earliest examples of pilgrimage routes is the Camino de Santiago, which has become one of the most thriving tourist offerings in Galicia, Spain (Santos, 2002). The Camino passes through five autonomous communities, spanning across 166 cities and towns, and encompasses over 1800 buildings of historical significance, representing the cultural heritage of the Middle Ages, particularly those from the Romanesque era. This is why the pilgrimage to Santiago de Compostela has received numerous accolades, such as being recognized as a European Cultural Route by the Council of Europe since 1987 and being listed as a United Nations Educational, Scientific and Cultural Organization (UNESCO) World Heritage Site since 1993. Despite the popularity of the Camino fluctuating throughout history, the pilgrimage has persisted over the years. Although it is challenging to determine the exact number of pilgrims, statistics are available from the Pilgrim's Office for those who request a Compostelana (certificate) upon completing the route. According to official data, the year with the highest number of visitors prior to the coronavirus disease 2019 (COVID-19) pandemic was 2019, which reached 347,000 pilgrims (surpassing the last Holy Year in 2010) (Oficina del Pelegrino, nd). Travel was restricted due to the pandemic to reduce the spread of the disease, so visitor numbers for Holy Year 2021 included the total numbers of pilgrims for 2 years, but even so, this number was considerably reduced by comparison with 2019. The year 2022 received the most pilgrims since available records began, with 438,000 (Oficina del Pelegrino, nd). In addition to the pilgrims that are counted, considerable numbers of individuals arrive annually in Santiago without passing through any official offices. These visitors choose to walk the Camino for a multitude of reasons, frequently not limited to religious motives.

The author interviewed pilgrims and pilgrims' organizations in 2019, walked the Camino in 2021 also interviewing pilgrims, and participated in the project promoted by Helsinki España (2021) 'Team Innovation' based on promoting the United Nations Sustainable Development Goals, tourism and human rights through working on innovative tourism products on the Camino de Santiago.

Among the testimonies collected, most of visitors were motivated by spiritual and/or personal transformation, but various levels of intensity were found. There are those who began the Camino with a clear spiritual motivation, a desire for personal transformation, to find themselves, to find God. There are also cases of pilgrims who start the Camino because they are looking for something, without knowing exactly what it is. However, having heard the testimony of other pilgrims who have changed their lives, they hope to have a similar experience.

In general, it can be said that pilgrims expect, directly or indirectly, to be transformed by the Camino; and most experiences at the end of the Camino are positive or incredibly positive. It is also true that there are no studies that have attempted to measure the extent of the transformation (i.e. whether it has a lasting effect over time). In fact, a considerable proportion of the pilgrims interviewed said that it was not the first time they had done the Camino, and many said that it was an experience they wanted to repeat in the future, so it could be that some kind of renewal of this experience or transformation is needed.

In a study by Aulet (2020b) of pilgrims on both the Camino de Santiago and the Ignatian Way or Camino Ignaciano, she also concluded that spiritual (not religious) motivations were particularly important, and that many of the pilgrims were already practising some kind of spiritual activity or exercise on a regular basis before going on the pilgrimage. For example, some of the pilgrims travel in an organized way, in groups, with their own trainers or spiritual guides who accompany them in activities such as meditation, yoga and others.

At the level of organization of the pilgrimage route, although some agencies and companies are offering packages with different activities, specific services along the route are lacking. Churches along the route are frequently closed (Fig. 5.1) and finding religious individuals to speak with can often be challenging. Volunteers are available in some instances, but the accompaniment they offer is often unorganized and unstructured. Upon reaching Santiago, attending the pilgrims' mass and following the fixed confessional schedule are viable options. Beyond this, overcrowding at some points along the Camino and the commercialization of certain sites may impact the experience for some pilgrims. However, it is worth noting that there are also those who feel the spiritual aspect falls short of their expectations.

Another of the pilgrimage routes in Spain is the Camino Ignaciano, that goes from Loyola

Fig. 5.1. Pilgrim praying in front of a closed church in the Camino de Santiago. Photo taken by the author.

(Basque Country) to the Holy Cave (Manresa, Catalunya) following the path of St Ignatius. The proposal was launched by the Jesuit community to prepare the big commemoration of the 500th anniversary of the transformation of the saint. Even though the numbers of visitors are considerably fewer compared with the pilgrims that follow the Camino (around 350 pilgrims/year), the proposal of this pilgrimage route is about transformation. It is pertinent to note that the Camino de Santiago and Camino Ignaciano share the same facilities and services. The route from Loyola to Manresa is identical, except for the direction of the pilgrimage. Unlike the Camino de Santiago, which proceeds from east to west, the Camino Ignaciano progresses from west to east (Abad Galzacorta and Guereño Ómil, 2016; Álvarez Orozco and Aulet, 2021).

One of the intriguing aspects of the life of St Ignatius is his personal metamorphosis. Ignatius was born into a noble family in Loyola in 1491. Following a career in the military, he sustained a leg injury from a cannonball in 1521, which required him to be bedridden. During this period, he read a few books about the lives of Christ and the saints, which deeply affected him. Therefore, having recuperated in 1522, he embarked on a pilgrimage to the Holy Land. His aim was to reach Barcelona to start his journey. During his pilgrimage, he visited several shrines including Nuestra Señora de Aránzazu and Montserrat. In Montserrat, remarkably close to Barcelona, he spent a night watching over the Virgin and conversing with a confessor. This significant moment in history involves the soldier's renunciation of his military life, symbolized by the abandonment of his sword. After leaving Montserrat, he received warnings that Barcelona was closed due to fears of the plague. He decided to remain and wait in Manresa, where he stayed in a Dominican convent and a hospice for the poor. During his stay, he intensified his prayers and engaged in penance, which led him to experience illumination and revelation. A transformation occurred without reaching the destination of his pilgrimage (the Holy Land), which he would attain years later (Tylenda, 2001).

One of St Ignatius's most important legacies are the Spiritual Exercises, a structured sequence of meditations and contemplations that he wrote based in his experience with the aim to assist others deepen their relationship with God (O'Brien, 2011). The Ignatian Way is not simply a route; it is offered as a path of transformation founded on the Spiritual Exercises. Therefore, the proposal is to walk the 650 km route in 4 weeks:

- The first stage runs from Loyola to Logroño and centres on the principle and foundation of one's own existence and the ultimate source of fulfilment.
- The second stage runs from Logroño to Zaragoza and focuses on calling upon the Eternal King to combat evil forces. The objective is to analyse the salvation history, with particular focus on Christ's birth, and illustrate how following His example can aid in the redemption of humanity.
- The third phase, spanning from Zaragoza to Fraga, concentrates on Christ's Passion and death. Thus, it is an apt time to re-evaluate the positive intentions established in the previous week and refine them through the experience of dying on the cross. It is like a difficult journey in the desert, which can also remind us of an initiatory death.
- The fourth stage, from Fraga to Manresa, focuses on the risen Christ appearing to the Virgin Mary as a symbol of a new beginning. This awakening marks the end of the journey where transformation takes place.

While it is true that individuals can walk at their own pace, many choose to do so with the guidance of instructors on the path. Iriberri and Lowney (2015) provide multiple examples of transformation. According to Álvarez Orozco and Aulet (2021):

> The spiritual impact of religious tourism on the Ignatian Way is evident in the transformative experience of pilgrims. Not everyone begins as a pilgrim; some embark as spiritual tourists concentrating on religious heritage, viewing it as a religious tourism experience (Cánoves, 2006). However, the journey transfigures them as they discern the objectives, and the spiritual quest starts to become meaningful as a sacred ritual (Liutikas, 2017). A spiritual journey involves a new path of self-exploration with emphasis on surrendering, accepting, and seeking true self-realisation and nature through spiritually, religiously, and culturally significant tourism.
>
> (Álvarez Orozco and Aulet, 2021, p. 39)

2. Social transformation

The gathering of people for religious purposes or the enrolment into pilgrimage routes can contribute to interreligious and intercultural dialogue. The social transformation refers to how the interaction between people in sacred sites can change their attitude. In tourism, encounters can take place in different contexts and depend on aspects such as the time when tourists and locals meet, the conditions (they share the same space as users, one is the customer and the other is the worker), the tourist season, cultural differences, etc. (Nash and Smith, 1991). Tourism should encourage encounter and dialogue between people of different origins, it being a useful tool to promote interculturality. Interculturality refers to developing relationships between disparate cultural groups, the equitable interaction among them, and the potential to establish fresh communal expressions (UNESCO, 2019). Panikkar (2004) explicitly highlights the concept of equitable interaction, as intercultural discussion can frequently be based on a situation of dominance from certain participants. Interactions between diverse cultures can sometimes result in subordination or the belief that one culture or idea is superior to another. This dynamic is particularly common in tourism, where unequal relationships between residents and tourists are not uncommon. Tourism may result in a novel manifestation of colonialism, where stereotypical images are propagated and localities become reliant on tourist revenue, which can alter and adulterate Indigenous conditions (Hall and Tucker, 2004; Jamal, 2019). In this case, intercultural dialogue becomes impracticable.

For fruitful dialogue to occur, one must explore the identity of the other, establish an exchange to enrich their own beliefs or traditions and become aware of the value and richness of diversity. Doing so leads to positive and beneficial sociocultural impacts of tourism for both visitors and residents. The experience of pilgrimage and the pilgrims' quest for inner peace can enhance a respectful attitude towards people and places, including local cultures, religions, traditions and heritage. This encounter between cultures should reinforce local cultures and heritage. Culture and cultural heritage play a key role in people's well-being, as well as bringing meaning to life (or travel), as affirmed by institutions and associations such as Europa Nostra. After researching different pilgrimage routes and interviewing those responsible for different shrines and temples, one of the conclusions is that pilgrims have the feeling that they never walk alone (Aulet, 2020b).

The studies conducted on the Camino indicate that pilgrims travel either alone or in groups, and some individuals travel solo but join an organized group (Aulet, 2020b). Throughout the stages of the route, a sense of goodwill and *communitas* arises, which fosters a commitment to helping one another. The personal transformation plus the sense of belonging to a bigger community of pilgrims can make them more aware and more respectful (Fig. 5.2).

A good example of how tourism can make a change in the social aspect is the company, MEJDI Tours. This company is a travel agency established in 2009 by a Palestinian and a Jewish person who shared a goal of using tourism to dispel stereotypes and encourage a culture of harmony and courtesy. As a result, they introduced the idea of a dual narrative and organized tours that were guided by two individuals showcasing distinctive perspectives. Guides play a vital role in facilitating personal and collective transformation of the group through active dialogue, which equips tourists with the necessary tools to achieve a deeper understanding. It is concerned with favourable interactions among individuals and societies, as another underpinning principle of this institution is respectable and upright tourism. Therefore, they endeavour to guarantee that the nearby communities are the principal recipients of the tourism industry (Sarah, 2020). This practice has also enabled them to coordinate religiously diverse journeys.

> MEJDI Tours brings the Torah, New Testament, and Qur'an to life with extraordinary interfaith group tours of Israel and Palestine unmatched by anything else. From classic Jewish, Christian, and Islamic destinations to MEJDI exclusives—homestays and home-cooked meals, discussions with religious leaders, peace builders, and so much more— MEJDI Tours works with group leaders to create unique interfaith group tours that not only educate, but connect travelers to the Holy Land and the people that live there.
>
> Only MEJDI Tours enables your group with the power to build intercultural bridges amongst

Fig. 5.2. Pilgrims walking the Camino. Photo courtesy of Guilherme Ferreira Rodrigues (used with permission).

> those that live in the Holy Land, strengthening your faith while forging lifelong bonds amongst participants.
>
> (MEJDI Tours, 2022)

3. Ecological transformation

Religious tourism has the potential for promoting more sustainable ways of travelling. Several sacred sites are based in natural environments, where a strong connection through symbolism is developed, as already mentioned in this chapter. It is not only about the environmental aspect, but also about providing meaningful spaces or to preserve the meaning and the spirit of the site (Shackley, 2006). The concept of spiritual sustainability also refers to the preservation of the symbolic and spiritual values of sacred spaces (Aulet and Duda, 2020). Interaction with natural spaces and sacred sites should be not only on a physical level but also in a deeper, spiritual context. This assertion is supported by a number of studies that have explored the spiritual and conscious dimension of experiencing natural places, such as wilderness areas, forests or the outdoors in general, where increased sensory perception and awareness lead to a meaningful experience (Sharpley and Jepson, 2011). Nature is a means to foster mindfulness through appreciative activities in the natural environment (Wang *et al.*, 2023), for example being high in the mountains is the most common source of spiritual feelings, conveying a sense of connectedness, smallness/insignificance in the face of the 'grandeur' of nature (Sharpley and Jepson, 2011).

The disruption of ecological balance, caused by the acceleration of current development models, has led to activities that focus on reconnecting with nature and the pursuit of inner harmony between human beings and the natural world (Bagri *et al.*, 2009). Pope Francis highlights the urgent need to care for our shared home, the Earth, in his encyclical *Laudato Si'* (Pope Francis, 2015). In fact, a clear connection between spirituality and consciousness and the present ecological crisis has been established by some authors. As early as 1969, Berger portrayed the current situation as one of uprootedness.

> Peter Berger has described development as the 'spreading condition of homelessness'. The creation of homelessness takes place both through the ecological destruction of the 'home' and the cultural and spiritual uprooting of peoples from their homes.
>
> (Mies *et al.*, 2014, p. 155)

We are heirs to a materialistic view that sees the Earth as a mere storehouse of resources and not as the Mother Earth of traditional Indigenous cultures. Despite this, there is still a 'spiritual geography' of the landscape, made up of unique, identitary, symbolic places, prehistoric or historic shrines, hermitages, pilgrimage routes, etc. with which many people still have strong ties. Many protected landscapes are significant to local communities: (i) they inspire respect and affection; (ii) they strengthen cultural identity; and (iii) they are essential for spiritual well-being. Most are related to sacred symbolism that is common to different religious traditions: elements such as mountains, stones, forests and trees or streams of water are considered sacred places (Eliade, 1965).

Nature and the environment offer a sense of inner peace and tranquillity that would explain, also, the increasing number of (young) people engaging in old pilgrimages such as the Camino de Santiago (Collins-Kreiner and Kliot, 2017) and new ones based on ecofeminism, goddesses or New Age (Fedele, 2013).

An interesting case to look at as an example is the Monastery of Poblet, in Catalunya, Spain. Poblet is a Cistercian monastery founded in 1150, whose monks have followed the *Rule of Saint Benedict* (St Benedict, 2008). It was closed in 1835 due to Spanish law but refounded in 1940 by Italian monks of the same Order. In 1991, the monastery was declared a World Heritage Site due to its austere and majestic architecture, the fortified royal residences, and the pantheon of the kings of Catalonia and Aragon (Fig. 5.3).

In 2007, the monastery began an ecological conversion process based on three axes (Aulet, 2020a):

1. Improve the management of all monastery facilities, buildings and properties by applying measures based on social and environmental justice criteria.
2. Inform and educate visitors interested in environmental values linked to spiritual principles.
3. Protect the quality of the rural and forest landscape, by improving its ecological integrity, silence and beauty.

To combat climate change, the primary focus is on reducing energy consumption, with an emphasis on substituting fossil fuels and other polluting energy sources with cleaner, renewable alternatives. Other measures involve improving water conservation, such as reducing household usage, harvesting rainwater, reusing grey water, and preventing pollution of surface and groundwater due to the application of non-organic cleaning agents or agricultural practices detrimental to the environment. The monastery's waste management system underwent improvement by implementing a selective waste-collection process, resulting in a significant reduction of non-organic waste that is subsequently recycled. In addition, all the organic fraction is utilized as fertilizer.

Fig. 5.3. Monastery of Poblet, Catalonia, Spain. Photo courtesy of Guilherme Ferreira Rodrigues (used with permission).

On the second aspect, various measures have been implemented over time to ensure a favourable reception for visitors. Despite the monastic lifestyle followed by the community in the monastery, visits are encouraged and visiting hours and an information centre are available. A downloadable audio-guide is obtainable during the visit, which assists guests in comprehending each room in the monastery and its significance. In 2021, an interpretation centre was inaugurated with an audiovisual presentation that emphasizes the monastery's spiritual aspects and its surroundings. The aim of this challenging endeavour is to raise visitors' awareness.

Additionally, the monastery played a key role in creating the Poblet natural protected area, which encompasses a vast woodland area extending beyond the monastery's confines. This measure has been crucial in preventing urban speculation and maintaining landscape identity (Torcal and Mallarach, 2008).

These measures form part of a sequence of modifications taking place at the premises of the Order. For instance, the General House of the Cistercian Order is currently engaged in several areas, including energy, water, waste management, hygiene products, food, mobility, garden maintenance, among others. According to Torcal (2003, p. 86): 'through these transformations, the Cistercian Order endeavours to follow in the footsteps of Pope Francis and contribute humbly'.

Summary

Since the COVID-19 pandemic, various health and lifestyle issues have arisen. Certain tourism academics have perceived the lockdown as an opportunity to shift towards a more sustainable model of tourism. None the less, upon scrutinizing tourism's progression over the previous year, there remains substantial room for improvement in ameliorating detrimental aspects of the industry. In this vein, religious travel and pilgrimages ought to serve as transformative and holistic experiences which foster more conscious modes of travel and living.

According to the WHO, health is defined as a state of complete physical, mental, and social well-being, not simply the absence of disease or infirmity. Spiritual health is also recognized as a dimension alongside physical, emotional and social health. Some health experts propose a new paradigm that demands a more holistic approach, emphasizing inclusive narratives, promoting social trust and fostering meaningful lives. Religious tourism and pilgrimage have the potential to play a significant role. Recently, Health Tourism Worldwide and the Wellness Tourism Association have released a wellness travel report. According to representatives in the industry, there is substantial growth potential in spirituality and nature (Wellness Tourism Worldwide, 2020).

As evidenced, visiting sacred sites while travelling holds immense potential for transformation as it involves a type of ritual. This shift is not only occurring at a personal level but also can contribute to the improvement of social and environmental realities. At the current time, there is an urgent need for such improvements but it is also a challenge. Observing younger generations connecting with pilgrimage routes is a positive aspect that can assist in creating more dedicated communities in the future.

This chapter demonstrates the connection between religious tourism and three types of transformation: (i) personal; (ii) social; and (iii) environmental. However, this connection must be further investigated and researched to propose changes in tourism that can result in a transformation of not only the tourism model but also our way of life. Instead of focusing solely on smart destinations, we should consider becoming wiser (religious) travellers.

As academics, it is important to consider a holistic view of reality. Interdependence is a crucial concept when attempting to comprehend the world. To enhance our comprehension, academics have segmented reality into various disciplines. This fragmentation is similar to how an art expert concentrates on individual details and techniques, in addition to the painting as a whole. However, it is crucial to bear in mind the complete picture. As proposed by Pritchard *et al.* (2011), academics should advance towards genuinely interdisciplinary models that provide diverse perspectives on reality and allow research to progress accordingly. Furthermore, academics require a transformation to tackle present challenges.

References

Abad Galzacorta, M. and Guereño Ómil, B. (2016) Las necesidades del peregrino Ignaciano: percepciones de una experiencia. *International Journal of Scientific Management Tourism* 2, 9–25.

Álvarez Orozco, M.A. and Aulet, S. (2021) Análisis del impacto espiritual del turismo religioso en la experiencia turística: el caso del Camino Ignaciano. *Rotur, Revista de Ocio y Turismo* 15(2), 24–44. DOI: 10.17979/rotur.2021.15.2.7580.

Augé, M. (1993) *Los no Lugares: Espacios del Anonimat: Una Antropología de la Sobremodernidad*. Gedisa, Barcelona, Spain.

Aulet, S. (2020a) Ecological reconversion to fight climate change: the case of the Monastery of Poblet. In: *Proceedings of Climate Change, World Heritage and Tourism. 10e séminaire de la Chaire UNESCO et du Réseay UNITWIN-UNESCO 'Culture, Tourisme, Développement'*. Université La Sorbonne, Paris, pp. 99–114.

Aulet, S. (2020b) Pilgrim's motivations: a theoretical approach to pilgrimage as a peacebuilding tool. *International Journal of Religious Tourism and Pilgrimage* 8(6), 60–75. Available at: https://arrow.tudublin.ie/ijrtp/vol8/iss6/7 (accessed 1 September 2023).

Aulet, S. and Duda, T. (2020) Tourism accessibility and its impact on the espiritual sustainability of sacred sites. *Sustainaiblity* 12(22): 9695. DOI: 10.3390/su12229695.

Aulet, S. and Sureda, M. (2019) The semantics of the sacred: a tool for interreligious dialogue. In: Vidal-Casellas, D., Aulet, S. and Crous-Costa, N (eds) *Tourism, Pilgrimage and Intercultural Dialogue. Interpreting Sacred Stories*. CAB International, Wallingford, UK, pp. 14–28.

Aulet, S., Altayó, J. and Vidal-Casellas, D. (2019) Managing visitors at sacred sites: the case of Montserrat. In: Griffiths, M. and Wiltshier, P. (eds) *Managing Religious Tourism*. CAB International, Wallingford, UK, pp. 47–64.

Bagri, S., Gupta, B. and George, B. (2009) Environmental orientation and ecotourism awareness among pilgrims, adventure tourists, and leisure tourists. *Tourism: an International Interdisciplinary Journal* 57(1), 55–68.

Beckstead, Z. (2021) On the way: pilgrimage and liminal experiences. In: Wagoner, B. and Zittoun, T. (eds) *Experience on the Edge: Theorizing Liminality. Theory and History in the Human and Social Sciences*. Springer, Cham, Switzerland, pp. 85–105.

Berger, P. (1969) *A Rumour of Angels: Modern Society and the Rediscovery of the Supernatural*. Doubleday and Co., New York.

Bimonte, S. and Faralla, V. (2012) Tourist types and happiness a comparative study in Maremma, Italy. *Annals of Tourism Research* 39(4), 1929–1950. DOI: 10.1016/j.annals.2012.05.026.

Bremer, T.S. (2006) Sacred spaces and tourist places. In: Timothy, D. and Olsen, D. (eds) *Tourism, Religion and Spiritual Journeys*. Routledge, Abingdon, UK, pp. 25–35.

Burckhardt, T. (2000) *Principios y Métodos del Arte Sagrado*. José J. de Olañeta, Palma de Mallorca, Spain.

Cánoves, G. (2006) Turismo religioso en Montserrat: Montaña de fe, montaña de turismo. *Cuadernos de Turismo* 18, 63–76.

Cohen, E. (1992) Pilgrimage centers: concentric and excentric. *Annals of Tourism Research* 19, 33–50. DOI: 10.1016/0160-7383(92)90105-X.

Collins-Kreiner, N. (2010) Researching pilgrimage. *Annals of Tourism Research* 37(2), 440–456. DOI: 10.1016/j.annals.2009.10.016.

Collins-Kreiner, N. and Kliot, N. (2017) Why do people hike? Hiking the Israel National Trail. *Tijdschrift Voor Economische En Sociale Geografie* 108(5), 669–687. DOI: 10.1111/tesg.12245.

Devereux, C. and Carnegie, E. (2006) Pilgrimage: journeying beyond self. *Tourism Recreation Research* 31(1), 47–56. DOI: 10.1080/02508281.2006.11081246.

Eliade, M. (1965) *Le Sacré et le Profane*. Éditions Gallimard, Paris.

Fedele, A. (2013) *Looking for Mary Magdalene: Alternative Pilgrimage and Ritual Creativity at Catholic Shrines in France*. Oxford University Press, Oxford.

Gandhi, M. (1964) *The Collected Works of Mahatma Gandhi*, Volume XII. The Publications Division, Ministry of Information and Broadcasting, Government of India.

Graburn, N. (1989) Tourism: the sacred journey. In: Smith, V.L. (ed.) *Hosts and Guests: the Anthropology of Tourism.* University of Pennsylvania, Philadelphia, Pennsylvania, pp. 21–36.

Greenia, G.D. (2018) What is pilgrimage? *International Journal of Religious Tourism and Pilgrimage* 6(2): 3. Available at: https://arrow.tudublin.ie/ijrtp/vol6/iss2/3 (accessed 1 September 2023).

Guénon, R. (1995) *Símbolos Fundamentales de la Ciencia Sagrada*. Paidós, Barcelona, Spain.

Hall, M. and Tucker, L. (2004) *Tourism and Postcolonialism*. Routledge, London.

Haq, F. and Wong, H.Y. (2009) Is spiritual tourism a new strategy for marketing Islam? *Journal of Management, Spirituality and Religion* 6(2), 141–156.

Helsinki España (2021) Proyecto TEAM INNOVATION 2021 "ODS, Turismo Sostenible y Derechos Humanos: Innovaciones y Trabajos del Futuro en el Camino de Santiago". Available at: https://helsinkiespana.org/proyecto-team-innovation-2021-ods-turismo-sostenible-y-derechos-humanos-innovaciones-y-trabajos-del-futuro-en-el-camino-de-santiago/ (accessed 14 November 2023).

Iliev, D. (2020) The evolution of religious tourism: concept, segmentation and development of new identities. *Journal of Hospitality and Tourism Management* 45, 131–140. DOI: 10.1016/j.jhtm.2020.07.012.

Iriberri, J.L. and Lowney, C. (2015) *El Camino Ignaciano*. Ediciones Mensajero, Bilbao, Spain.

Jackowski, A. and Smith, V.L. (1992) Polish pilgrim-tourists. *Annals of Tourism Research* 19, 92–106. DOI: 10.1016/0160-7383(92)90109-3.

Jackson, R.H. and Hudman, L. (1995) Pilgrimage tourism and English cathedrals: the role of religion in travel. *The Tourist Review* 50(4), 40–48. DOI: 10.1108/eb058206.

Jamal, T. (2019) *Justice and Ethics in Tourism*. Routledge, Abingdon, UK.

Liutikas, D. (2017) The manifestation of values and identity in travelling: the social engagement of pilgrimage. *Tourism Management Perspectives* 24, 217–224. https://doi.org/10.1016/j.tmp.2017.07.014

MacCannell, D. (1999) *The Tourist: a New Theory of the Leisure Class*. University California Press, Oakland, California.

MEJDI Tours (2022) Jerusalem: Dual Narrative Tour. For people who believe that both sides of the story matter. Available at: https://mejditours.com/day-tours/ (accessed 1 September 2023).

Mies, M., Shiva, V. and Salleh, A. (2014) *Ecofeminism*. Bloomsbury Publishing, Sidney, Australia.

Nash, D. and Smith, V.L. (1991) Anthropology and tourism. *Annals of Tourism Research* 18(1), 12–25. DOI: 10.1016/0160-7383(91)90036-B.

O'Brien, K. (2011) *The Ignatian Adventure: Experiencing the Spiritual Exercises of St. Ignatius in Daily Life*. Loyola Press, Chicago, Illinois.

Oficina del Pelegrino (nd) Estadísticas antiguas del Camino de Santiago. Available at: https://oficinadelperegrino.com/estadisticas/ (accessed 13 September 2023).

Olsen, D.H. (2019) The symbolism of sacred space. In: Vidal-Casellas, D., Aulet, S. and Crous-Costa, N. (eds) *Tourism, Pilgrimage and Intercultural Dialogue: Interpreting Sacred Stories*. CAB International, Wallingford, UK, pp. 29–42.

Olsen, D.H. and Timothy, D.J. (eds) (2022) *Routledge Handbook of Religious and Spiritual Tourism*. Routledge, Abingdon, UK.

Ostrowski, M. (2000) Pilgrimages or religious tourism? *Peregrinus Cracoviensis* 10, 53–61.

Otto, R. (1965) *Lo Santo: Lo Racional y lo Irracional en la Idea de Dios*. Alianza Editorial, Madrid.

Panikkar, R. (2004) *Pau i interculturalitat, una Reflexió Filosòfica*. Proa, Barcelona, Spain.

Pardellas, X. (2005) *Turismo Religioso: O Camiño de Santiago*. Universidade de Vigo, Vigo, Spain.

Pigem, J. (2015) *Intelligència Vital. Una Visió Postmaterialista de la Vida i la Consciència*. Kairos, Barcelona, Spain.

Pope Francis (2015) *Laudato Si'*. Edizioni p emme, Rome.

Pritchard, A., Morgan, N. and Ateljevic, I. (2011) Hopeful tourism: a new transformative perspective. *Annals of Tourism Research* 38(3), 941–963. DOI: 10.1016/j.annals.2011.01.004.

Raj, R. and Griffin, K.A. (eds) (2015) *Religious Tourism and Pilgrimage Management: an International Perspective*. CAB International, Wallingford, UK.

Ries, J. (1989) *Lo Sagrado en la Historia de la Humanidad*. Encuentro, Madrid.

Rinschede, G. (1992) Forms of religious tourism. *Annals of Tourism Research* 19, 51–67. DOI: 10.1016/0160-7383(92)90106-Y.

Sánchez Sánchez, A. and Fernández Herrera, A. (1996) *Dimensiones de la Educación Para la Paz: Teoría y Experiencia*. Eirene, Granada, Spain.

Santos, X.M. (2002) Pilgrimage and tourism at Santiago de Compostela. *Tourism Recreation Research* 27(2), 41–50. DOI: 10.1080/02508281.2002.11081219.

Sarah, A.A. (2020) *Crossing Boundaries: a Traveler's Guide to World Peace*. Berrett-Koehler Publishers, Oakland, California.

Schnell, T. and Pali, S. (2013) Pilgrimage today: the meaning-making potential of ritual. *Mental Health, Religion and Culture* 16(9), 887–902. DOI: 10.1080/13674676.2013.766449.

Schulz, E.K. (2005) The meaning of spirituality for individuals with disabilities. *Disability and Rehabilitation* 27(21), 1283–1295. DOI: 10.1080/09638280500076319.

Severino Croatto, J. (2002) *Experiencia de lo Sagrado y Tradiciones Religiosas*. Verbo Divino, Estella, Spain.

Shackley, M. (2006) Empty bottles at sacred sites: religious retailing at Ireland's national shrine. In: Timothy, D. and Olsen, D. (eds) *Tourism, Religion and Spiritual Journeys.* Routledge, Abingdon, UK, pp. 94–103.

Sharpley, R. (2009) Tourism, religion and spirituality. In: Jamal, T. and Robinson, M. (eds) *The SAGE Handbook of Tourism Studies*. Sage, London, pp. 237–253.

Sharpley, R. and Jepson, D. (2011) Rural tourism: a spiritual experience? *Annals of Tourism Research* 38(1), 52–71. DOI: 10.1016/j.annals.2010.05.002.

St Benedict (2008) *The Rule of St Benedict* Edited and translated by White, C. Penguin Classics, London.

Timothy, D. and Olsen, D. (2006) *Tourism, Religion and Spiritual Journeys.* Routledge, Abingdon, UK.

Torcal, L.M. (2003) La transformació ecològica de la Casa General de l'Ordre Cistercenc. *Questions Vida Cristiana* 276, 77–86.

Torcal, L.M. and Mallarach, J.-M. (2008) The monastic landscape of Poblet: a place where spirituality, culture and nature join hands. In: Mallarach, J.-M. (ed.) *Protected Landscapes and Cultural and Spiritual Values. Volume* 2. International Union for Conservation of Nature (IUCN), GTZ and Obra Social de Caixa Catalunya, Barcelona, Spain, pp. 80–91.

Turner, V. and Turner, E.L.B. (1989) *Image and Pilgrimage in Christian Culture.* Columbia University Press, New York.

Tylenda, J.N. (2001) *A Pilgrim's Journey: the Autobiography of Ignatius of Loyola*. Ignatius Press, San Francisco, California.

United Nations Educational, Scientific and Cultural Organization (UNESCO) (2019) *International Decade for the Rapprochement of Cultures (2013–2022)*. Available at: https://en.unesco.org/internationaldecaderapprochementofcultures (accessed 1 September 2023).

van Gennep, A. (2019) *The Rites of Passage*, 2nd edn [first edition 1909, in French]. Translated by Vizedom, M.B. and Caffee, G.L. University of Chicago Press, Chicago, Illinois.

Vukonic, B. (1996) *Tourism and Religion*. Pergamon Press, London.

Wang, S., Blasco, D., Hamzah, A. and Verschuuren, B. (2023) Tourists and 'philosophers': nature as a medium to consciousness and transcendence in spiritual tourism. *Annals of Tourism Research* 99: 103543. DOI: 10.1016/j.annals.2023.103543.

Wellness Tourism Worldwide (2020) *4WR: Wellness for Whom, Where and What? Wellness Tourism 2020 Full Report*. Available at: https://www.globalwellnesssummit.com/wp-content/uploads/Industry-Research/Global/2011-wellness-tourism-worldwide-wellness-for-whom.pdf (accessed 14 November 2023).

Willson, G.B., McIntosh, A.J. and Zahra, A.L. (2013) Tourism and spirituality: a phenomenological analysis. *Annals of Tourism Research* 42, 150–168. DOI: 10.1016/j.annals.2013.01.016.

6 Change from Within: Pilgrimage as a Method of Personal and Political Integration in Response to Crisis

Stephen F. Haller*
Wilfrid Laurier University, Brantford, Ontario, Canada

Abstract

The chapter critically evaluates modern pilgrims who no longer view themselves as sinners in need of redemption. To varying degrees, they are: (i) spiritually alienated; (ii) psychologically distressed; or (iii) suffering a loss of meaning in their lives. A substantial number of modern pilgrims feel the dissonance between their self and their society. According to philosophers Kierkegaard and Plato, the solution to alienation and a divided self is an inner transformation of character rather than a political solution. Travel and pilgrimage can spark this transformation, for they are conducive to peak experiences, which connect one with existence itself, and to reflection on one's individual role in community and global challenges. Pilgrimage can heighten one's awareness of one's connection to other pilgrims, to communities and to nature. This chapter further explores the concept of 'change from within' as a deeply personal response to the uncertainties of our times, emphasizing that its value should not be dismissed based on narrow consequentialist judgements. Instead, it proposes that such inner transformation can play a vital role in addressing personal, social and political challenges in the context of crisis.

Introduction

Spiritual reflection is often triggered when one is faced with a sense of a divided self. When events in our lives, in our culture, in our politics do not unfold as all our previous experience might have led us to assume that they would, we sense that something is out of balance between our personal goals and the communities we live in. We seek to find a way to live, as an individual, in a world where familiar patterns of events are disrupted, and the trust in institutions to address global problems is seriously eroded. The climate is changing in undesirable ways; irrational wars seem to erupt for no reason; biodiversity loss and extinction threaten the stability of the biosphere; and global pandemics are an omnipresent threat. One might despair, as the Spanish philosopher of religion, Unamuno, once wrote, and believe that they are 'sailing, perhaps without a set course, on an unstable and sinkable raft' (Unamuno, 1954 [1921], p. 262).

Our consciences ask each of us whether there is anything about our lives that we could do in reaction to the situation. The deviation during events from our former expectations leads us to review the choices we make, and the connection between our individual choices and the collective good. Some may try out programmes, or religions and lifestyles that are novel to them. One might turn to Buddhism, while another commits to a vegan lifestyle.

*shaller@wlu.ca

DOI: 10.1079/9781800623675.0006

There is much disagreement about what to do, and what politics should be supported. The world offers no agreed-upon foundations to stand on when contemplating the complex situation. Everyone is faced with questions that cannot be evaded, and that no one else can decide for them. This is a spiritual contemplation about integrating oneself with their actions in the world. Rosemont, following the philosopher Ludwig Wittgenstein, defines a spiritual experience as a feeling of absolute safety and a sense of 'fully belonging ... to those that have preceded us, those in whose midst we live, and those who will follow us' (Rosemont, 2001, p. 31). He argues that the world's religions offer many different routes to this same sense of unity with the world, its past and future. They provide direction on 'how to live an integrated life' (Rosemont, 2001, p. 30).

Dante's *Inferno* can be interpreted in terms of this struggle between self and society (Sayers, 1949). Sayers (1949) explains that the poem is an allegory about an *individual* soul in a corrupt *society* of Florence. Today, we search for a meaningful, empathetic, practical response about how to respond to global crises. Rosemont argues that we will need both political change and religious commitment (Rosemont, 2001, p. 30). However, an individual could give up hope of finding political solutions. If so, then an individual might wonder whether there is any point to their own individual efforts to confront the situation; or whether those efforts would be futile. Dante's solution is an inner transformation of character rather than a political solution (Fig. 6.1).

This chapter will survey a few philosophers' views on how an individual's reaction to crises often results in a divided self: that is, a self where one's personal values are not coordinated with their community behaviour. Next, the chapter will explain that political solutions to this problem are difficult to achieve in practice because of 'community-action problems' where consequential reasoning leads to a failure to make sense of the benefits of cooperation. The best response to this problem is to find, like Dante, an inner transformation where one achieves a unity within oneself by doing the right thing simply because it is what you value, and not because one has calculated positive expected results. Once again, a survey of a few philosophers who defend this idea will be explained. Finally, the chapter will suggest that modern-day pilgrimage is a suitable means for inner transformation. Modern pilgrims have divided selves, and, to varying degrees, are: (i) spiritually alienated; (ii) psychologically distressed; or (iii) suffering malaise, or a loss of meaning in their lives. The experience of pilgrimage is conducive to inner self-focus, as well as participation in a community. It is an effective way to find harmony between self and society.

Crises in Various Times Have Created Divided Selves

The Norwegian philosopher, deep ecologist and mountaineer, Naess considered the uncertain times of his generation, the second half of the 20th century, and wrote that the 'approaching ecocatastrophe' can make people 'sad' and 'passive' (Naess, 2008, p. 125). This, he thinks, is the result of a lack of 'maturity and integration' (Naess, 2008, p. 125). To be 'integrated' is to find a way to synchronize one's personal desires with our actions in the world. A mature, integrated person would 'partake in the life of the community ... with joy, and within the framework of his or her value priorities' rather than from an external sense of duty (Naess, 2008, p. 123). To illustrate the divided self, he describes many of his fellow environmentalists. While they might speak articulately about their academic beliefs, which value the natural world as a source of beauty, joy and meaning, their lifestyles do not always include much joyful time and experience in nature. Thus, he suspects, there is an inconsistency in what they say and what they do. They do not 'walk the talk'. Their personal and political selves are divided. They do not show, by their life example, that they are whole and unified in their engagement with the world.

Fromm tried to diagnose the ills of the mid-20th century, such as war and authoritarianism, and collected them all under the term 'alienation' (Fromm, 1962, pp. 26, 31, 49, 56). He argued that this alienation was rooted in the 'chains of illusion' that keep a person identified with 'objects outside' oneself, and hence enslave a person to only part of oneself, for instance 'money, [or] power' (Fromm, 1962, p. 56). For

Midway upon the journey of our life / I found myself within a forest dark, /
For the straightforward pathway had been lost.

Inf. I, lines 1–3

Fig. 6.1. Gustave Doré's 1857 illustration for Dante's *Inferno*. From Wikimedia Commons, the free media repository. Available at: https://commons.wikimedia.org/wiki/Category:Gustave_Dor%C3%A9_-_Inferno (accessed 20 November 2023).

example, as alienated consumers, or sellers, we relate to the world in one way only (Fromm, 1962, p. 51). Our possessions distract us from facing deeper questions.

Thus, both authors, Naess (2008) and Fromm (1962), argue the need for integration of the total personality. This integration will occur when one's work, and life, is *its own end*, and not merely a means to something external to them. He hopes that, and explains how a person might 'overcome the split between himself and the world outside of him in order to arrive at the experience of unity and oneness with his fellow man and with nature' (Fromm, 1962, p. 174). This will require a recognition of the facts of the world, and an acceptance of the limitations of one's own psychological makeup. None the less, even within these determined constraints, one can find the freedom to relate with the world in a life-affirming way. Most people do not make a choice at all, he laments. Instead of choosing between life or death, they distract themselves by keeping busy with external concerns (Fromm, 1962, p. 175).

The early 19th-century Danish philosopher of religion, Kierkegaard, also emphasized the need to avoid distractions and evasions, which keep one from achieving self-unity by making difficult choices about existential issues. He writes: 'For he who is not himself a unity is never really anything wholly and decisively; he only exists in an external sense' (Kierkegaard, 1938, p. 184). Kierkegaard's answer to this problem is expressed in his maxim to 'will one thing', and this central concept of inner transformation will be discussed in more detail in section 'Change from Within' later in this chapter.

'Change from within' or 'change from without'?

Koestler, in his *Yogi and the Commissar* (1945), described the dilemma of choosing how to improve things in our world as a choice between 'change from within' or 'change from without'. The 'Yogi' in Koestler's metaphor, believes that one can improve things only by 'individual effort from within' and not by 'legislation' (Koestler, 1945, p. 10). However, Koestler rejects this path of change from within because it leads to passivity in the face of evil. The 'Commissar', on the other hand, is one that believes that the only way to effect change is a 'reorganization of the system' from 'without' (Koestler, 1945, p. 9). The commissar gets results through top-down political solutions and force. Koestler also rejects this path because authority is too often abused. Unfortunately, he argues, there can be no workable compromise between these two poles of the saint and the revolutionary (Koestler, 1945, p. 10).

Limit situations, choices and identity

Koestler's (1945) ambiguous conclusion (described in the previous section), however, is of no help to those who must choose one of these paths as a means of healing the divided self. Some choices are existential argue Kierkegaard (1944) and Sartre (Detmer, 2008). That is, they are not merely about efficiency, or facts, or means-ends calculations; but rather, they are choices that determine our very identity, and what kind of persons we choose to be. They are about taking responsibility for a decision and not merely letting things happen.

Existential psychologist and philosopher, Jaspers, wrote about decisions in the face of a crisis that he called a 'limit' (or 'border') situation (Bakewell, 2016). These are the momentous decisions one faces when living outside the boundaries of ordinary existence. Sarah Bakewell describes them as: 'Moments when one finds oneself constrained ... by what is happening, but at the same time pushed by these events towards the limits or outer edge of normal experience' (Bakewell, 2016, p. 82). Jaspers was thinking of the 'limit situation' of his friends living under Nazi rule. Early on, people were faced with life-and-death decisions of whether to flee the country or stay a bit longer to see what happens. One's natural psychological tendency, he argues, is to avoid the decision and continue as if everything were normal for as long as possible. People cannot quite bring themselves to believe that a disaster is happening. Jaspers suggests that they did not expect these events, but rather expected things to go on as they always had in their previous experience. Jaspers contends that they think they can tolerate it until it is over – hoping things will improve. Unfortunately, this response of inertia is not what the situation under Nazism demanded. Some people saw the coming dangers, and, if they had the means, fled Germany and Austria right away, realizing 'that life could not continue unaltered' (Bakewell, 2016, p. 86). They were 'open' to what the situation demanded.

Humanity is now in a global 'limit situation' with respect to the global climate and its planetary life-support systems. Wars, pandemics, species extinction and the rise of authoritarian politics are among the contributors to the crisis we are in today. This is testing our moral identities and demanding each of us to open our minds to the revealed realities of our situation. Just as in Jaspers' analogous context about the rise of Nazism (Bakewell, 2016), recent warnings about the scope of the problem are difficult to take in. It is hard to absorb the fact that it is happening, even if one knows, intellectually, that it is. Life experience leads many people to expect things to go on as they always have before. Many people cannot grasp the proportions of the crisis; and, if they do, they might not see the urgency to act. Many people evade the question with vague hopes of some technical solutions that will allow us to continue our lives unchanged.

Community versus Individual Action Problems

As suggested, above, to respond to this problem might need both political and religious inner transformation. However, the political solutions can be impeded by our divided selves. The consistency one seeks between personal desires and social responsibilities is often frustrated by a deep philosophical problem known as the 'community-action problem'. Community-action problems are those where the self-centred actions of many individuals result in consequences that, paradoxically, make everyone in society, including those very same individuals, worse off. It is difficult to escape this self-defeating dilemma because the consequentialist logic in favour of self-centred behaviour is strong.

Community-action problems can be found in Hobbes' important political theory of the 'social contract' described in his *Leviathan* (Hobbes, 1968 [1651]). His starting assumption is that human nature is egoistic and motivated solely by self-interest. Everyone attempts to get as much benefit as possible for themselves. Given these drives of human nature, the only way to make sense of duties to others is in terms of a social contract, where people can be motivated to cooperate with others for the benefit of the community *only* in cases where they can perceive that it is in their own enlightened self-interest that calculates the mutual benefits that would come back to themselves. In these cases, cooperation with others is necessary to avoid a competitive life of all against all – which would make life worse for everyone. However the incentive to work with others evaporates – if one can get away with it.

Today, this type of problem is researched mathematically using something known as 'game theory' (Axelrod, 1984). Game theory explores decision making in the context of other people making decisions. In many situations, one's decision will depend upon what other people do. The counter-intuitive insight gained from game theory problems is that it is *never rational* for a self-interested individual to cooperate with others. The thinking is expressed in terms of a dilemma.

On the one hand, *if* everyone else *is* already cooperating for mutual benefit, then one's own personal choice to not cooperate will not make any significant difference to the overall end result, but it will provide a big benefit to that individual who has opted out. Other individuals will reason the same way and conclude that there is no point to their own cooperation. Thus, there will be no cooperation, and everyone loses. Conversely, *if* many people are *not* cooperating, then there is no point in my cooperating either, since the mutual benefits will not materialize for *anyone*, and one's own personal efforts will be a wasted effort.

When everyone runs through this same logical reasoning, then no one sees a reason to cooperate with others, and this results in an outcome that is less desirable for all than the outcome that could be achieved through cooperation. Again, the paradoxical result is that the self-centred actions of many individuals result in consequences that make everyone, including those very same individuals, worse off. Many social problems, including war, traffic congestion, family squabbles about chores, nuclear arms races, vaccination hesitancy and environmental degradation, can be interpreted as resulting from this kind of failure to cooperate (Heath, 2001). These are the kinds of problem that create the alienation between self and society that was described above.

The political solution to this problem, provided by Hobbes, is a 'social contract' that *forces* everyone into cooperation. In Hobbes' solution, the enforcement by a political leader is critical. Rational, self-interested individuals will only cooperate if they can be *guaranteed* that others will cooperate as well. Hobbes notes, famously, that 'Covenants, without the sword, are but Words' (Hobbes, 1968 [1651], p. 223).

The problem for us in the 20th century, of course, is that such enforcement is often lacking, or impossible, in many real-life situations. One can fine a citizen of a particular country for littering, for example, but there is no international police force capable of enforcing a no-litter policy on the world's oceans. A government can try to 'force' a reduction in carbon gas emissions by taxing citizens on their use of fossil fuels. However, there is no international police force guaranteeing that other nations will do the same. And if they don't, then what would be the point of personal 'sacrifices' if it makes little effective difference to the fate of the Earth's climate? Thus, most people find it impossible to behave in socially responsible ways, that they themselves might agree with, when others are not doing so as well.

Several *indirect* methods of enforcing the social contract may *sometimes* fill the gap. For example, it often pays to have a good reputation for caring about others. Interactions among people who know, and frequently interact with each other, face the same cooperation dilemmas, with the same people, repeatedly. In such contexts, it pays, for purely self-interested reasons, to develop a good reputation as a community-minded actor, so as not to be excluded from future cooperative activities.

Other indirect methods of enforcement may have developed through evolution by natural selection (Axelrod, 1984). It seems that incentives to cooperate have been internalized in humans and other primates. For example, cooperation can be ensured through means of sociobiological enforcement mechanisms, including guilt, aggression and friendship.

But in the context of a global crisis, such local, indirect enforcement mechanisms are absent and/or ineffective. There is no global police force ensuring that all people behave justly; and it is difficult to feel personally connected to distant, abstract, global problems. In the absence of enforcement, we seem doomed to self-destruction by individuals unwilling to risk being 'suckered' by others who aim to 'free-ride'. It does not seem 'rational' to make any sacrifice of one's own well-being if it really makes little or no difference to effectively solving a problem – and it will not if enough others are not also doing their share.

Given this reasoning, it is difficult to know how to respond to the world in a way that achieves our goal of finding harmony and integration between our personal actions and community responsibilities. Instead, people end up feeling alienated, fragmented and divided from society.

Change from Within

The wisdom of philosophers from previous generations can guide our response to this dilemma of a divided self in our own uncertain times. It was argued above that the best reaction to alienation and uncertainty is to seek a perspective that confirms a consonance between our personal beliefs and actions, with our commitments to the communities to which we belong. In our life choices one should seek to regain a sense of harmony between our personal selves and our world. Remember that it is this sense of seeking consistency between personal beliefs and actions, and one's social responsibilities and place in the community, which is a spiritual transformation providing wholeness and connection with the world outside ourselves. It might restore the sense of 'fully belonging' that Rosemont (2001, p.31) refers to (quoted in the 'Introduction' above), when defining spiritual experiences.

Plato (1974), Fromm (1962) and Kierkegaard (1938) reject external political solutions to these problems and defend the 'internal' attitude rejected by the logic of game theory. All three authors, Plato (1974), Fromm (1962) and Kierkegaard (1938), argue that one should do the right thing, the just thing, for its own sake, and not because of the consequential ends in mind. Like Fromm, who argued, as explained above, that individuals can overcome feelings of 'alienation' by resisting the 'illusion' of identifying with 'objects outside' themselves (Fromm, 1962, pp. 26, 31, 49, 56), Plato and Kierkegaard also argue that one should be concerned with the workings of the internal 'soul', rather than consequences in the external world. In game theory language, they are urging that one should go ahead and do the right thing, regardless of whether this makes one a sucker in the community cooperation game. There are bigger issues afoot – like what Kierkegaard might call facing 'eternity' (Kierkegaard, 1938, p. 198).

The basic outline of this dilemma between self-interested action and social duties was described in Plato's *Republic*, where Socrates is challenged to defend the idea that one should act justly, fairly and virtuously – for that sake alone – regardless of any rewards or disadvantages to oneself that might result from that action (Plato, 1974). It was assumed, in Plato's dialogue, that most people would think otherwise. Why would anyone care about anything more than their own well-being, and that of those in their own immediate circle of friends and family? Socrates, in Plato's dialogue, responds to this challenge by arguing that there is no actual conflict between self-interest and duty. A 'virtuous' soul will always do the just thing, and that the community-directed interests of justice are not inconsistent with the self-directed interests of possessing a balanced soul. The bottom line is the simple assertion that those who 'do the right thing' for its own sake, without regard to rewards and costs, live better lives because they have more 'harmonious' souls. They live with a sense of well-being that comes when one is satisfied that they have not behaved unjustly.

Fromm, too, argued that a fulfilled, psychically healthy person will do 'what is right and good' not from obedience to external rules, but rather, internally 'by developing a taste' for it (Fromm, 1962, p. 176). Like Plato, he realized that just action must be motivated from within.

Kierkegaard also focuses on the individual's responsibility for their own choices, and not on calculations of cost and benefit. He is a philosopher of religion who asks his readers direct and demanding questions, such as 'What kind of life do you live?' (Kierkegaard, 1938, p. 182) and whether you can defend your life against 'eternity' (Kierkegaard, 1938, p. 198). Ask yourself, he urges, whether you are living a life 'of honor' (Kierkegaard, 1938, p. 196).

He summarizes this philosophy in his famous aphorism that one should live 'willing only one thing', and by this he means that we should always be conscious that each of us is responsible and accountable for our own choices (Kierkegaard, 1938, pp. 198, 182).

It is tempting, for most people, to find ways to evade responsibility for their actions by 'hiding in a crowd' (Kierkegaard, 1938, pp. 185, 192). We seek excuses like 'everyone else is doing it'; or 'my little contribution to the problem won't really make a difference in the long run'. This is the message one learns from game theory; namely, that there is no point to one's personal just actions if everyone else is evading their responsibility. 'So,' says Kierkegaard, one 'grows cowardly in the face of—number' (Kierkegaard, 1938, p. 190). This self-deception is what Kierkegaard calls 'double-mindedness' or 'divided mind' (Kierkegaard, 1938, p. 200); that is, trying to have it both ways, but with no consistency between belief and behaviour. Such a divided mind would then act as if the problem were outside oneself, and thus not one's personal responsibility.

Instead, one should base one's decisions and everyday actions on ethical principles of right and wrong, rather than the end results (Kierkegaard, 1938, p. 199). One can never know the results anyway – for they will always be uncertain. It is about believing that if one lives honourably or spiritually, then *whatever results* from that life will be honourable. If one is treating everyone with respect; if one is living up to their capabilities; if there is consistency between one's beliefs and behaviours; then a person is integrated, unalienated and undouble-minded. As argued above, this can be described as a spiritual goal of inner transformation that results in being at ease with one's place in the world (Rosemont, 2001, p. 31).

An example of a divided self making decisions in the face of a crisis

Consider one of the many global problems, mentioned in the first paragraph of the 'Introduction', which are creating the sense of uncertainty and alienation. One can find many ways to avoid changing one's lifestyle and attempt to reduce one's own greenhouse gas emissions and decrease one's personal 'carbon footprint'. It is easy to find advice about the most effective lifestyle changes, such as driving less, avoiding aeroplane flights and eating less red meat. It is just as easy to find excuses not to do these things. One might argue, as in game theory, that there is no point to an individual's efforts to cut back on their own emissions – if most others on the planet are not also do the same thing. The concentration of greenhouse gases in the global atmosphere will rise, and the ecocatastrophe will arrive, despite your efforts. If so, then why be the only 'sucker'? Thus, one excuses one's personal petrol (gasoline) usage by explaining that one will give up one's own automobile when China stops building more coal-fired energy plants.

One might rationalize, further, that one's own personal aeroplane travel is acceptable because one has calculated that one's whole-lifespan carbon footprint is less than that of others.

There are scientific and ethical reasons for acting against climate change, given the uncertainty of the magnitude of the risks involved (Haller, 2002). Here are added reasons of one's internal reflection and the desire for internal harmony. It is the attempt to achieve inner transformation that unites one's internal and external existence, rather than being a divided self.

Pilgrimage as a Vehicle for Finding Self/World Unity

Just how does pilgrimage and spiritual reflection help us decide between inner transformation or political action during periods of personal alienation and uncertainty? It does so by providing experiences that connect people with the world, and which unify their divided selves. Thus, religious experience involves both the personal and the communal. For example, prayer is a spiritual experience, in the sense that it is an attempt to find an insight into one's own interior soul. One is not necessarily concerned with its effectiveness in physically changing events in the external world – as in a prayer that one's favourite sports team wins the championship game. Rather, it is an 'inner communion' (James, 1958, p. 401). It is an attempt to find peace within oneself, and between our self and the world. And the activity of prayer has value, regardless of whether one's prayers 'come true' in the external world (Fig. 6.2).

Many modern pilgrims, to varying degrees, are: (i) spiritually alienated; (ii) psychologically distressed; or (iii) suffering a loss of meaning in their lives. Their identity feels alienated and divided – in the way explained by Fromm (1962), Naess (2008) and Kierkegaard, (1938) above. Pilgrimage, and some other forms of travel, can be a means for making oneself whole again. It can do this because it is often a way of producing certain kinds of experiences, like those described by psychologist Maslow (1970) as 'peak experiences', which encourage a higher degree of inner reflection than occurs in everyday life. Pilgrims ask themselves the questions that Kierkegaard says we must ask if we are to face 'eternity' (Kierkegaard, 1938, p. 198). Pilgrimage often embeds one in a community of others facing the same

Fig. 6.2. Ciraqui on the Camino de Santiago. Painting by the author.

thoughts and challenges. The experience of pilgrimage often widens the pilgrim's focus from their own individual worries and concerns, to include empathy for other pilgrims, and consider their own place in the world (Fig. 6.3).

What Maslow called 'peak experiences' are at the 'core' of every religion (Maslow, 1970, p. 19). He describes these peak experiences using the same terms, of unity and integration, used by other thinkers quoted earlier in this chapter. He writes that peak experiences provide one with 'a clear perception of the universe as a whole ... and that one has his place in it' (Maslow, 1970, p. 59). Quoting William Blake, Maslow describes these experiences as involving a 'unitive consciousness' (Maslow, 1970, p. 68). Such experiences 'can be so profound and shaking ... that it can change the person's character ... forever after' (Maslow, 1970, p. 59). The central experience is one of 'ego transcendence' (Maslow, 1970, p. 62). Such profound experiences are not easily achieved in one's everyday life. They often require a sustained consciousness or activity, and 'long, hard work' (Maslow, 1970, pp. xv–xvi).

Fig. 6.3. Burguette on the Camino de Santiago. Painting by the author.

Pilgrimage provides just such fertile ground to get this job done.

Other language that describes these transformational experiences refers to the 'sublime'. Sublime experiences are those that are both beautiful and terrifying at the same time, such as thunderstorms, oceans, sunrises and stargazing. These experiences make one feel insignificant against the awareness of the vast, uncontrolled forces and distances in the universe. Yet, at the same time, the perceiver knows that they are not insignificant, because they are creating this experience. What results is a religious feeling of being lifted out of one's individual ego and feeling connected to all existence.

Schweitzer had this experience while paddling, at night, through a herd of hippos on a river in West Africa. Fearful for his life, he had the epiphany of awareness of all living beings experiencing this same fear. He described it as 'a mystically ethical oneness with existence' (Schweitzer, 1923, p.253). The event led him to develop his philosophy of 'reverence for life'. Environmentalist Leopold felt the sublime when, as a game manager, he shot a wolf and watched the 'fierce green fire' die in its eyes (Leopold, 1981). This experience of death changed his relationship with wolves and their environment. He then developed his holistic 'land ethic' which emphasized a respectful acknowledgement of the role of wolves in an entire ecosystem.

Pilgrims have this experience of awe when confronted with sacred locations, cathedrals, final destinations, early morning dew and many other things. Similarly, they too feel a sense of connection to all things outside their own personal ego. This diminishing of personal ego is what Rosemont had in mind by defining religious experience in terms of 'belonging' (Rosemont, 2001, p. 31). He expands on this idea with a familiar description of religious experience: 'This sense is a feeling of being a part of, at one with, something larger than ourselves, ... something which will endure after us' (Rosemont, 2001, p. 33).

One's 'divided self' can be unified by seeking 'connection' with the world outside oneself. Desser's (2003) essay 'Only Connect' argues that some kinds of travel promote the experience of connection, while other kinds of travel are 'missing' something. For example, after describing the extraordinary comforts and luxuries to be experienced at the Bellagio Hotel in Las Vegas, she concludes that these experiences lack 'essential value' and 'depth' because they are fundamentally imitations of the real thing (Desser, 2003, p. 275). While the indoor gardens at the Bellagio are spared no expense in their re-creation of the seasons, they remain 'disconnected' from a functional ecosystem (Desser, 2003, p. 274). While there are endless seductions of music and gourmet restaurants in the hotel, they all manage to 'disorient' the hotel visitors, rather than engage them. The hotel guests are crowded together but are 'in no way connected' (Desser, 2003, p. 275). Borrowing the title phrase 'only connect' from Forster's (1910, p.195) *Howards End* to supply the title of her essay, she argues that what is missing from the luxury hotel experience is an experience of *connection* with things that matter.

Life becomes less fragmented, she argues, if one 'connects to experience fully' (Desser, 2003, p. 279). The examples she supplies of complete, unmediated experience include: (i) backpacking in the woods; (ii) being awed by Chartres Cathedral where she felt 'the totality of presence' (Desser, 2003, p. 280); (iii) attending the theatre; and (iv) being fully present in front of works of art and feeling 'their terrible beauty' (Desser, 2003, p. 281).

It is this experience of 'connection' that can resolve a divided self into one that finds unity with the external world. Desser writes of these experiences that: 'I felt connected, related, a part of something greater than myself. I felt at home in the universe' (Desser, 2003, p. 279). It is one of the purposes of pilgrimage to undergo this transformation. To relate to the world in a less fragmented way.

Summary

There are no half measures for Kierkegaard (1938). This is a difficult idea to absorb. He suggests that one should be concerned with one's own choices and responsibilities, rather than solely on political solutions. Game theory purports to show that political solutions to community-action problems are the only rational reaction to them. But that solution may never come. Instead, an individual can be motivated to do the right thing because it seems right, not because it will be in their personal self-interest. Kierkegaard argues that one cannot truly live with an evasion of personal responsibility when making life choices. One should stop trying to maximize efficiency in personal decision making. Do the right thing anyway, even if you cannot see by your calculations that your individual, small action will solve the problem. Do it anyway. Be 'earnest' (Kierkegaard, 1938, pp. 178, 182).

Koestler (1945) described the dilemma of choosing how to improve things in our world as a choice between 'change from within' or 'change from without' (meaning political solutions). Kierkegaard (1938), Plato (1974), Naess (2008) and Fromm (1962) defend the importance of individual choices and responsibilities, rather than deferring to solely political solutions. *Change from within* is a personal response to uncertain times and should not be dismissed merely because it might be 'ineffective' in terms of consequentialist calculations. Pilgrimage is one way to aid this inner transformation because it is conducive to inner reflection and experiences of connection with others and with existence itself.

References

Axelrod, R. (1984) *The Evolution of Cooperation*. Basic Books, New York.

Bakewell, S. (2016) *At the Existentialist Café*. Alfred A. Knopf, Toronto, Canada.

Desser, C. (2003) Only connect. In: Lightman, A. (ed.) *Living with the Genie*. Island Press, Washington, DC, pp. 273–286.

Detmer, D. (2008) *Sartre Explained:From Bad Faith to Authenticity*. Open Court Publishing Company, Chicago, Illinois.

Forster, E.M. (1910) *Howards End*. Alfred A Knopf, New York. Available at: https://archive.org/details/howardsend002321mbp/page/n233/mode/2up (accessed 20 November 2023).

Fromm, E. (1962) *Beyond the Chains of Illusion*. Simon and Schuster, New York.

Haller, S.F. (2002) *Apocalypse Soon? Wagering on Warnings of Global Catastrophe*. McGill-Queen's University Press, Montreal, Canada.

Heath, J. (2001) *The Efficient Society*. Penguin Books, Toronto, Canada.

Hobbes, T. (1968) [1651] Of the causes, generation, and definition of a common-wealth. In: *Leviathan*. Penguin Books, New York, pp. 223–228.

James, W. (1958) *The Varieties of Religious Experience*. Penguin Books (Mentor),Toronto, Canada.

Kierkegaard, S. (1938) *Purity of Heart Is To Will One Thing*. Translated by Steere, D.V. Harper & Row, New York.

Kierkegaard, S. (1944) *Either/Or*. Translated by Swenson, D.F. and Swenson, L.M. Princeton University Press, New York.

Koestler, A. (1945) *The Yogi and the Commissar and other Essays*. Jonathan Cape, London.

Leopold, A. (1981) Thinking like a mountain. In: *A Sand County Almanac*. Oxford University Press, Oxford, pp. 137–141.

Maslow, A. (1970) *Religions, Values, and Peak Experiences*. Penguin Books, New York.

Naess, A. (2008) The place of joy in a world of fact. In: Drengson, A. and Devall, B. (eds) *The Ecology of Wisdom*. Counterpoint, Berkeley, California, pp. 123–132.

Plato (1974) *Plato's Republic*. Hackett Publishing, Indianapolis, Indiana.

Rosemont, H. Jr (2001) *Rationality and Religious Experience*. Carus Publishing, Peru, Illinois.

Sayers, D.L. (1949) Alighieri, Dante (1949/1472) *The Divine Comedy*. Translated by Sayers, D.L. Penguin Books, New York.

Schweitzer, A. (1923) The ethic of reverence for life. In: Naish, J. (translator) *Civilization and Ethics*. A. & C. Black, London, pp. 251–276.

Unamuno, M. (1954) [1921] *The Tragic Sense of Life*. Dover Publications, Mineola, New York.

7 Moving Pilgrimage into the Virtual Realm: Understanding the Advantages and Complexities of Immersive Technologies

Mujde Bideci[1]* and Caglar Bideci[2]
[1]*Canterbury Christ Church University, Canterbury, UK;* [2]*University of the West of Scotland, London campus, UK*

Abstract

This chapter examines and investigates immersive technologies, including mixed reality (MR), virtual reality (VR) and augmented reality (AR), which have seen rapid development, offering innovative solutions that enhance interactions for visitors. While the impact of immersive technologies on enhancing the visitor experience has been well studied, there has been little attention given to the pilgrimage tourism experience in current literature, despite its significant role in the global tourism economy and its ability to attract millions of travellers annually. To better understand how visitors respond to the use of immersive technologies and address this gap, this chapter presents a conceptual framework by examining the overall role of immersive technologies in the pilgrimage experience, considering cognitive, emotional, spatial and social behavioural aspects. The chapter also discusses the challenges and limitations associated with implementing immersive technologies in pilgrimage tourism. It concludes with recommendations for the future of pilgrimage tourism and practical implications based on these findings.

Introduction

With the upsurge of coronavirus disease 2019 (COVID-19), immersive technologies such as mixed reality (MR), augmented reality (AR) and virtual reality (VR) have emerged as a means of enhancing the visitor experience by enabling them to engage with the past, present and future at tourism destinations in innovative ways that combine physical and virtual environments. These technologies are highly effective marketing tools (Wedel *et al.*, 2020) for the tourism economy, particularly in the context of contemporary tours (Lee *et al.*, 2017). While immersive technologies bring more affordable and customized experiences to people according to their needs, they also interact with users by enriching the degree of engagement and persuasion and eliciting vivid physical and emotional responses (Riva *et al.*, 2004). Despite the potential benefits and the growing interest in immersive technologies in the context of tourism, our knowledge of their conceptual boundaries and application possibilities for pilgrimage experiences is lacking. Most existing research on the topic has focused on the positive and negative effects of immersive technologies or conceptual and theoretical approaches, with little attention given to their impact on the overall pilgrimage experience. For example Wei *et al.* (2019) examined the beneficial effects of VR

*mujde.bideci@canterbury.ac.uk

DOI: 10.1079/9781800623675.0007

and AR on visitors' experiences, while Merkx and Nawijn (2021) considered the detrimental effects, and Jung and tom Dieck (2018) investigated the conceptual and theoretical approaches that have advanced marketing and heritage conservation. Recent studies have also investigated how immersive technologies can help researchers comprehend visitors' behavioural intentions (Tsai, 2020), especially in museum contexts (Errichiello *et al.*, 2019).

Pilgrimage and religious retreats have become increasingly popular around the world as a response to the growing interest in spirituality and their ability to enhance visitors' multisensorial experiences while also boosting local businesses. In this regard, tourism practitioners have developed specialized packages and experiences for people seeking spiritual fulfilment. These spiritual tourism packages include visits to sacred sites, participation in religious ceremonies and meditation retreats. Recently immersive technology-driven experiences have been added to this list, as they offer a unique way for travellers to immerse themselves in rituals, heritage and history, cultures and traditions with or without the opportunity of their being present. However, there is a need to examine the boundary conditions of the relationship between immersive technologies and behavioural outcomes in religious tourism to better understand their impact on visitors' experiences. Moreover, there has been limited exploration of the role of mental imagery with sensory inputs in this context, and studies on the emotional effects of immersive technologies such as VR, AR and MR have focused on consumers' immediate experiences rather than their nostalgic or ancestral bond reminiscences.

To bridge these knowledge gaps, this chapter will examine the overall role and aspects of using immersive technologies, investigating how the functional components of these devices influence pilgrimage experiences in terms of developing cognitive, emotional, spatial and social behavioural responses. Specifically, the study aims to explore the role of immersive technology (MR, AR, VR) applications in authentic pilgrimage practices and their impact on visitor experiences. Additionally, the findings will enhance our understanding of the complexities, especially in the engagement and sense of presence, which refers to the psychological feeling of being in a virtual environment via MR, AR and VR technologies. This understanding can support both managerial decision making and academia for future research.

Pilgrimage Tourism Experience

Tourism experience is an evolving 'socially constructed' term (Tussyadiah and Fesenmaier, 2009) that is influenced by various meanings, dynamics and stages (Bideci and Bideci, 2021b) throughout travel. Drawing on the well-established concepts of experience economy (Pine II and Gilmore, 1998) and service quality literature (Parasuraman *et al.*, 1988), research on tourism experience has expanded to meet the growing need for understanding visitors' journeys and their motivations, and also cognitive and emotional stages in order to create memorable (Larsen and Mossberg, 2007), authentic and transformative experiences (Kirillova *et al.*, 2017). Journeys for religious purposes are among the most widespread and ancient movements in human history (Kaelber, 2006). According to Morinis (1992), there are six different pilgrimage journeys based on the state of feeling. These are: (i) devotional pilgrimage (honouring the 'one' they believed in and testifying the traditions); (ii) instrumental pilgrimage (praying for something); (iii) normative pilgrimage (predetermined for specific seeking); (iv) initiatory pilgrimage (a journey for transformation of oneself); (v) obligatory pilgrimage (following religious purposes); and (vi) wandering pilgrimage (for personal merit and satisfaction). In this chapter, however, the motivations for undergoing a pilgrimage are considered to be twofold: (i) for religious reasons; or (ii) for secular reasons.

Traditional pilgrimage, as a form of religious travel, involves embarking on a journey for religious reasons that often include specific rituals, such as seeking forgiveness, engaging in prayer, improving health, being alone, meditation and visiting sites where miraculous events have occurred (Timothy and Boyd, 2003). Those who undertake these journeys are known as pilgrims and desire to live memorable, unique and authentic experiences once or a few times in their life. They wish to focus on the original identity of a place and its emotional connections, in contrast to the negative consequences of mass tourism and its rapid mobility and the homogenization of demand (Lewicka, 2011).

In this Instagrammable world, the pilgrimage tourism experience encompasses not only religious reasons but also secular movements with the motivation of immersion in cognitive stages, including nostalgia by dreaming beyond reality and appreciating history (Bideci and Bideci, 2023). It should be also noted that the fantasies associated with the stories at destinations (Chi and Chi, 2022) are also a big player in pilgrimage tourism. Pilgrimage tourism transcends a mere journey to a sacred place, offering an opportunity to establish a profound connection with the destination's culture, history and between both residents and visitors. Moreover, the pilgrimage has been an integral part of human culture for centuries (e.g. the Kumano Kodō in Japan or the medieval Via Francigena in Italy), representing an ancient practice that involves embarking on a spiritual or religious journey to a sacred site from which emerges inner transformative aspects. According to Kirillova *et al.* (2017), the transformative experience, without being secular or pious, can impact individuals, effectively while enabling them to connect with their inner selves and fostering a sense of community. This transformation also deepens visitors' or pilgrims' understanding of spirituality and degree of faith while they reflect upon their values and beliefs and their understanding of the world around them.

From a secular point of view, Bond *et al.* (2015) posit that worshipping or praying is no longer the primary motivation for visiting a pilgrimage site. People are motivated to visit pilgrimage sites for different purposes, such as: (i) having familiar connections to a site such as historical bonds (e.g. visiting their ancestors' graves); (ii) due to their interest in a site's association with celebrities of their time (e.g. Thomas Beckett or Elvis Presley); (iii) because of a site's architectural values (e.g. Antoni Gaudí's Sangria Família – the largest unfinished Catholic church in the world – in Barcelona or Catedral de Santa María de la Sede in Seville, Spain); or (iv) to see historic artworks such as stained glass (e.g. Mosque-Cathedral of Cordoba, Spain) (Hughes *et al.*, 2013). However, visitors can still have spiritual feelings that are influenced by the atmosphere of the archaeological structure and the rituals inside performed by others (Bideci and Bideci, 2021b). This gives broader insight regarding the societal trends towards secularization in which people seek cultural experiences (such as in Notre Dame de Paris in France), and connections to the pilgrimage heritage sites but are less engaged in traditional religious practices. Therefore, travelling to sacred sites has become a popular leisure activity for people from diverse cultures, religions and nationalities. Visiting a pilgrimage site is not only limited to individuals with religious beliefs, but also those who do not adhere to a specific faith, or are not interested in fulfilling religious obligations but follow traditional routes and visit sacred pilgrimage sites out of curiosity about rituals. In this sense, religion-based or curiosity-based movements have become a significant contribution to the tourism economy that caters for discerning visitors' demand that seeks unique experiences.

The growing demand for pilgrimage tourism has also influenced religious rituals, related activities and transformed old traditions and archetypes into new and innovative forms of travel within the context of new religious movements and leisure. This popularity has led to a reinterpretation of the experience at sacred sites, which is a crucial factor in travel patterns (Damari and Mansfeld, 2016) for both researchers and practitioners. In this sense, more research is needed on pilgrimage tourism and related experiences to better understand the behaviours of visitors derived from different motivations and their impacts on the tourism economy. As a result, we approach pilgrims and visitors from a marketing perspective. We therefore categorize visitors as either *passive receivers* driven by a desire to remember their ancestral past and seeking to sense the ambience of the place; or *active seekers* motivated by a quest for authenticity in their experiences and having a personal bond, seeking spirituality in their experiences at the pilgrimage heritage tourism destinations (Bideci and Bideci, 2021a). In line with current trends, religious sites have adapted their structure to attract a wide range of visitors who have various motivations; and these sites offer diverse events, especially immersive technology-based applications, at or outside the sites, to enrich interactivity and increase visitor engagement.

Immersive Technologies

Immersive technologies deliver interactivity with engagement and enjoyment (Fan *et al.*, 2022) within a system of an inclusive and extensive illusion of reality to the users and enhance interactivity and user experience (Raptis *et al.*, 2018) in an immersive way. As they offer different services to users because of their unique structures, MR, AR and VR are the various forms of immersive technologies that provide and offer users enriched participation through cognitive (Flavián *et al.*, 2019) and physical immersion with all senses with a realistic (Flavián *et al.*, 2021) and high-fidelity preview. In particular, the interaction – through the digital realm – has a key role in improving users' habits and behaviours by providing easy access to the unknown or inaccessible world and enriching the connections between the user and the world in question from various aspects (Barrera and Shah, 2023). For this reason, despite traditional services, immersive technologies can offer simulated versions of the past and future of places, sites or destinations by creating a mental imagery phase to seek out the stories and possibilities of presence.

While exploring the unknown with these technologies is attractive understanding the differences between realities remains complicated. To respond to the complexities of realities, the 'reality–virtuality continuum' introduced by Milgram and Kishino (1994) has been used by researchers to categorize the diverse range of realities and virtuality. This classification spans from authentic environments to entirely virtual ones, located at the two ends of the continuum (as illustrated in Fig. 7.1). **Real environments** refer to the physical reality that surrounds us, including real-life scenes observed directly or indirectly, such as through a vivid video display. In contrast, **virtual environments** are entirely computer-generated environments that highlight objects that do not exist in the real world. Users interact with these environments in real time through a technological interface (Flavián *et al.*, 2019). In this reality continuum, extended reality (XR) is an umbrella term for MR, VR and AR, and has been defined as a bridge between reality and a coexisting-simulated world (Hoyer *et al.*, 2020). This allows people to communicate and interact in real time in the online-to-offline digital realm (Jung *et al.*, 2016) effectively and dynamically (Flavián *et al.*, 2019) and provide live immersive experiences (Berg and Vance, 2017) with added value (Patrício *et al.*, 2011).

To further expand on this, the continuum also includes AR and VR that lie between the two end points. AR adds virtual elements to the real world with 3D modelling on images or videos, allowing users to interact with both simultaneously (Azuma *et al.*, 2001). VR – a computer-generated and 3D-simulated environment – is defined as 'a fully synthetic world that may or may not mimic the real world' (Loureiro *et al.*, 2020, p. 2) and therefore has the potential of reducing the physical inconvenience of places (Robaina-Calderín *et al.*, 2023) by using head-mounted display tools (HMDs). MR, on the other hand, combines the physical and digital worlds by using a 3D context such as HoloLens (a head-mounted MR device) to create a unique environment where physical and virtual objects interact with each other in real time (Milgram and Kishino, 1994). The categorization provided by Milgram and Kishino's (1994) reality continuum offers a useful framework for researchers investigating immersive technologies in various contexts such as museums (Errichiello *et al.*, 2019), gaming (Lin *et al.*, 2018a) or health (Freeman *et al.*, 2017; Lin *et al.*, 2018b), and a comprehensive understanding of differences between real and virtual environments and objects. Moreover, the combination of MR, AR and VR as XR tools provides a sense of new experience-based perceptions (Eimer, 2004), embodiment where users can feel deeply connected with the 3D environment (Flavián *et al.*, 2019), and an emotional bond that

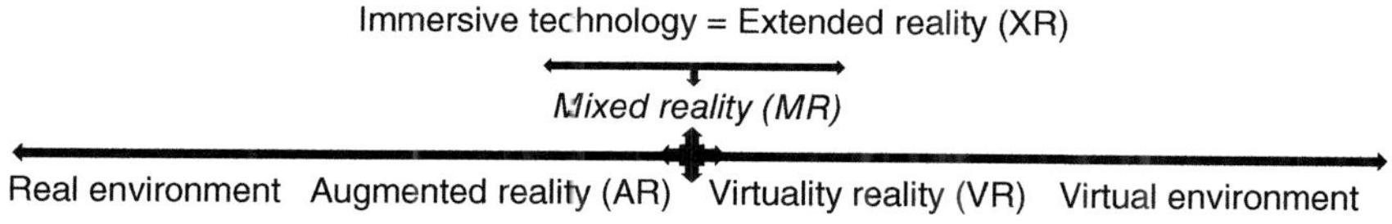

Fig. 7.1. Reality–virtuality continuum. Adapted from Milgram and Kishino, 1994.

can lead to positive behavioural actions such as loyalty (Lemon and Verhoef, 2016), hedonism (Mirghadr *et al.*, 2018) and sense of presence (Tussyadiah *et al.*, 2018), which leads to a unique, authentic and memorable experience.

Applications of Immersive Technologies in Pilgrimage Tourism

Recently, Robaina-Calderín *et al.* (2023) explained the term immersion as a multifaceted form of action inherently tied to individual experience and psychological presence in VR environments. Hence, they have found that immersion in the experience of tourist destinations has a positive effect on behavioural and psychological engagement. In terms of pilgrimage tourism, immersive technologies support this multifaceted form of immersion and can present creative, interactive and memorable experiences and shape the future of new religious movements by offering cognitive, social, emotional and spatial experiences for visitors by opening new horizons for the tourism economy (Fig. 7.2).

For instance, *cognitively* visitors can gain interactive and immersive learning experiences and learn about the geography, culture and history of the pilgrimage heritage sites in a more engaging way while being present at the site, or even without being physically there. In addition, these technologies can help visitors to get various skills such as: (i) increasing attention; (ii) creating a collective memory (e.g. memorabilia of lived experiences through virtual souvenirs such as pilgrim scallop shells); (iii) facilitating decision making (e.g. a virtual discussion board for sustainable actions); (iv) attracting foreign language learning; or (v) understanding cultural sensitivities which are simulated by real-life scenarios. Enabling virtual exploration and learning about a site without requiring a physical presence, immersive technologies make it particularly advantageous for those who are unable to travel or for sites that are challenging to access. Moreover, this approach helps to safeguard the site by limiting the number of physical visitors and mitigating any harm caused by tourism, thereby contributing to its preservation. VR tools create a connection close to reality as they mediate and tailor the experience according to the individual's needs and desires (e.g. Bec *et al.*, 2019; Fan *et al.*, 2022). According to Yuqiang *et al.* (2017), visitors to Lincoln Cathedral, an example of early English Gothic-style architecture, have restricted access to the cathedral due to its sensitive structure (narrow staircases and the sensitive fabric needs protection). However, visitors can experience the whole

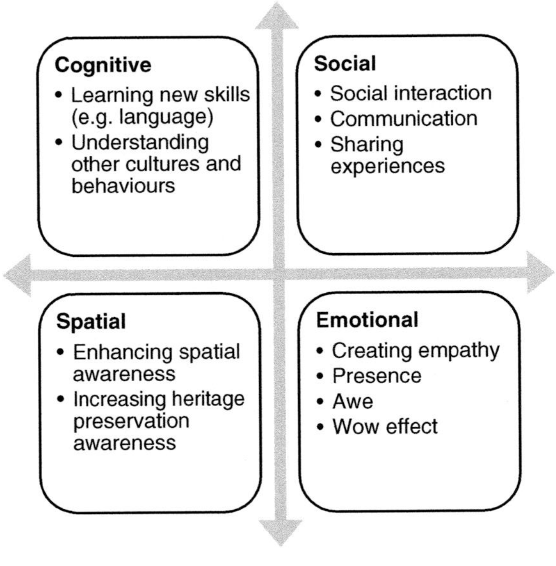

Fig. 7.2. Immersive technologies experiences in pilgrimage.

building with a VR application in order to explore and feel their presence in the cathedral. This also makes the experience accessible to a much wider audience more interactively and engagingly. At the same time, it provides essential information that can help to deepen visitors' understanding of the history and significance of the site, enhancing their sensory feelings and making it more meaningful.

Socially, while pilgrimage is an individual inner journey and is mostly preferred as a solo way to find inner peace, interactions between pilgrims create value for a memorable experience along the way (Casais and Sousa, 2020). Immersive technologies can increase social interaction and communication among visitors by allowing them to share their experiences, even if they are physically located in separate places or physically present at the site with no direct connections. In this way, virtual communities can be created where visitors or pilgrims can connect via these technologies.

Emotionally immersive technologies can evoke an intense sense of emotions and presence and create empathy while allowing visitors or pilgrims to experience the perspective of the place, its history and its spiritual ambience. Moreover, as Hinsch *et al.* (2020) state, visiting a destination with one or other form of immersive technology (e.g. AR) can enhance the emotional experience related to feelings such as awe and the wow effect. Religious retreats, especially, are a common source of awe using AR.

Finally, *spatially* immersive technologies can enhance the spatial awareness of visitors and allow them to explore a site with its unique features in a more detailed and interactive way, including the ability to move around the space. It can also be used to increase awareness towards the conservation of pilgrimage heritage sites and enhance the simulation of architectural design patterns in pilgrimage sites which are difficult or impossible to recreate in real life due to the sensitivity of the place or restricted access (e.g. Temple Mount in Jerusalem).

Immersive technologies can facilitate an interactive experience of pilgrimage tourism according to individuals' needs and capacities. For example, individuals can create their own unique pilgrimage journey by deciding the visit duration and the destinations, and therefore customize their experience by using immersive tools to address their spiritual, physical and emotional needs. In practice, a person can use a VR headset with applicable software to create a personal pilgrimage to a location that holds personal significance for them, such as a childhood home that helps them to live nostalgic feelings (e.g. VR Pilgrim – a VR application to visit Greek local monasteries), or a place of spiritual significance to their family (e.g. Labbaik VR – an immersive Umrah application). Moreover, they can offer an opportunity to visit sensitive and significant locations with no restrictions, such as the Dunhuang Cave 61, the largest Buddhist remains at the Mount Wutai mural, Mogao Grottoes, in the borderland of Central Asia and China. Immersive technology offers users an interactive exploration of Vatican City's landmarks using a VR tool (Vatican VR, 2023). In addition to well-known destinations like St Peter's Basilica and the Sistine Chapel, visitors can take virtual tours of various religious and cultural sites within the Vatican, including the Vatican Museums. A realistic and vivid 3D representation can be accessed by users through this innovative technology (Zheng *et al.*, 2022). Moreover, it is possible for users to wander virtually through St Peter's Basilica, marvelling at the intricate details and even seeing the renowned dome without leaving their home or sofa. An exceptional experience is created when the application captures the grandeur and presence of these sacred spaces of the Holy See, Vatican City.

Another example is the Hajj, which is an annual Islamic pilgrimage to Mecca, considered one of Islam's five pillars. Each year, millions of Muslims from around the world travel to Mecca to perform the Hajj rituals (Tagliacozzo and Toorawa, 2016). Due to the outbreak of the COVID-19 pandemic and the resulting travel restrictions, the Saudi Arabian government developed a VR-based Hajj application. This innovative solution allows individuals to perform the Hajj rituals virtually, offering a more accessible and inclusive option for Muslims who may be unable to travel physically to Kaaba (in Mecca), the most sacred site in Islam. The experience comprises a 360° view of the Kaaba and a guided tour of the Hajj rituals. Immersive technology applications have been developed by other countries, such as the German Muslim3D (2023), the Canadian HajjiVR (2023) and the 'Jerusalem VR' experience, developed by the

Jerusalem Development Authority and Western Wall Heritage Foundation (Virtual Reality Experience in Jerusalem, 2023). The experience allows individuals to explore the Old City of Jerusalem, a holy site for Jews, Christians and Muslims, by means of VR. It includes virtual tours of historical and religious sites such as the Western Wall, the Church of the Holy Sepulchre and the Dome of the Rock. The experience also includes narration and historical context to provide a richer and more informative experience for pilgrims.

Understanding the Complexities of Immersive Technologies

Based on the above applications of immersive technologies in various pilgrimage sites, researchers have explored how immersive technologies affect the intangible aspect of the visitor experience. On the one hand, the technology-based pilgrimage demands extensive exploration and understanding due to its enduring nature for both researchers (e.g. Hill-Smith, 2009, 2011; Mu *et al.*, 2023) and practitioners. Considering the opportunities of applying immersive technologies in pilgrimage heritage sites, practitioners are especially interested in and have seen increased funding for creative experience designs in such places and historic sites. In contrast to 'cyberpilgrimage' (Hill-Smith, 2009, 2011), which is based on web pages and second-life Hajj (i.e. the ritual of Hajj in VR form) or pilgrimage experiences like virtual pilgrimage to Ireland's Croagh Patrick (MacWilliams, 2013), the emerging immersive technologies cater for current market trends. They provide 'technology-driven, hedonism-oriented, and immersive personalised' experiences (Guo *et al.*, 2021, p. 1030) through narratives of multisensory interaction including touch, sight and hearing. In addition to the above-mentioned aspects, psychological dimensions: (i) enhance visitors' immersion (Agrewal *et al.*, 2020); (ii) provide involvement with emotions (McLean and Barhorst, 2022); (iii) increase the positive sense of presence (Rodríguez-Ardura and Martínez-López, 2014); and (iv) embody social and emotional presence of visitors or pilgrims (Fan *et al.*, 2022). Moreover, a mental imagery phase is delivered via various types of equipment such as HMDs, that may be tethered or standalone, and controllers, or for an advanced experience tools such as sensors, haptic bodysuits, immersive domes and immersive sound and light installations. Allal-Chérif (2022) stated that visiting 'intelligent' cathedrals, which use immersive technologies as a part of pilgrimage activity that serves to 'collective imagination' (p. 1), can help historic reconstruction and provide replicas of living one's faith. Furthermore, the utilization of immersive technologies in pilgrimage heritage sites allows visitors to experience a sense of being present virtually but close to reality, feel emotions with values, and travel in time and space with no limits according to their own capacity. These technologies help to create engaging narratives that connect with visitors, enabling them to go beyond their reality and immerse themselves in a unique cognitive experience (Bec *et al.*, 2019).

On the other hand, understanding the sensory aspects of visitors that shape their experiences during their travels needs more research (Freidus and Caro, 2021). This is especially the case since our understanding of how immersive technologies have affected their experience in this context has remained insufficient in the current literature and knowledge is limited. Moreover, while immersive technologies hold promise for current developments, they also pose questions about future potential and possible opportunities for use, as well as the challenges this may present. For example, by simulating the sights, sounds and even emotions associated with these sacred sites, further research is needed to explore the specific contributions of immersive technologies to create memorable pilgrimage experiences. Empirical studies can provide a better understanding of how they can enhance the pilgrimage tourism journey without compromising reality, or whether immersive technologies offer pilgrims unique opportunities to deepen their connection and understanding of the pilgrimage experience.

Although there are many potential benefits to using immersive technologies in pilgrimage tourism, there are also potential implications and concerns. Along with the tourism sector, immersive technologies have been extensively used in education (Merchant *et al.*, 2014), entertainment (Lin *et al.*, 2018b), healthcare (Freeman *et al.*, 2017), retail (Bonetti *et al.*, 2018) and research (Meißner *et al.*, 2019). However, researchers and practitioners have not yet clearly defined the boundaries of those new realities, technologies and related experiences (Flavián *et al.*, 2019). One of the most significant concerns is the potential for technology to replace the physical experience of pilgrimage altogether. The act of physically travelling to a location and interacting with other pilgrims (Casais and Sousa, 2020), and experiencing the culture, hi-story (its history and story) and traditions of the places or destinations are critical parts of the pilgrimage experience. Therefore, the use of immersive technologies should be viewed as a supplement to, rather than a replacement for, the physical experience of pilgrimage. Another concern is the degree and limitation of commercializing the visitor experience through the implementation of immersive technologies in the pilgrimage heritage sites. Recently, Alegro *et al.* (2023) examined how VR is used as a versatile destination marketing strategy and how these technologies allow visitors to virtually explore a location before embarking on a physical journey. As part of their study they delved into the challenge of monetizing immersive technologies in user experiences and their potential implications. According to findings, VR experiences are provided free of charge, offering a glimpse of the destination's attractions. Moreover, VR serves as an extension of the experience, particularly in the interpretation, enhancing visitors' understanding of the cultural and historical significance of a place. From a different point of view, the progress of VR technology can eventually replace physical travel, and the pilgrimage experience may become a substitute service sold for its value instead of a real journey. This raises concerns about the potential loss of authentic interactions and personal connections that physical travel can provide. Parallel to these challenges, the use of MR, VR and AR technologies to recreate sacred sites and locations can be transformed into a way to monetize the experience of pilgrimage, turning it into a commodity that can be bought and sold. This could have negative implications for the pilgrimage's spiritual and cultural significance, reducing it to a commercial transaction rather than a transformative and meaningful experience journey.

In the context of cultural tourism destinations, with the progressive acceleration of immersive technology there will be a gradual rise in visitors' expectations for authenticity in their travel experience (Shehade and Stylianou-Lambert, 2020). However, the use of immersive technologies in cultural and pilgrimage places may bring about important ethical considerations, particularly around issues of representation and authenticity. For example, in their study at Mount Tai (China), Zhu *et al.* (2023) found that despite the fact that AR can enhance and enrich tourism experiences, they draw attention to earlier research that presents limited authenticity in virtual tourism. In this sense, two questions need to be responded to before executing the technology-driven experience: (i) who has the right to create a virtual representation of the sacred site?; and (ii) how can this representation be created? Shehade and Stylianou-Lambert (2020) also add that AR designers should prioritize enhancing the representation of tourism products to ensure a captivating and immersive experience for visitors, augmenting their perception of authenticity. Therefore, the balance between authenticity and its own customized representation is a prominent issue to address to ensure that an optimal immersive experience will be provided for visitors and pilgrims.

Beyond these discussions, we need to ask also about the appropriateness of presenting technology-driven experiences to recreate a location that is sacred and holy for a particular community or culture without their consent or input. There are other concerns regarding the possibility of misrepresentation and distortion of the cultural and historical importance of the location, particularly when the technology is developed by individuals or organizations that lack a deep understanding of that specific culture but are keen on generating income. Nevertheless, more studies are needed to comprehend how immersive technologies can transport pilgrims to virtual sacred sites by allowing them to participate in rituals, explore historical events, connect with the spiritual essence of the place, and how destination managers can use the optimum potential of these technologies to increase visitors' engagement. It is essential to carefully consider how these technologies are implemented to ensure they enhance the authenticity and reverence of the pilgrimage experience. Balancing the use of immersive technologies with preservation of the sacred and respect for religious traditions is crucial. Moreover, understanding the impact of immersive technologies on the cognitive, emotional, spatial and social dimensions of the pilgrimage experience requires further research and exploration.

To address these concerns, the solution is to ensure the use of immersive technologies in pilgrimage is seen as a supplement or support to, not a replacement of, an actual pilgrimage journey. It is important to acknowledge the cultural and spiritual significance of physical pilgrim rituals and ensure that the use of technology will not detract from that experience, while also protecting the common spiritual and heritage value of the places. Creating standards and guidelines to promote knowledge of local communities and protect jobs for cultural experts can be important implementations. In line with this approach, communities or governance bodies can work collaboratively and ensure the technology is accurate and beneficial while promoting ethical representation, authenticity, and historical and cultural significance of the pilgrimage heritage sites and providing equal and inclusive access to the pilgrimage experiences for all.

Summary

Immersive technologies such as VR, AR and MR have the potential to enhance the pilgrimage tourism experience in terms of cognitive, emotional, spatial and social aspects. While the practical implementation of these technologies is still relatively new to the tourism economy, there is an increasing supply among experienced designers and demand for immersive experiences among visitors (Bogicevic *et al.*, 2019). In particular, the role these immersive technologies have in enhancing mental creativity, the nostalgia effect and accessibility makes them attractive for all stakeholders in the pilgrimage tourism sector. In the meantime, it is important to better understand their potential positive impacts and drawbacks on the pilgrimage experience, as researchers explore the possibilities and limitations of the level of immersion and interactivity that these technologies provide (Zhu *et al.*, 2023). It should be noted that these technologies can enhance the feeling of being present at the pilgrimage heritage site, while they also create a sense of detachment from the physical world. Therefore, it is also important to comprehend the negative impacts on the level of cognitive, emotional, social and spatial interactions between pilgrims, and the degree of disconnection from the pilgrimage site's cultural and historical significance, to develop a meaningful and fulfilling pilgrimage experience and find the balance between immersion and involvement. Another critical factor to consider is the accessibility of immersive technology to all pilgrims and visitors who may be unable to go to pilgrimage destinations due to physical, financial or other limitations such as health due to post-COVID-19 conditions. Immersive technology might provide an alternative way for these people to participate in the pilgrimage experience. However, it is important to ensure that these technologies will not create a division between those who can physically travel to the place and those who cannot.

Overall, the application of immersive technologies continues to evolve in pilgrimage heritage sites and inevitably they become more widespread especially because of secular movements. By understanding the benefits and limitations of these technologies, researchers and practitioners can work towards designing a more inclusive and enriching pilgrimage experience for all. Future research can also explore how to increase access to these technologies for all stakeholders including experienced service designers, destination managers and visitors. Therefore, researchers and practitioners should seek how immersive technologies can be used more effectively to enhance cognitive, emotional, social and spatial dimensions along with pilgrims' spiritual experience. To achieve this, the applicable methods to incorporate immersive technologies into the pilgrimage tourism experience should be chosen by integrating traditional pilgrimage practices, while protecting the authenticity and cultural significance of those heritage sites. Moreover, while the peak experience is targeted to pilgrimage journeys, how those technologies affect pilgrims' behaviour during and after their pilgrimage practices should be investigated, especially for pilgrims or secular visitors from different religious and cultural backgrounds. Finally, how these technologies can be designed and implemented in a culturally sensitive, ethical and sustainable way should be taken into consideration.

References

Agrewal, S., Simon, A.M.D., Bech, S., Bærentsen, K.B. and Forchammer, S. (2020) Defining immersion: literature review and implications for research on audiovisual experiences. *Journal of the Audio Engineering Society* 68(6), 404–417.

Alegro, T., Turnšek, M., Špindler, T. and Petek, V. (2023) Introducing Amazon Explore: a digital giant's exploration of the virtual tourism experiences. *Journal of Tourism Futures.* https://doi.org/10.1108/JTF-02-2022-0072

Allal-Chérif, O. (2022) Intelligent cathedrals: using augmented reality, virtual reality, and artificial intelligence to provide an intense cultural, historical, and religious visitor experience. *Technological Forecasting and Social Change* 178: 121604.

Azuma, R., Baillot, Y., Behringer, R., Feiner, S., Julier, S. and MacIntyre, B. (2001) Recent advances in augmented reality. *IEEE Computer Graphics and Applications* 21(6), 34–47.

Barrera, K.G. and Shah, D. (2023) Marketing in the metaverse: conceptual understanding, framework, and research agenda. *Journal of Business Research* 155: 113420.

Bec, A., Moyle, B., Timms, K., Schaffer, V., Skavronskaya, L. and Little, C. (2019) Management of immersive heritage tourism experiences: a conceptual model. *Tourism Management* 72, 117–120.

Berg, L.P. and Vance, J.M. (2017) Industry use of virtual reality in product design and manufacturing: a survey. *Virtual Reality* 21, 1–17.

Bideci, M. and Bideci, C. (2021a) Designing a tourist experience for numen seekers. *Qualitative Market Research: an International Journal* 24(5), 704–725.

Bideci, M. and Bideci, C. (2021b) Framing the visitor experience in sacred places. *TQM Journal* 34(3), 377–397.

Bideci, M. and Bideci, C. (2023) Back in time with immersive heritage tourism experience: a study of virtual reality in archaeological sites. In: Ferrer-Rosell, B., Massimo, D. and Berezina, K. (eds) *Information and Communication Technologies in Tourism 2023: Proceedings of the ENTER 2023 eTourism Conference, January 18–20, 2023*. Springer Nature, Cham, Switzerland, pp. 312–317.

Bogicevic, V., Seo, S., Kandampully, J.A., Liu, S.Q. and Rudd, N.A. (2019) Virtual reality presence as a preamble of tourism experience: the role of mental imagery. *Tourism Management* 74, 55–64.

Bond, N., Packer, J. and Ballantyne, R. (2015) Exploring visitor experiences, activities and benefits at three religious tourism sites. *International Journal of Tourism Research* 17(5), 471–481.

Bonetti, F., Warnaby, G. and Quinn, L. (2018) Augmented reality and virtual reality in physical and online retailing: a review, synthesis and research agenda. In: Jung, T. and tom Dieck, M. (eds) *Augmented Reality and Virtual Reality: Empowering Human, Place and Business*. Springer, Cham, Switzerland, pp. 119–132. https://doi.org/10.1007/978-3-319-64027-3_9

Casais, B. and Sousa, B. (2020) Heterogeneity of motivations and conflicts in pilgrim-to-pilgrim interaction: a research on the way of Saint James. *Tourism Management Perspectives* 36, 100748.

Chi, O.H. and Chi, C.G. (2022) Reminiscing other people's memories: conceptualizing and measuring vicarious nostalgia evoked by heritage tourism. *Journal of Travel Research* 61(1), 33–49.

Damari, C. and Mansfeld, Y. (2016) Reflections on pilgrims' identity, role and interplay with the pilgrimage environment. *Current Issues in Tourism* 19(3), 199–222.

Eimer, M. (2004) Multisensory integration: how visual experience shapes spatial perception. *Current Biology* 14(3), R115–R117.

Errichiello, L., Micera, R., Atzeni, M. and Del Chiappa, G. (2019) Exploring the implications of wearable virtual reality technology for museum visitors' experience: a cluster analysis. *International Journal of Tourism Research* 21(5), 590–605.

Fan, X., Jiang, X. and Deng, N. (2022) Immersive technology: a meta-analysis of augmented/virtual reality applications and their impact on tourism experience. *Tourism Management* 91: 104534.

Flavián, C., Ibáñez-Sánchez, S. and Orús, C. (2019) The impact of virtual, augmented and mixed reality technologies on the customer experience. *Journal of Business Research* 100, 547–560.

Flavián, C., Ibáñez-Sánchez, S. and Orús, C. (2021) The influence of scent on virtual reality experiences: the role of aroma-content congruence. *Journal of Business Research* 123, 289–301.

Freeman, D., Reeve, S., Robinson, A., Ehlers, A., Clark, D. *et al.* (2017) Virtual reality in the assessment, understanding, and treatment of mental health disorders. *Psychological Medicine* 47(14), 2393–2400.

Freidus, A. and Caro, L. (2021) Mobilizing affect in the search for self-transformation: a case study of volunteer transformation in orphanage tourism. *Annals of Tourism Research* 86: 103084.

Guo, K., Fan, A., Lehto, X. and Day, J. (2021) Immersive digital tourism: the role of multisensory cues in digital museum experiences. *Journal of Hospitality & Tourism Research* 47(6), 1017–1039.
HajjiVR (2023) Learn the rites of Hajj and Umrah in virtual reality! Available at: https://hajjivr.com/ (accessed 10 May 2023).
Hill-Smith, C. (2009) Cyberpilgrimage: a study of authenticity, presence and meaning in online pilgrimage experiences. *Journal of Religion and Popular Culture* 21(2): 6.
Hill-Smith, C. (2011) Cyberpilgrimage: the (virtual) reality of online pilgrimage experience. *Religion Compass* 5(6), 236–246.
Hinsch, C., Felix, R. and Rauschnabel, P.A. (2020) Nostalgia beats the wow-effect: inspiration, awe and meaningful associations in augmented reality marketing. *Journal of Retailing and Consumer Services* 53: 101987.
Hoyer, W.D., Kroschke, M., Schmitt, B., Kraume, K. and Shankar, V. (2020) Transforming the customer experience through new technologies. *Journal of Interactive Marketing* 51(1), 57–71.
Hughes, K., Bond, N. and Ballantyne, R. (2013) Designing and managing interpretive experiences at religious sites: visitors' perceptions of Canterbury Cathedral. *Tourism Management* 36, 210–220.
Jung, T. and tom Dieck, M.C. (2018) The future of heritage tourism: perspectives of augmented reality. *Journal of Travel Research* 57(8), 1054–1067.
Jung, T., tom Dieck, M.C., Lee, H. and Chung, N. (2016) Effects of virtual reality and augmented reality on visitor experiences in museum. In: Inversini, A. and Schegg, R. (eds) *Information and Communication Technologies in Tourism 2016: Proceedings of the International Conference in Bilbao, Spain, February 2–5, 2016*. Springer, Cham, Switzerland, pp. 621–635.
Kaelber, L. (2006) Paradigms of travel: from medieval pilgrimage to the postmodern virtual tour. In: Timothy, D. and Olsen, D. (eds) *Tourism, Religion and Spiritual Journeys*. Routledge, Abingdon, UK, pp. 49–63.
Kirillova, K., Lehto, X. and Cai, L. (2017) Tourism and existential transformation: an empirical investigation. *Journal of Travel Research* 56(5), 638–650.
Larsen, S. and Mossberg, L. (2007) The diversity of tourist experiences. *Scandinavian Journal of Hospitality and Tourism* 7(1), 1–6.
Lee, I.J., Chen, C.H. and Su, C.Y. (2017) App based souvenirs and entry tickets: a new means of enhancing post visit memories: a case study from Taiwan. *Tourism Management Perspectives* 24, 177–185.
Lemon, K.N. and Verhoef, P.C. (2016) Understanding customer experience throughout the customer journey. *Journal of Marketing* 80(6), 69–96.
Lewicka, M. (2011) Place attachment: how far have we come in the last 40 years? *Journal of Environmental Psychology* 31(3), 207–230.
Lin, J.H.T., Wu, D.Y. and Tao, C.C. (2018a) So scary, yet so fun: the role of self-efficacy in enjoyment of a virtual reality horror game. *New Media and Society* 20(9), 3223–3242.
Lin, Y.C., Chen, Y.P., Yien, H.W., Huang, C.Y. and Su, Y.C. (2018b) Integrated BIM, game engine and VR technologies for healthcare design: a case study in cancer hospital. *Advanced Engineering Informatics* 36, 130–145.
Loureiro, S.M.C., Guerreiro, J. and Ali, F. (2020) 20 years of research on virtual reality and augmented reality in tourism context: a text-mining approach. *Tourism Management* 77, 104028.
MacWilliams, M.W. (2013) Virtual pilgrimage to Ireland's Croagh Patrick. In: Dawson, L.L. and Cowan, D.E. (eds) *Religion Online: Finding Faith on the Internet*. Routledge, New York, pp. 223–238.
McLean, G. and Barhorst, J.B. (2022) Living the experience before you go ... but did it meet expectations? The role of virtual reality during hotel bookings. *Journal of Travel Research* 61(6), 1233–1251.
Meißner, M., Pfeiffer, J., Pfeiffer, T. and Oppewal, H. (2019) Combining virtual reality and mobile eye tracking to provide a naturalistic experimental environment for shopper research. *Journal of Business Research* 100, 445–458.
Merchant, Z., Goetz, E.T., Cifuentes, L., Keeney-Kennicutt, W. and Davis, T.J. (2014) Effectiveness of virtual reality-based instruction on students' learning outcomes in K-12 and higher education: a meta-analysis. *Computers and Education* 70, 29–40.
Merkx, C. and Nawijn, J. (2021) Virtual reality tourism experiences: addiction and isolation. *Tourism Management* 87, 104394.
Milgram, P. and Kishino, F. (1994) A taxonomy of mixed reality visual displays. *IEICE Transactions on Information and Systems* 77(12), 1321–1329.
Mirghadr, L., Torabi Farsani, N., Shafiei, Z. and Hekmat, M. (2018) Identification of key components of visitor education in a museum. *Museum Management and Curatorship* 33(3), 223–234.

Morinis, A. (1992) *Sacred Journeys: the Anthropology of Pilgrimage*. Greenwood Press, Westport, Connecticut.

Mu, R., Nie, Y., Cao, K., You, R., Wei, Y. and Tong, X. (2023) Pilgrimage to pureland: art, perception and the Wutai mural VR reconstruction. Preprint. arXiv preprint arXiv:2304.07511. Available at: https://www.researchgate.net/publication/370070733_Pilgrimage_to_Pureland_Art_Perception_and_the_Wutai_Mural_VR_Reconstruction (accessed 1 November 2023).

Muslim3D (2023) Embark on a wonderful journey. Available at: https://muslim3d.io/ (accessed 9 May 2023).

Parasuraman, A., Zeithaml, V.A. and Berry, L.L. (1988) SERVQUAL: a multiple-item scale for measuring consumer perceptions of service quality. *Journal of Retailing* 64(1), 12–40.

Patrício, L., Fisk, R.P., Falcão e Cunha, J. and Constantine, L. (2011) Multilevel service design: from customer value constellation to service experience blueprinting. *Journal of Service Research* 14(2), 180–200.

Pine II, B.J. and Gilmore, J.H. (1998) Welcome to the experience economy. *Harvard Business Review* 76(4), 97–105.

Raptis, G.E., Fidas, C. and Avouris, N. (2018) Effects of mixed reality on players' behaviour and immersion in a cultural tourism game: a cognitive processing perspective. *International Journal of Human-Computer Studies* 114, 69–79.

Riva, G., Waterworth, J.A. and Waterworth, E.L. (2004) The layers of presence: a bio-cultural approach to understanding presence in natural and mediated environments. *Cyberpsychology and Behavior* 7(4), 402–416.

Robaina-Calderín, L., Martín-Santana, J.D. and Muñoz-Leiva, F. (2023) Immersive experiences as a resource for promoting museum tourism in the Z and millennials generations. *Journal of Destination Marketing and Management* 29, 100795.

Rodríguez-Ardura, I. and Martínez-López, F.J. (2014) Another look at 'being there' experiences in digital media: exploring connections of telepresence with mental imagery. *Computers in Human Behavior* 30, 508–518.

Shehade, M. and Stylianou-Lambert, T. (2020) Revisiting authenticity in the age of the digital transformation of cultural tourism. In: Katsoni, V. and Spyriadis, T. (eds) *Cultural and Tourism Innovation in the Digital Era: Sixth International IACuDiT Conference, Athens 2019*. Springer, Cham, Switzerland, pp. 3–16.

Tagliacozzo, E. and Toorawa, S.M. (eds) (2016) *The Hajj: Pilgrimage in Islam*. Cambridge University Press, Cambridge.

Timothy, D.J. and Boyd, S.W. (2003) *Heritage Tourism*. Pearson Education, London.

Tsai, S.P. (2020) Augmented reality enhancing place satisfaction for heritage tourism marketing. *Current Issues in Tourism* 23(9), 1078–1083.

Tussyadiah, I.P. and Fesenmaier, D.R. (2009) Mediating tourist experiences: access to places via shared videos. *Annals of Tourism Research* 36(1), 24–40.

Tussyadiah, I.P., Wang, D., Jung, T.H. and tom Dieck, M.C. (2018) Virtual reality, presence, and attitude change: empirical evidence from tourism. *Tourism Management* 66, 140–154.

Vatican VR (2023) Ultimate virtual reality tour vatican 2023. Available at: https://vatican-tickets.tours/vatican-tours/vatican-virtual-reality-tour/ (accessed 3 May 2023).

Virtual Reality Experience in Jerusalem (2023) A look into the past. Virtual reality experience. Available at: https://english.thekotel.org/attractions/a-look-into-the-past/ (accessed 2 May 2023).

Wedel, M., Bigné, E. and Zhang, J. (2020) Virtual and augmented reality: advancing research in consumer marketing. *International Journal of Research in Marketing* 37(3), 443–465.

Wei, W., Qi, R. and Zhang, L. (2019) Effects of virtual reality on theme park visitors' experience and behaviors: a presence perspective. *Tourism Management* 71, 282–293.

Yuqiang, B., Niblock, C. and Bonenberg, L. (2017) Lincoln Cathedral interactive virtual reality exhibition. In: Çağdaş, G., Özkar, M., Gül, L.F. and Gürer, E. (eds) *Future Trajectories of Computation in Design – 17th International Conference, CAAD Futures 2017, Istanbul, Turkey, July 12–14, 2017, Proceedings*. Springer, Cham, Switzerland, pp. 585–595.

Zheng, C., Chen, Z., Zhang, Y. and Guo, Y. (2022) Does vivid imagination deter visitation? The role of mental imagery processing in virtual tourism on tourists' behavior. *Journal of Travel Research* 61(7), 1528–1541.

Zhu, C., Fong, L.H.N. and Gan, M. (2023) Rethinking the consequences of postmodern authenticity: the case of a world cultural heritage in augmented reality. *Current Issues in Tourism* 26(4), 617–631.

8 A Global Turn? Pandemic 'Virtual Pilgrimage' and the Politics of COVID-19 in the UK

Eleanor O'Keeffe*
National Centre for Social Research, London, UK

Abstract

This chapter investigates how virtual pilgrimage reached a new level of societal and academic recognition during the coronavirus disease 2019 (COVID-19) pandemic and the politics that informed and animated these processes. Virtual pilgrimage is often considered a global and globalizing force, defying national borders and democratizing engagement. Evidence from a data-driven approach to understanding the scope and character of virtual pilgrimage growth during the COVID-19 pandemic complicates these interpretations. Based on a study of ritual adaptation in the UK, the argument assembles social media and media analysis to investigate virtual pilgrimage trends and developments, demonstrating the responsiveness of virtual pilgrimage forms to the (national) politics of COVID-19. The evidence presented here identifies two contemporaneous operations: (i) the dominance of nationally grounded institutional digital infrastructures over virtual pilgrimage production to meet particularly pressing challenges after COVID-19; and (ii) the transformation of pilgrimage discourses, as they became invigorated by the identity politics of COVID-19. The chapter thus explores virtual pilgrimage in all its 'virtualities' – as a genre of digital culture and social media discourses – as a valve for the pandemic's anxieties and tensions in the UK.

Introduction

Virtual pilgrimage reached a new level of societal and academic recognition during the coronavirus disease 2019 (COVID-19) pandemic. While cyber or 'virtual pilgrimage' is as old as the Internet (older, indeed, medieval historians will argue) the sudden rise in popularity of 'virtual' pilgrimages during a period of extensive social and travel restrictions created headlines across religious and secular media. 'Virtual pilgrimage' relayed a positive narrative about social life in the pandemic – a parable of technological ingenuity and the endurance of human connection in the face of crises. Its particularities and prevalence during the pandemic have piqued scholarly interest and enquiry into these forms of digital culture (e.g. Shinde, 2022). Research has ranged from identifying and exploring the range and variation in outputs under COVID-19 (Bailey, 2022a, pp. 192–193) to examination of the theological ramifications of productions emerging from geopolitical and demographic shifts that COVID-19 initiated in Christian and Muslim contexts (Dowson, 2020; Niu, 2023). Some have argued that 'DIY pilgrimage' (facilitated by smart technologies)

*eleanor.okeeffe@natcen.ac.uk

 Religious Tourism and Globalization. D. Liutikus and R. Raj (eds)
DOI: 10.1079/9781800623675.0008

met global social needs under lockdowns (Barush, 2020) or generated global engagement, inclusivity and 'communitas' (Dunn-Hensley, 2020; Dowson, 2021). Whether COVID-19 sparked a global turn in pilgrimage through smart and communications technologies has emerged as an important question for future research (Dowson and O'Keeffe, 2023).

The realities of the various guises of virtual pilgrimage under COVID-19, and their impacts, are still unclear. We are far from comprehending the nature and significance of these practices during the pandemic from a macro perspective and charting how an array of structural, geopolitical and national pressures influenced them. Assessing COVID-19 adaptation, whether this fostered inclusivity or founded new global communities in pilgrimage, is extremely difficult. This is partly due to the methodological challenge of analysing any macro shifts in dispersed and hyper networked digital culture. But the study of 'cyber' pilgrimage (in all manifestations) is still an emerging field of scholarship (see Hill-Smith, 2011, pp. 239–240). Some research has catalogued virtual pilgrimage outputs through empirical research, identifying commonalities and iterating frameworks (MacWilliams, 2002). However, most pre-pandemic research into virtual pilgrimage analyses unique domains of virtual pilgrimage action, qualitatively and often through ethnographic approaches. This collection of studies, when examined empirically, reveals the contours of virtual pilgrimage on the web before 2020, but the uncertainty about pre-pandemic activity makes assembling a 'before' and 'after' picture difficult.

The relative nascence of the field makes it harder to comprehend all aspects of virtual pilgrimage under COVID-19. 'Virtual pilgrimage' scholarship has largely promoted individualistic or agency-centred perspectives, proposing that the democratizing aspects of web 2.0 undercut institutional power structures. Studies foreground 'experience' as a metric of authenticity, for instance the democratization of pilgrim identity and ritual experience through smart and connective technologies (van der Beek, 2019; Polus and Carr, 2021), or the grassroots use of such forms in diasporic communities (Helland, 2007). While virtual forms of pilgrimage satiated more individual or communal needs under COVID-19, as Barush (2020) has argued, there are also more complex issues of power and authority to contend with, considering the widespread governmental regulation of society and the heightened inequalities of social capacity. Research has previously reflected on the making of virtual pilgrimage, and its collaborative and infrastructural foundations, from a critical and reflective insider perspective (e.g. Clarke, 2020; Bergeron and Green, 2021). There are few (albeit stand out) analyses of how communications technologies create new structures of authority (Golan and Martini, 2018). Such considerations are important to consider for the COVID-19 context, which brought virtual pilgrimage into the mainstream.

This chapter reports findings from a data-driven approach to understanding the scope and character of virtual pilgrimage during the pandemic. It draws together a range of social media and media data to fashion some macro-level analysis of virtual pilgrimage developments during 2020. The pandemic provided an unprecedented opportunity to study the variation evident in these digital practices, while also examining their relationship to the political economy and what infrastructures (knowledge, digital, institutional) resourced them. The evidence presented here suggests the alignment of pandemic virtual pilgrimage production to institutional and cultural digital infrastructures, and the continuance of institutional power over form, meaning and 'communitas', which mattered in the pandemic's initial phases. It thus questions whether digital communications tools in pilgrimage (or for other uses) were inherently democratizing, inclusive, or, indeed, truly globalizing during the pandemic (Beaverstock *et al.*, 2023); in this, it echoes findings of *in situ* studies that a significant effect of the global pandemic has been to increase local and national pilgrimage practices (Mroz, 2021; Bailey, 2022b).

The research applied a national lens to study how pilgrimage conducted, or channelled, the nationalizing and globalizing forces instigated or heightened by the pandemic, such as the growth in social media or Internet use, increased state regulation of the lives of citizens and restrictions on travel. Although pilgrimage is often considered as the quintessential mode of transcendence over the nation state, it lays bare tensions between identities grounded in patrimonial claims over heritage and transnational networks and modalities of being (Geary, 2014). Virtual pilgrimage bears even greater assumptions of such transcendence. Analyses emphasize individualist, almost libertarian, possibilities for conducting religious travel on the web; they assess the authenticity of such rituals through subjectivities, investigating experience, how individuals use these forms, and their motivations (Hill-Smith, 2011). Investigating whether similar tensions exist (between global and national networks and identities) in virtual realms as they do *in situ* practices is, thus, important to establish. Case studies of national contexts illustrate how globalizing forces (corporate, technological, geopolitical) congregate differently in particular sites and spaces, recognizing that personal and collective significances and identities within social life may remain primarily local or 'glocal' (Griffin, 2015). This chapter considers virtual pilgrimage impacts within a national framework, as far as is possible, to explore how these tensions resolved in this genre of digital culture.

The findings presented in this chapter are drawn from research undertaken during the pandemic. In 2020, the body responsible for the allocation of public funding in the UK, the UK Research and Innovation (UKRI) Funding Service, issued an emergency call for funding to study the social impacts of the COVID-19 pandemic. UKRI devolved a small portion of this resource to the Arts and Humanities Research Council (AHRC), which fast-tracked 77 new projects that would study the pandemic's various and multifaceted impacts in real time. Recognizing that ritual 'does social work', *British Ritual Innovation under COVID-19* (Edelman *et al.*, 2021), hereafter referred to by its abbreviated title, BRIC-19, examined changes in ritual life occasioned by societal lockdowns. This included investigating the impacts of 'digital innovation' on the UK's faith and belief communities and the ways in which ritual itself adapted to meet the needs (social, existential) of wider society. Virtual pilgrimage became a significant subject of this research initially because of the considerable media attention that virtual, cyber or 'e-pilgrimage' captured in the pandemic's early months, which drew project focus alongside other prominent ritual shifts online (e.g. Tadie, 2020).

That the trajectories of virtual pilgrimage appeared to resonate with broader findings relating to the growth of spirituality in the face of crisis made it an important subject to pursue beyond the headlines in religious and secular media (Bentzen, 2021). But the BRIC-19 case study first investigated, according to the project's central aims, engaging with emerging debates about the pandemic in social and political science and religious studies, such as the inequalities of infrastructure, social and community capacities under crisis, resilience, citizenship and the place of faith and belief in UK society. It was these research concerns, rather than those already established for virtual pilgrimage by scholars in the fields of digital religion, pilgrimage or communications studies, which guided how the case study proceeded. The research was conducted as a two-stage enquiry. First, it aimed to locate and gather an evidence base to test and analyse claims to growth. Second, it sought to investigate the nature and character of growth, to see where it was most evident, and how it related to community capacity.

This study exemplifies how pandemic-impact research intersects with scholarly concerns in this field and how it can enhance and contribute to these areas of scholarship. Particularly, this chapter has taken considerable inspiration from the work of Simon Coleman (2021), who has invited scholars to embrace the multivalence of contemporary pilgrimage and its intersections with other 'articulations' of social and cultural life. Coleman (2021) acknowledges the methodological difficulties of such a project, which became all too clear in this research considering the intricate ways in which pilgrimage becomes entangled with web 2.0. For that reason, I was not prescriptive about which points of intersection between pilgrimage and digital culture should be studied and did not propose a definition for 'cyber pilgrimage', as scholars have done (Hill-Smith, 2011). I have embraced as many of the 'virtualities' of pilgrimage as the data presented, which have included both individual and communal ritual practices and discourses, although I do not claim that the data can be seen as an objective or exhaustive record of all of these.

Evidence and Data: Twitter as an Archive of Communication Flows

Generating macro-level insights to examine changes in ritual practice during the pandemic is fraught with difficulty. The route adopted in this research has relied on collecting a large corpus of UK-based tweets to investigate and analyse the major changes in practice and discourse. The Twitter data was generated by Pulsar Platform, an audience insights agency based in the UK with paid access to Twitter's application programming interface (API). Pulsar congregated tweets based on keyword searches of Twitter from January 2019 to 31 December 2020, which allowed for the comparison of the pandemic period with the previous year. The central search (under generic pilgrimage keywords – pilgrimage, #Pilgrim, Pilgrimage, pilgrim) produced a large corpus of tweets (n 82,429). This was supplemented with database research (Nexis UK) into UK-based media, including national and local media, magazine outputs and religious press. This provided one macro-level view of changes in the UK.

There are caveats to acknowledge regarding the representative nature of this dataset. Twitter engagement in the UK is skewed towards men, younger sections of the population, and managerial/professional occupations, although usage has a wider age and occupational diversity than many expect (Sloan, 2017). Despite what could be considered as relatively diverse user population for a social media platform, there is still little demographic data to appreciate how pilgrimage users engage with Twitter. Social media sites (SMS) are foundational to meaning making and identity forming in pilgrimage, with users networking across platforms to benefit from their distinct affordances (van der Beek, 2018, 2019). How this manifested in the UK, and the range of practices UK Twitter might collect, is unclear. Pulsar's functionalities also influenced the representativeness of the data. The search facility drew on geolocation data and/or user biographical information to establish a 'UK-based' dataset with some validity. Yet the resources of the research team, along with the author's limitations, limited the search to English-language tweets only. This automatically excluded any geolocated post in languages other than English, which are used by many residents, citizens and visitors to the UK. The data captures, then, a normative and distinct set of pilgrimage uses and discourses.

Yet Twitter – more than other social media platforms – centralizes the multi-directional, networked communication flows routed by pilgrimage in all its 'virtualities'. The Twitter data assembled digital artefacts created from the slippage between *in situ* and digital practices at an individual level, acknowledging that these may not be fully representative. It also captured outputs from the social media activity embedded into the touristic, heritage and commercial infrastructures that create or contribute to pilgrimage practice. These entities, groups or institutions are likely to employ the social media platform for outreach, public engagement and brand development (Kidd, 2011). Examining the traces left on social media from individual and collective activity, therefore, can also identify what forms of virtual pilgrimage became adopted in the pandemic, and to whom this genre of digital culture mattered most. Twitter's reputation and status (at least in 2019–2020) as a marketplace of ideas and pseudo public sphere enhanced its utility. The dataset provided the basis to examine pilgrimage discourses, and how pilgrimage 'tropes' (Coleman, 2021) interfaced with (or were invigorated by) the politics and debates created by the sociopolitical dynamics created by the pandemic in the UK. As such, the Twitter data is employed as an archive of communication flows established through pilgrimage action and public discourse.

Exploring the many interfaces between pilgrimage and ritual, social and political operations embedded in this archive required a multi-methods approach. First, the data has been used to chart patterns of engagement (Fig. 8.1), tabulate salient topics (Table 8.1) across 2019 and 2020, and as an archive of user content. These concurrent processes generated comparative data on user activity and impacts of digital innovations, as well as uncovering shifts in meaning within specific social and political contexts through empirical study. Second, building on its connections to the event-driven apparatus of virtual pilgrimage activity, the Twitter data provided the basis for a new dataset (referred to as the *virtual pilgrimage dataset* henceforth) to catalogue the study of virtual pilgrimage events and communal activities during 2020. This incorporated additional data drawn from Nexis UK media databases to ensure comprehensive coverage of activities, with Twitter serving as the primary source. This virtual pilgrimage dataset yielded insights into the occurrences of virtual pilgrimage production by month (Fig. 8.2), their representational qualities and destinations (Fig. 8.3) and facilitated analysis of their technological formats (Fig. 8.4). Finally, the

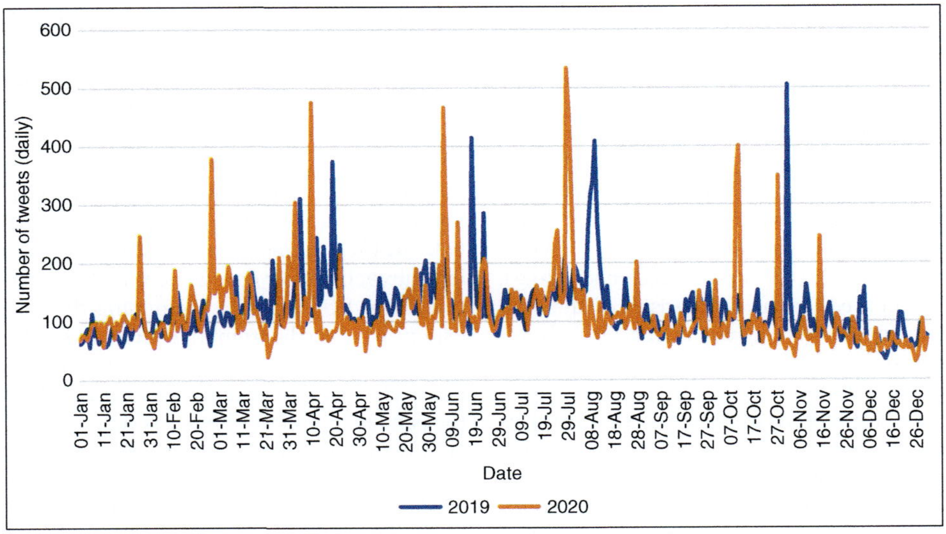

Fig. 8.1. Engagement trending patterns on Twitter, 2019–2020. Based on plotting daily data from n 42,826 (2019) and n 39,603 (2020) tweets generated by Pulsar Platform.

Table 8.1. Salient topics on Twitter, 2019–2020: a comparison of most frequent keyword instances 2019–2020. Based on keyword searches (NVivo) of a corpus of n 42,826 (2019) and n 39,603 (2020) tweets generated by Pulsar Platform.

	Year	
Topic	2019	2020
Pandemic-related words lockdown; COVID-19; COVID; C19; coronavirus; pandemic; social distance	256	5398
Words relating to technology virtual; online; website; Facebook; Instagram; YouTube; podcast	426	3652
Words relating to Islam Mecca; Hajj; Muslim; Muslims; umrah; #Hajj; #Hajj2020; Islam; #prophetmuhammad; Eid; Muhammed	6114	8741
Words relating to history history; #history; nuns; medieval; #medieval	1346	2311

Twitter data, which included public user biographical information, also provided a resource to study 'pilgrimage influencers' (Fig. 8.5) – the 387 users who posted more than ten times on pilgrimage topics.

This chapter can only outline and discuss these findings in the most general of terms. The first section compares the engagement and user trends between 2019 and 2020, tracing the major impacts of COVID-19, particularly rapid digital acculturation. The second section works within the framework of *controversies* and *identities*, exploring the separate ways in which technology and the global context of the pandemic reconfigured the relationship between pilgrimage and identity, through the study of salient pilgrimage topics and qualitative, in-depth and empirical examination of the content of the tweets and their social and political contexts. This section highlights the ways in which national-level contexts and politics intersected with tropes and discourses of pilgrimage and contributed to the reshaping of civic identities during the time of social transition and crisis.

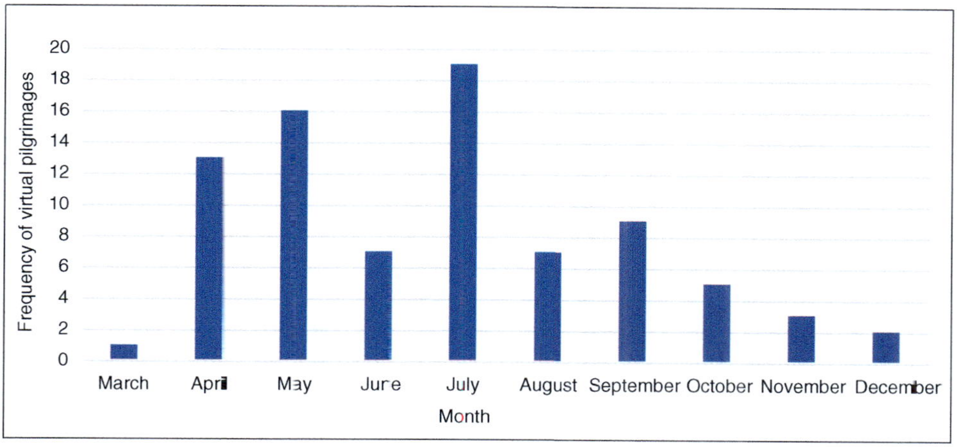

Fig. 8.2. Virtual pilgrimage events by month, 2020. Based on a study of 81 events identified in the 'virtual pilgrimages dataset'.

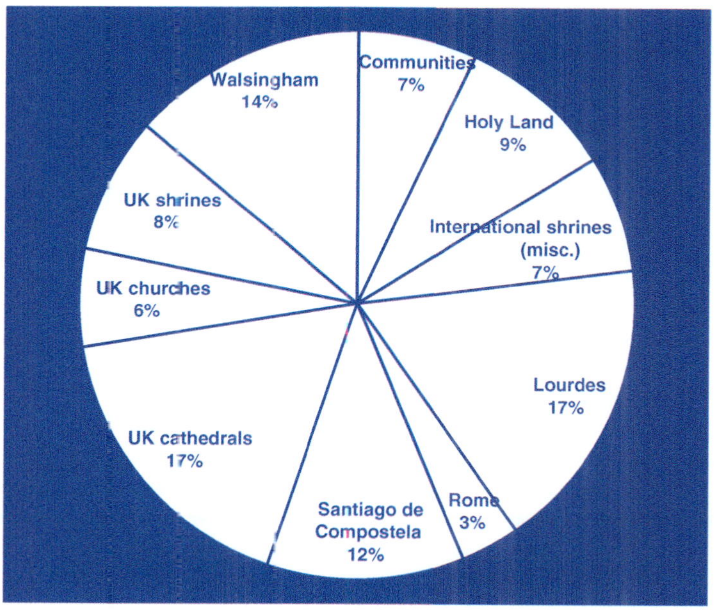

Fig. 8.3. Destinations of virtual pilgrimage events, 2020. misc., miscellaneous. Based on a study of 81 events identified in the 'virtual pilgrimages dataset'.

The Pandemic and its (Dis)continuities: the Impact and Extent of Digital Practices

From research that emphasizes the flourishing of pilgrimage practices in the UK during COVID-19 (Bailey, 2022a; O'Keeffe, 2023b), how digital acculturation can contribute to individual ritual creativity (Barush, 2020), the rise in social media and online use during the pandemic (Ofcom, 2022) and how the Internet fostered spiritual responses to the challenges of the global crisis (Bentzen, 2021), one might have hypothesized growth in social media engagement with

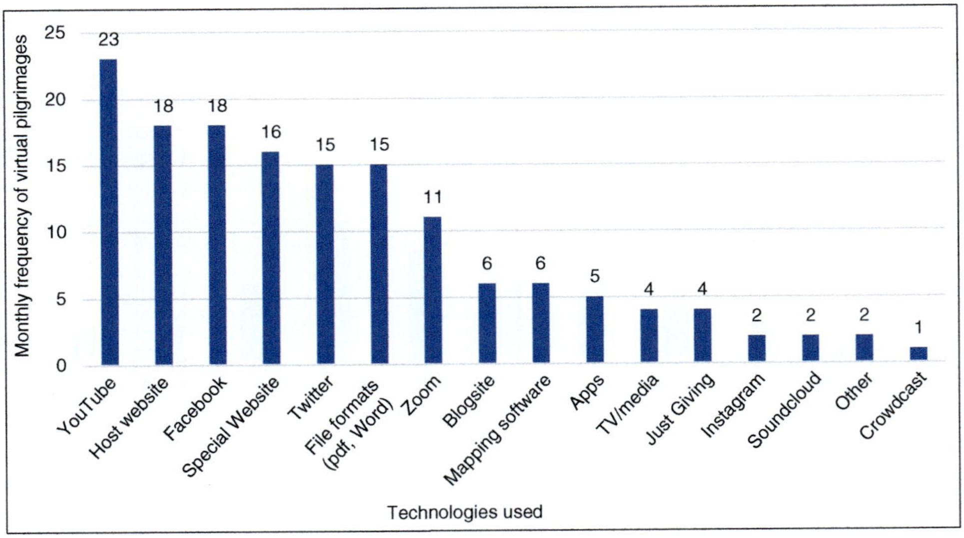

Fig. 8.4 Technologies used in virtual pilgrimage events (2020). Based on a study of 81 events identified in the 'virtual pilgrimages dataset'.

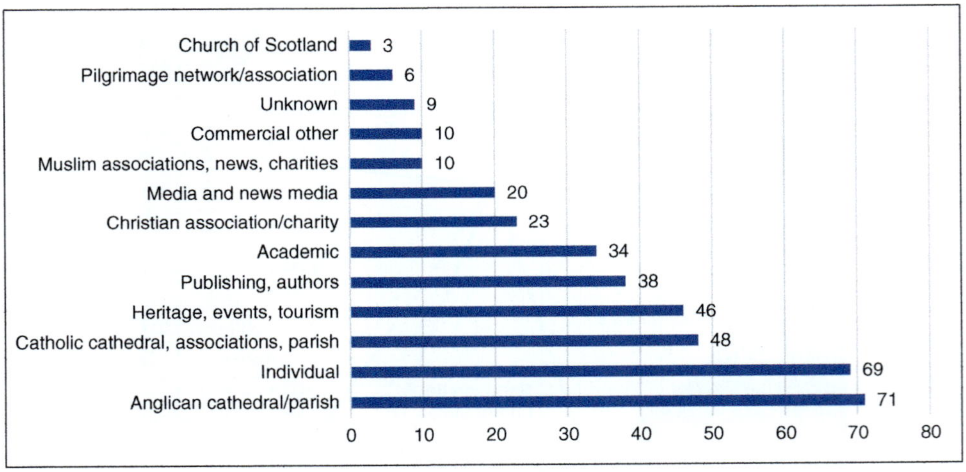

Fig. 8.5. Pilgrimage influencers (2020). Based on analysis of users who tweeted more than ten times using the pilgrimage keywords (n 387).

pilgrimage topics during the pandemic. The Twitter data tempers these expectations: 2020 saw a slight decrease (7%) in Twitter activity employing pilgrimage keywords. In the wider context of pandemic conversation, pilgrimage also constitutes an exceedingly small part of national-level Twitter discourse. For instance, the number of UK tweets (80,000 or so) captured under a 'pilgrimage' search are lower than the 1.3 million tweets captured by Pulsar under the keyword search for 'memorial' for the same period (O'Keeffe, 2023a). Despite widespread and rapid growth in online use in the UK during the pandemic, there were 10% fewer users employing the main pilgrimage keywords on Twitter in 2020 compared with the previous year.

These patterns indicate the disruptive influence of COVID-19 and the national and global travel restrictions that ensued in the pandemic, which many predicted would present enormous challenges to religious tourism (Raj and Griffin, 2020). Indeed, the lowest engagement over the 2 years came on 23 March 2020, the day that the UK's first national lockdown and social distancing restrictions were announced. However, while activity may be lower overall in 2020 when compared with the previous year, engagement patterns demonstrate that pilgrimage was still a living topic of concern during the pandemic's first months. Peaks in the 2020 data are slightly more frequent and more pronounced, suggesting a higher prevalence of controversy or trending even if daily engagements run low. The substance of conversation content is, however, clearly different, especially the introduction of pandemic terminology, for which there is no precedent 2019 (Table 8.1). This includes references to the pandemic and COVID-19, along with terminology relating to the social restrictions in the UK ('lockdown', 'social distance'). These words or short phrases often coincide with words describing rupture: 'closed', 'postponed' and 'cancelled'. In fact, 'cancelled' appears as frequently as 'journey' (a central trope of post-secular pilgrimage) in the 2020 data, with one-third of these references concentrated in the first few weeks of the pandemic.

The next most significant qualitative difference between the 2019 and 2020 data relates to digital innovations and adoption. References to 'virtual' pilgrimage, meeting 'online' or other social media platforms increased tenfold in the 2020 data compared with the previous year, despite overall fall in engagement. The timing of these references shows the rapid and temporally bounded nature of adaptive activity in the first few months of the pandemic. Usage of these terms rose steeply in the first weeks after March 2020, but became pronounced from the start of May and are most common in the month of July. This finding corresponds with the *virtual pilgrimage dataset*, which identified 81 UK-based, communal online pilgrimages held between April and December 2020. Most of these events occurred within the initial months of the pandemic and particularly the summer (Fig. 8.2).

The virtual pilgrimage dataset facilitates further study of these genres. Scholarly attention has focused on two poles of virtual pilgrimage outputs: (i) the individual as primary locus for virtual pilgrimage action; or (ii) the large shrine-based 'mega-events'. However, the pandemic triggered a considerable productive effort between these two scales and created an adaptive new genre of virtual pilgrimage activity. This was Christian, with 77% of these events coming from the work of Catholic or Church of England cathedrals or diocesan networks or Christian pilgrimage associations. Much of this can be seen as a rapid, crisis-driven response, quickly assembled in the wake of the first national lockdown (23 March–1 June 2020). Regardless of their theological or spiritual efficacy, which was not discussed in the social media engagement, their publicity material positioned them as consolations for visits that could not take place, or placeholders for future visits – they represented important continuity markers. Most productions relied on simple, inexpensive software tools, which associations, communities and others coordinated in different ways to realize a sort of 'virtual' experience that could be experienced collectively: (i) video conferencing platforms; (ii) PowerPoint slides; and (iii) videos (in some cases, pre-recorded prior to engagement campaigns) uploaded on to YouTube (Fig. 8.4).

These outputs did not come close to achieving record levels of participation seen in the pandemic's 'mega-event' online pilgrimages, nor did they draw the same remarked-upon global engagement. Many did not attempt to broadcast to or entice global networks – their main purpose was to meet specific community needs and participation levels would be known only to those who led them. In cases where greater accessibility became possible or was encouraged, such as those publicized on YouTube, the resulting views, comments or likes were limited in number and especially when compared to the engagement levels publicized for 'mega-events' such as Lourdes. For instance, Canterbury Cathedral's virtual pilgrimage in November and December 2020 achieved its highest viewing figures for the first of its six weekly episodes (2991 views) but averaged under 2000 views and 87 likes for each overall. There were very few comments to reveal the textures of experience or to evidence the kinds of global communities that congregated through these digital means (Canterbury Cathedral, 2020).

Several factors drove adaptation, including theological, institutional and infrastructural considerations. Anglican and Catholic cathedrals, for instance, are the most significant single producers within the virtual pilgrimage dataset and hold substantial stakes in the conversation about pilgrimage. Analysis of 387 'pilgrimage influencers' (defined as users who tweeted over ten times on pilgrimage topics), responsible for a quarter of the Twitter content, shows that pilgrimage collates an array of interests and affiliations, reflecting the multifaceted nature of pilgrimage activity (Fig. 8.5). Pilgrimage on Twitter has touristic and commercial dimensions and is heavily influenced by publishing, as well as arts and heritage sectors and interests. But institutional actors, Catholic and Anglican, feature prominently in this analysis. This encompasses accounts from diocesan networks and cathedrals, and those of individual representatives of both churches at parish level. 'Virtual pilgrimage' has certainly provided a new and important stratum of interlocution, supplying another arena for authority and mediation, which cathedral public engagement teams, cathedral managers and digitally resourced and skilled parish priests or networks are clearly entering (Golan and Martini, 2018).

Virtual pilgrimage capabilities reflected strategic and infrastructural investment that has happened over the past two decades, particularly in the Church of England. Most importantly, the 'heritagization' of UK cathedrals – the reframing and repurposing of the materiality of religious spaces as historic and heritage sites – has not only made pilgrimage a good topic for cathedral managers to position religious sites in the modern world but has introduced the organizational development of communications and social media skills aligned with the heritage sector (Kidd, 2016; Bowman and Sepp, 2019; Coleman, 2019). Additionally, the Church of England has cultivated a wholesale development of digital outreach capabilities to 'support digital evangelism and discipleship' at the level of the parish engagement (Church of England, nd, 2018a, b). Given a context where 'caminoization' (Bowman and Sepp, 2019) in the UK has re-signified rural churches as heritage landscapes, it is unsurprising that parish-level actors have also seen pilgrimage as a productive means for social reach during a time when congregations were already dwindling.

The dilemmas and challenges faced by faith and belief groups and institutional religion, which 'heritagization' and 'caminoization' had aimed to counter, became far worse in the pandemic. Lockdowns put a spotlight on the role of religion in society, sparking concerns of the freedom of religion overall (Cranmer and Pocklington, 2020; Robinson, 2020). It was not solely the closure of places of worship, whose doors were forced shut until 10 June, and the increased reliance on digital methods of outreach, which spurred virtual pilgrimage production (Church of England, 2020). Inequalities in digital infrastructure certainly prevented many churches conducting virtual 'tours' or pilgrimages online (Centre for the Study of Christianity and Culture, 2021). As important as digital skills and capacity was how Anglican and Catholic churches managed labour resources. Neither institution embraced the Coronavirus Job Retention Scheme (known as the 'furlough' scheme), where the state subsidized a portion of employees' salaries if employers closed shop, as it undermined social importance and the 'keyworker' role of the churches and its employees (Sherwood, 2020; *Church Times*, 2021). While the heritage and arts sectors, therefore, were heavily furloughed throughout the pandemic (Francis-Devine *et al.*, 2021), it is likely that the rates of furlough in both churches were far less, with a possible resultant effect on digital outputs (Clark, 2021).

These political and infrastructural factors influenced virtual pilgrimage under COVID-19 and heavily influenced their representational characteristics. Previous research on virtual pilgrimage has noted how online and digital methods tend to remove pilgrimage from a 'post-secular' iteration focused on journeying and refocus attention on spatial dynamics. As Kratchovil (2021) has recently argued, 'the spatiality of pilgrimage ... is always political' and the refocus of pilgrimage efforts online expresses 'a diversification of political claims about what the Church is and how it should relate to the modern world of today'. In Kratchovil's (2021) analysis of the geopolitics of Catholic spatial constructions in virtual pilgrimage within SMS, representational constructions convey two central and contrasting political readings: focus on the destination, he argued, projects a more 'conservative' idea of pilgrimage and sacrality of place, as opposed to more 'progressive' ideas that focus on movement and, particularly, landscapes.

This study demonstrates that the politics that infused and animated virtual pilgrimage were as much national as geopolitical in the context of COVID-19. Analysing locations in the virtual pilgrimage dataset, the endless possibilities that web 2.0 provided in fact had little sway on the content or directions of 'journeying' of these virtual pilgrimage outputs. The conservative 'destination' focus – on shrines and sites – was pronounced and the importance of national framings is clear: 45% of destinations are to UK sites. The heritagization of UK cathedrals and parish churches is evident through the focus on the materiality and interiors and the fact that they were sometimes assembled from the libraries of pre-pandemic public engagement campaigns. Virtual pilgrimage production related fundamentally to how Anglican and Catholic institutions had communicated their significance to UK contemporary society in the period before the pandemic. The imperative to use such methods to underscore this significance only increased during COVID-19.

Controversies and Identities: the Contemporary Meanings and Politics of UK Pandemic Pilgrimage Discourse

During the pandemic, it became common for analyses of virtual pilgrimage to read innovations through a lens of continuity between past and present. In fact, a distinct strand of Twitter's discourse about pilgrimage and specifically virtual pilgrimage under COVID-19 belongs to public history initiatives that draw connections between the 'virtual' practices of medieval convents and the re-appropriations of pilgrimage online in response to COVID-19 (e.g. Boyle, 2020). The 'medievalizing' of virtual pilgrimage, and pilgrimage as a topic, is demonstrated by the *salient topics data*. There are striking parallels between this tendency during COVID-19 and the aftermath of World War I, when creators of collective memorial culture turned to medievalism to heal wounds from the war's social rupture (Goebel, 2007). Yet, regardless of how fair the comparison between the past and the present may be, or however comforting the narrative, it inadvertently conceals the contingencies and presentism of pandemic virtual pilgrimage and its producers. It may also unwittingly cement a normalization of Christian spirituality in pilgrimage, which does not reflect contemporary realities of pilgrimage under COVID-19. This section discusses these relevancies and the politics of pandemic pilgrimage discourse in 2020.

Analysis of pandemic pilgrimage discourse on Twitter underscores that pilgrimage is evolving, multivalent and contemporary (Coleman, 2021). Meanings associated with pilgrimage, or that which pilgrimage crystallizes, have been heavily shaped by the nation's politics under COVID-19. The collapse of global travel and the politicized restrictions on domestic travel in the UK enlivened pilgrimage as part of national political discourse, particularly because regulations centred on movement and journeying. Inhibition of movement was one of the most significant public health regulations enacted during the UK's pandemic. The 'stay at home' orders (often referred to as 'lockdowns') mandated that individuals could only leave homes with a 'reasonable excuse', which included exercising alone, caring for others or travelling to access work or public services. These regulations could be enforced by the police with fines of varying levels. The scope and nature of these lockdown laws changed throughout the pandemic and remain one of the more contentious and heavily debated aspects of the pandemic in the UK (Barber *et al.*, 2022). These laws invoked fundamental issues such as the state's relationship to its citizens and their rights in this new legal framework.

In this setting, 'pilgrimage' (in its broadest definition of a 'meaningful journey') acquired a set of political usages as a means of both contesting and defending the regulations and those who broke them. In May 2020, the Prime Minister's special advisor, Dominic Cummings, was heavily criticized for travelling from his home in Islington, London, to Barnard Castle in County Durham (covering over 250 miles) during a time of restrictions on domestic movement. The news triggered a flurry of media attention centred on this north-eastern tourist spot. Cummings defended his actions as a necessary journey for the well-being of his young son and family as he struggled with his own health, but the scandal undermined the faith of the British public in the lockdown regulations and trust in the UK government (Fancourt *et al.*, 2020). The incident had other unforeseen consequences, including an increase in tourist visits to Barnard Castle, as media attention from the Cummings affair incidentally highlighted its medieval heritage and local amenities (Sherwood, 2022).

Amid the controversies that erupted on Twitter over Cummings' breach of the rules, pilgrimage became deployed with pointed irony, a gauge for both the absurdity of the regulations and the hypocrisy of their instigators. Users employed lofty language of 'pilgrimage' to mock Cumming's defence of necessity and obligation. Individuals who later visited the north-east site tweeted photos of their 'pilgrimage to Barnard Castle', referencing the significance of the episode in the life of the UK and playing on the medieval association of pilgrimage and the site. Opponents of the government were not the only parties prone to engaging pilgrimage for these purposes. Supporters of the Conservative government used the same tactics against their adversaries. For instance, the son of former Labour leader Neil Kinnock was publicly criticized for travelling to London to visit his father; he also defended his actions as 'essential travel'. One user posted: 'why are the left screeching themselves hoarse over Dominic Cummings but did not utter a single squeak over Kinnock doing a pilgrimage to London during national lockdown?' Pilgrimage discourse, therefore, became entangled with debates and challenges to lockdown regulations.

A more profound politicization of pilgrimage revolved around the relationships between the state and citizens from multi-faith and ethnic minority communities. In the UK, as elsewhere, the pandemic has illuminated the extent of social health inequalities and aggravated discrimination that British Muslims face in the media and from wider society (Poole and Williamson, 2021). Pilgrimage was not the only (or most significant) frontier to defend Muslim rights to equal citizenship, but it became an important one, partly because lockdown regulations imposed blanket restrictions on ritual and communal events, such as weddings, funerals and festivities. Such injunctions had an uneven impact on the UK population. The inequalities in ritual provision became exacerbated in the autumn of 2020 when, to deflect public criticism for instigating an unpopular second national lockdown, the UK government deployed rhetoric of 'saving Christmas', satiating needs of majority white voters to observe this overwhelmingly secular festival (e.g. Vaughan, 2020). In a context where other faiths had either cancelled, or been denied access to, important rituals such stark disparities in treatment sent a 'powerful symbolic message' about Christian norms and citizenship in the UK (De Marco, 2020).

These controversies constitute an important distinguishing feature of the 2020 data. Words related to Islamic identity and observance of the Hajj increase 54% when compared with the previous year, despite an overall drop of 7% in user activity. Some of this increase took place in the preliminary stages of the pandemic, driven by what are sometimes thinly veiled racially charged concerns of superspreading at the Hajj. But the most prominent spikes in this discussion occurred during two periods in the first few months of the pandemic. The first period coincided with the opening of the 'downsized' or 'socially distanced' Hajj in late July 2020, which created a visible peak in the dataset. Announcement of Hajj planning and provisions drew considerable interest from Muslims and non-Muslims, who congregated around the spectacle of social distance conveyed through the images of deserted and distanced shrines – images that were frequently retweeted and repurposed by media agencies (Fig. 8.6). The second period of increase occurred in the Autumn of 2020, when debates over what festivities could be deemed worthy of 'saving' became pronounced after the then-Prime Minister Boris Johnson's promise to the British public.

Both periods of topic growth relate to the proactive 'IslamoCovid' British Muslim response to the COVID-19 pandemic in the UK (Fig. 8.7). British Muslims used the Islamic adherence to regulations to articulate a positive identity in relation to the COVID-19 regulations, particularly through social media (Ouassini *et al.*, 2022). Some of this activity was driven by Muslim anti-racist associational action in the autumn of 2020, during the controversies sparked by the government's rhetoric. Memes created from the words of prophet Muhammad 'at his farewell

Fig. 8.6. Examples of tweets concerning the socially distanced Hajj 2020.

Fig. 8.7. Examples of 'IslamoCovid' Twitter responses.

pilgrimage' advocating racial equality, protesting the rampant Islamophobic social media in the pandemic's first wave, as well as the unfairness in treatment of minority ethnic communities from the UK state (Awan and Khan-Williams, 2020). Adil Ray OBE, a prominent Muslim public figure in the UK, articulated this sentiment in a tweet that went viral in December: 'Never let it be said that minorities don't put Britain first'. These discussions about pilgrimage on Twitter, specifically those concerning the Muslim practices of the Hajj, became a platform for forging solidarity to defend the rights of British Muslims and demonstrate their civic responsibility (Maravia, 2020).

The prevalence of this discourse indicates that, irrespective of the significance of Christian and 'post-secular' trends, an increasing portion of the UK's pilgrimage discourses are created through Islamic pilgrimage practices. In this context, Twitter presented a means to crystallize social solidarity, even a sense of 'communitas' through retweets and other engagements, around an inclusive vision of UK society. This became more pronounced when a report by anti-racist organization Hope not Hate revealed the extent of Islamophobia in the UK Conservative party (Hope not Hate, 2020). But such solidarity also exacerbated divisions as much as it mobilized to heal them, engaging in the kind of public shaming widely reported in the pandemic (Cooper *et al.*, 2022). In October 2020, 'socialist' account @ToryFibs, in a tweet that gained the most traction in that later period, used the willing 'sacrifice' of 'English Muslims' to shame the working and lower middle-class white majority voters of Johnson's appeal. Those who had foregone meaningful religious obligations for the greater good exposed the selfish behaviour of others who clung to empty identity-based cultural tastes, which the tweet derided in class-laden terms ('dry Turkey' and 'fizz'). Pandemic pilgrimage discourse thus erupted through the political and social fissures of UK society, conducting the politics of citizenship under COVID-19.

Summary

This chapter has endeavoured to catalogue and periodize genres of virtual pilgrimage created in the pandemic. It has examined the ways in which some communities and institutions in the UK employed pilgrimage, virtually – as an event and as social media discourse – in the immediate aftermath and what turned out to be the initial phase of the pandemic. It shows that, analogous with other contexts and the broader history of the pandemic, infrastructures mattered (Shinde, 2022). Institutions with resources, capacity and strategic interest were more able to deploy these forms and did much to move virtual pilgrimage from a futuristic and still niche mode of religious tourism into the mainstream (Dunn-Hensley, 2020). But this important action was most available and most useful for Anglican and Catholic communities in the first few months, more so than 'post-secular' associations or charities. While there is no comparable national study that could balance or confirm these findings, this chapter certainly demonstrates the need to take the nation state seriously in future study of this genre of digital culture.

This chapter has challenged the view that virtual pilgrimage was necessarily a globalizing or inclusive socializing force in the pandemic, demonstrating its embeddedness in state–society relations and the fraught, and sometimes divisive, politics it reinforced. During the pandemic, pilgrimage in the UK became a valve for anxieties about the role of the state and its regulation of citizens. It functioned too as a locus for social justice. The suppression of global travel and the regulation of domestic movement made ritual obligation and faith adherence issues for the national polity. While it is possible to see the solidarity enabled through pilgrimage as inclusive, these operations, the events and discourses produced were heavily demarcated by race, religion and politics. Virtual pilgrimage is yet another example of digital culture where web use, in and of itself, cannot substitute for intentional application of resources. Congregating people in new ways and building inclusivity, which many welcomed as a de facto effect of virtual pilgrimage, does not flow simply through technology. A truly inclusive global 'communitas' online may be illusory, but it is most likely to become a reality through a resourced and collaborative effort between representatives from across faith and social divides, software developers and creatives, charities, academia and heritage (Clarke, 2020).

References

Awan, I. and Khan-Williams, R. (2020) Coronavirus, fear and how Islamophobia spreads on social media. Report. Independent members of the Anti-Muslim Hatred Working Group. Available at: https://anti-muslim-hatred-working-group.home.blog/2020/04/20/coronavirus-fear-and-how-islamophobia-spreads-on-social-media/ (accessed 1 June 2023).

Bailey, A.E. (2022a) Pilgrimage during Covid-19: impacts, adaptations, and recovery. *International Journal of Religious Tourism and Pilgrimage* 10(1), article 16. https://doi.org/10.21427/hjtc-4m56

Bailey, A.E. (2022b) Micro-pilgrimages: a new post-secular trend? *Religions* 13: 665.

Barber, S., Brown, J. and Ferguson, D. (2022) Coronavirus: the lockdown laws. House of Commons Library Research Briefing. Available at: https://commonslibrary.parliament.uk/research-briefings/cbp-8875/ (accessed 27 June 2023).

Barush, K. (2020) As coronavirus curtails travel, backyard pilgrimages become the way to a spiritual journey. *The Conversation*, 10 August. Available at: https://theconversation.com/as-coronavirus-curtails-travel-backyard-pilgrimages-become-the-way-to-a-spiritual-journey-143518 (accessed 23 August 2022).

Beaverstock, J.V., Cohen, R., Rogers, A. and Vertovec, S. (2023) Covid-19 and *Global Networks*: reframing our understanding of globalization and transnationalism. *Global Networks* 23(1), 9–13. https://doi.org/10.1111/glob.12425

Bentzen, J.S. (2021) In crisis, we pray: religiosity and the COVID-19 pandemic. *Journal of Economic Behavior & Organization* 192, 541–583. https://doi.org/10.1016/j.jebo.2021.10.014

Bergeron, S.J. and Green, R.S. (2021) Exploring the Henro Michi: student-centred field research in Japan as a component of the Virtual Shikoku Pilgrimage project. *The Geography Teacher* 18(3–4), 121–126.

Bowman, M. and Sepp, T. (2019) Caminoisation and Cathedrals: replication, the heritagisation of religion, and the spiritualisation of heritage. *Religion* 49(1), 74–98. https://doi.org/10.1080/0048721X.2018.1515325

Boyle, M. (2020) No novelty here: Virtual pilgrimage. To be a pylgryme, 7 April. Available at: https://maryrboyle.wordpress.com/2020/04/07/no-novelty-here-virtual-pilgrimage/ (accessed 27 June 2023).

Canterbury Cathedral (2020) Virtual Becket Pilgrimage – Episode One: Beginning the journey (the Nave). Canterbury Cathedral, 28 November. Available at: https://www.youtube.com/watch?v=UEyvdJTo2nY (accessed 14 June 2022).

Centre for the Study of Christianity and Culture (2021) *Churches, COVID-19 and Communities: Experiences, Needs and Supporting Recovery*. Report. The Centre for the Study of Christianity and Culture, University of York, York, UK.

Church of England (nd) Church of England Digital Labs – a hub of resources, news and events to support digital evangelism and digital discipleship in your church. The Church of England. Available at: https://www.churchofengland.org/resources/church-england-digital-labs (accessed 14 August 2023).

Church of England (2018a) An introduction to the next steps in Church of England Digital Labs. The Church of England, 17 July. Available at: https://www.churchofengland.org/resources/digital-labs/blogs/introduction-next-steps-church-england-digital-labs (accessed 14 August 2023).

Church of England (2018b) Our top five tips to develop an effective church digital strategy. The Church of England, 31 August. Available at: https://www.churchofengland.org/resources/digital-labs/blogs/our-top-five-tips-develop-effective-church-digital-strategy (accessed 14 August 2023).

Church of England (2020) Church of England to close all church buildings to help prevent the spread of coronavirus. The Church of England, 24 March. Available at: https://www.churchofengland.org/news-and-media/news-and-statements/church-england-close-all-church-buildings-help-prevent-spread (accessed 14 August 2023).

Church Times (2021) Michaelmas ordinations from vocation to vacation: clergy on furlough. *Church Times*, 8 October.

Clark, H. (2021) Examining the end of the furlough scheme. House of Commons Library, 15 November. Available at: https://commonslibrary.parliament.uk/examining-the-end-of-the-furlough-scheme/ (accessed 14 August 2023).

Clarke, C.A.M. (ed.) (2020) *The St. Thomas Way and the Medieval March of Wales: Exploring Place, Heritage, Pilgrimage*. Arc Humanities Press, Leeds, UK.

Coleman, S. (2019) On praying in an old country: ritual, replication, heritage, and powers of adjacency in English cathedrals. *Religion* 49(1), 120–141. https://doi.org/10.1080/0048721X.2018.1515326

Coleman, S. (2021) *Powers of Pilgrimage: Religion in a World of Movement*. New York University Press, New York.

Cooper, F., Dolezal, L. and Rose, A. (2022) *COVID-19 and Shame: Political Emotions and Public Health in the UK*. Bloomsbury Academic, London.

Cranmer, F. and Pocklington, D. (2020) The impact of the COVID-19 pandemic on the exercise of religion in the United Kingdom. *Revista General de Derecho Canónico y Derecho Eclesiástico del Estado* 54: 36.

De Marco, G. (2020) Selectively Saving Christmas? Practical Ethics, 21 December. Available at: http://blog.practicalethics.ox.ac.uk/2020/12/selectively-saving-christmas/ (accessed 23 August 2022).

Dowson, R. (2020) A discussion of the practical and theological impacts of COVID-19 on religious worship, events and pilgrimage, from a Christian perspective. *International Journal of Religious Tourism and Pilgrimage* 8(7), article 5. https://doi.org/10.21427/ytdq-3s12

Dowson, R. (2021) Virtual pilgrimage to the Holy Land during the pandemic: contemplating the face of Christ with Mary – the Magdala Experience. Paper presented at 'Pilgrimage in the midst of Pandemic: Virtual Conference', William and Mary College, Williamsburg, Virginia, USA, January 2021. YouTube. Available at: https://youtu.be/x0Yx42OZoH3 (accessed 23 August 2022).

Dowson, R. and O'Keeffe, E. (2023) Virtual pilgrimage – an agenda for future research. In: Shinde, K. and Cheer, J. (eds) *A Research Agenda for Religious Tourism*. Edward Elgar Publishing, Cheltenham, UK, (in press).

Dunn-Hensley, S.M. (2020) Virtual pilgrimage in a time of pandemic: lessons from the Shrine of Our Lady of Walsingham. *International Journal of Religious Tourism and Pilgrimage* 8(7), article 13. https://doi.org/10.21427/ncaj-y766

Edelman, J., Vincent, A., Kolata, P. and O'Keeffe, E. (2021) *British Ritual Innovation under COVID-19*. Manchester Metropolitan University, Manchester, UK. Available at: https://tinyurl.com/bric19 (accessed 30 October 2021).

Fancourt, D., Steptoe, A. and Wright, L. (2020) The Cummings Effect: the relationship between controversial political actions, trust and behaviours during the COVID-19 pandemic. Available at: https://discovery.ucl.ac.uk/id/eprint/10108759/3/Fancourt_The%20Cummings%20effect-%20politics%2C%20trust%2C%20and%20behaviours%20during%20the%20COVID-19%20pandemic_AAM.pdf (accessed 28 June 2023).

Francis-Devine, B., Powell, A. and Clark, H. (2021) Coronavirus Job Retention Scheme: statistics. House of Commons Library, 23 December. Available at: https://researchbriefings.files.parliament.uk/documents/CBP-9152/CBP-9152.pdf (accessed 2 December 2023).

Geary, D. (2014) Rebuilding the navel of the Earth: Buddhist pilgrimage and transnational religious networks. *Modern Asian Studies* 48(3), 645–692. https://doi.org/10.1017/S0026749X12000881

Goebel, S. (2007) *The Great War and Medieval Memory: War, Remembrance and Medievalism in Britain and Germany 1919–1940*. Cambridge University Press, Cambridge.

Golan, O. and Martini, M. (2018) Digital pilgrimage: exploring Catholic monastic webcasts. *The Communication Review* 21(1), 24–45. https://doi.org/10.1080/10714421.2017.1416795

Griffin, K. (2015) The globalisation of pilgrimage tourism? Some thoughts from Ireland. In: Griffin, K. and Raj, R. (eds) *Religious Tourism and Pilgrimage Management: an International Perspective*, 2nd edn. CAB International, Wallingford, UK, pp. 57–78.

Helland, C. (2007) Diaspora on the electronic frontier: developing virtual connections with sacred homelands. *Journal of Computer-mediated Communication* 12(3), 956–976. https://doi.org/10.1111/j.1083-6101.2007.00358.x

Hill-Smith, C. (2011) Cyberpilgrimage: the (virtual) reality of online pilgrimage experience. *Religion Compass* 5(6), 236–246. https://doi.org/10.1111/j.1749-8171.2011.00277.x

Hope not Hate (2020) Conservative Party Inquiry on all forms of Racism including Islamaphobia. Hope not Hate Submission of Evidence (report). September. Available at: https://hopenothate.org.uk/wp-content/uploads/2020/09/conservative-party-investigation-islamophobia-2020-09-v5.pdf (accessed 2 December 2023).

Kidd, J. (2011) Enacting engagement online: framing social media use for the museum. *Information Technology and People* 24(1), 64–77. https://doi.org/10.1108/09593841111109422

Kidd, J. (2016) *Museums in the New Mediascape: Transmedia, Participation, Ethics*. Routledge, Abingdon, UK.

Kratochvíl, P. (2021) Geopolitics of Catholic pilgrimage: on the double materiality of (religious) politics in the virtual age. *Religions* 12(6): 443. https://doi.org/10.3390/rel12060443

MacWilliams, M.W. (2002) Virtual pilgrimages on the Internet. *Religion* 32(4), 315–335. https://doi.org/10.1006/reli.2002.0408

Maravia, U. (2020) Rationale for suspending Friday prayers, funerary rites and fasting Ramadan during COVID-19: an analysis of the fatawa related to the coronavirus. *Journal of the British Islamic Medical Association* 4(2), 1–6.

Mroz, F. (2021) The impact of COVID-19 on pilgrimages and religious tourism in Europe during the first six months of the pandemic. *Journal of Religion and Health* 60(2). 625–645.

Niu, S. (2023) Virtual hajj as a response to demographic and geopolitical pressures. *Contemporary Islam* 17, 95–108. https://doi.org/10.1007/s11562-023-00512-1

Ofcom (2022) *Online Nation: 2022 Report*. 1 June. Ofcom, London.

O'Keeffe, E. (2023a) Assessing the impact of the pandemic on rituals of memorialization in the UK – a data-driven approach. Paper presented at the *6th International Symposium of the Death Online Research Network 6*, 1 June 2023. Northumbria University, Newcastle, UK.

O'Keeffe, E. (2023b) Finding data pathways through the 'pandemic pilgrimage boom': embedded-like research, COVID-19, and the British Pilgrimage Trust. *International Journal of Religious Tourism and Pilgrimage* 11(4), article 8.

Ouassini, A., Ouassini, N. and Amini, M. (2022) The 'IslamoCovid' response: British Muslim mobilization(s) at the intersection of COVID-19 and Islamophobia. *Journal of Media and Religion* 21(4), 193–206. https://doi.org/10.1080/15348423.2022.2147752

Polus, R. and Carr, N. (2021) The role of communication technologies in restructuring pilgrimage journeys. *International Journal of Religious Tourism and Pilgrimage* 9(5), article 5. https://doi.org/10.21427/MG4M-E412

Poole, E. and Williamson, M. (2021) How racist narratives about Muslims in the British press were reconfigured during the initial peak of COVID-19. London School of Economics (LSE) British Politics and Policy blog, 7 September. Available at: https://blogs.lse.ac.uk/politicsandpolicy/press-reporting-muslims-covid19/ (accessed 28 June 2023).

Raj, R. and Griffin, K.A. (2020) Reflecting on the Impact of COVID-19 on religious tourism and pilgrimage. *International Journal of Religious Tourism and Pilgrimage* 8(7), article 2. https://doi.org/10.21427/8f91-6z16

Robinson, T. (2020) *Effect of the Covid-19 pandemic on Freedom of Religion or Belief*. Debate Pack CDP 0146 (2020). House of Commons Library, 20 November. UK Parliament, London.

Sherwood, H. (2020) Catholic church leader criticises Covid worship restrictions in England. *The Guardian*, 1 November. Available at: https://www.theguardian.com/world/2020/nov/01/catholic-church-leader-criticises-covid-worship-restrictions-england (accessed 27 June 2023).

Sherwood, H. (2022) Visitors flock to Barnard Castle and other historic sites in England. *The Guardian*, 28 January. Available at: https://www.theguardian.com/culture/2022/jan/28/barnard-castle-visitor-numbers-english-heritage (accessed 16 August 2023).

Shinde, K. (2022) Smart heritage tourism destinations in India: rhetoric to reality. Paper presented at the *3rd UNWTO Conference on Smart Destinations*, 21–23 November 2022, Valencia, Spain.

Sloan, L. (2017) Who tweets in the United Kingdom? Profiling the Twitter population using the British Social Attitudes Survey 2015. *Social Media + Society* 3. https://doi.org/10.1177/2056305117698981

Tadie, S. (2020) Online world pilgrimage 'Lourdes United' transcends borders and health restrictions. *National Catholic Register*, 16 July. Available at: https://www.ncregister.com/news/online-world-pilgrimage-lourdes-united-transcends-borders-and-health-restrictions (accessed 28 June 2023).

van der Beek, S. (2018) New pilgrim stories. Narratives, identities, authenticity. *Yearbook for Ritual and Liturgical Studies* 34, 67–72. https://doi.org/10.21827/5bfeaada9f1c7

van der Beek, S. (2019) Post-Camino storytelling: navigating transformation in a digital age. In: Hoondert, M. and van der Beek, S. (eds) *Ritual in a Digital Society* (pp. 65-79). *Netherlands Studies in Ritual and Liturgy* No. 21. Institute for Ritual and Liturgical Studies, Protestant Theological University and Institute for Christian Cultural Heritage, University of Groningen, Groningen, the Netherlands. Available at: https://www.pthu.nl/irilis/Publications/nsrl21-ritual-in-a-digital-society-hoondert-and-van-der-beek.pdf (accessed 28 June 2023).

Vaughan, R. (2020) Boris Johnson is working on plans to 'save Christmas' and allow families to celebrate together after a 'difficult year'. inews.co.uk, 17 November 2020. Available at: https://inews.co.uk/news/politics/boris-johnson-plans-save-christmas-covid-restrictions-england-lockdown-762969 (accessed 28 June 2023).

9 Religious Tourism Landmarks in Lebanon: Gigantism and Oversized Christian Pilgrimages Sites

Nour Farra Haddad*
Lebanese University, Beirut, Lebanon

Abstract
This chapter discusses and explores the constantly evolving sacred geographies in Lebanon, which testify to the importance of religious Christian and Muslim markers in the landscape. In the late 20th century and early 21st century, the development of gigantic Christian projects, including large saint statues and crosses, has been observed. This trend is related to the development of religious tourism, but it is also connected to several other factors, such as the concerns of some Lebanese Christians feeling threatened by a growing Muslim population and increasing fanaticism. The chapter addresses this phenomenon of developing large Christian pilgrimage projects all over the country, its touristic and social identity dimensions, and its impact on local host communities. The selected cases illustrate that the negotiations of spatial and social identities in a context of interreligious coexistence are ongoing projects in the post-war period. These projects are also significantly contributing to the development of religious tourism, which, in turn, supports local communities' economies.

Introduction

Lebanon is located at the extreme east of the Mediterranean Sea, sandwiched between Syria and Israel/Palestine. What sets this small country apart from others in the region is not only its geography and climate but also its religious diversity, with 18 different communities.

The war in Lebanon (1975–1990) and the post-war period seriously damaged the Lebanese economy, as well as its infrastructure, and devastated the tourism industry. Since 1991, Lebanon has been struggling for development, but tourism has yet to return to its pre-war levels, mostly due to insecurity and rising political and sectarian tensions. Planning strategies have demonstrated that niche tourism diversification, such as religious tourism, can lead to sustainable development.

Constantly evolving, sacred geographies bear witness to the importance of Christian and Muslim religious markers in the Lebanese landscape. In the late 20th and early 21st centuries, we observed the development of gigantic Christian projects featuring huge saint statues and crosses. This trend is related to the development of religious tourism, but also it is connected to other considerations, such as the perception of some Lebanese Christians feeling threatened by a growing Muslim population and increasing fanaticism.

Oversized statues and religious monuments are erected all around the world for various purposes, as instruments of: (i) the expression of

*nour@neosl.com

DOI: 10.1079/9781800623675.0009

faith; (ii) communication; and (iii) control. They reflect the need of the faithful to materialize their beliefs and trust in their communication with the divine, but they also serve as symbols against secularism or in support of causes. For promotional purposes, project initiators often use superlative qualifications. For example, news reports highlight the construction of the world's largest Marian statue in the Philippines (Mauro, 2021) or one of the largest Marian statues in the world in Croatia (Catholic News Agency, 2017). Many international rankings also list the largest statues or crosses worldwide. For example, the Cross of All Nations in Qanat Bakish (Lebanon) ranks fourth, following the tallest cross in the world (150 m – the Cross of the Valley of the Fallen) in Spain, the second tallest one (the Memorial Cross of Dambana) in the Philippines, and the third tallest one (the Third Millennium Cross of Coquimbo) in Chile (Miratico, 2016). Regarding statues, Lebanon cannot compete with the tallest statue in the world located in China, the Spring Temple Buddha, standing at 153 m in Henan Province. However, Lebanon's tallest statue, St Charbel (Faraya), at 24 m in height, still ranks well compared to the iconic Christ the Redeemer in Brazil (30 m tall).

This chapter will address the phenomenon of developing gigantic Christian pilgrimage projects, its touristic and social identity dimensions, and the involvement and impact on local host communities. This study represents an attempt to understand the use of territorial religious markers such as crosses and huge statues. In this perspective, the aim is to examine the complexity of coexistence in Lebanon and national identity issues in the context of developing religious tourism. A particular focus of interest is understanding the involvement of local host communities in these projects, as well as the multidimensional perspective of this phenomenon (economic, social and environmental).

The selected cases illustrate that the negotiations of spatial and social identities in a context of interreligious coexistence are ongoing projects in post-war Lebanon. They also significantly contribute to the development of religious tourism and support the local communities' economies.

Seven religious destinations were selected as case studies after visiting more than 15 sacred sites with colossal monuments or statues. Special attention is given to some of these projects intended to serve as landmarks located all over Lebanon. The monuments take various forms, including crosses, statues, rosaries and sanctuaries. The development of these sites also varies, as some have already been completed and are open to visitors, while others are still under construction.

Field trips were organized to each location at least three times, with hours of on-site participant observation and several interviews with the local host community. The fieldwork was conducted by the author between February 2018 and March 2019. In total, over 84 hours of participant observation and 25 interviews were conducted during the investigations, involving research notes and photos. The qualitative anthropological method used combined semi-structured interviews and participant observation. The interview guidelines consisted of a dozen open-ended questions in Arabic.

The chapter analyses the motivations behind these constructions and the impact of social, political and economic dimensions. Describing these initiatives will help us understand both the motivations behind the development of these projects and the public perception of these religious space markers. They will also have an impact on the landscape and the sustainable development of religious tourism.

Historical Background of Lebanon's Religious Tourism

Christianity first spread in the regions corresponding to nowadays Lebanon, Syria, Jordan and Iraq, and throughout history, Oriental Christians would play a key role in the area on various levels. From the Muslim conquest, in the 7th century, Lebanon's history of Islamic–Christian relations has been marked by many tensions, clashes and episodes of war but never overriding the dialogue and conviviality between these communities.

Throughout history, people of all religions have built monuments, shrines and statues to depict their deities, and, over the years, the world has seen the construction of ever taller and bigger statues. While being sacred places, some religious monuments stand out as mind-blowing due to their colossal size and amount of sacrifice needed to construct them. Constantly evolving, sacred geographies mark out the Lebanese landscape with Christian and Muslim markers.

Lebanon has an immense potential for religious tourism. It is a part of the Holy Land and as well, a land of Holiness with thousands of religious sites spread all over its territory and many vocations enabling the canonization of many saints. An abundance of oratories, churches, chapels, monasteries, mosques and shrines testify to the importance of the piety of Christians and Muslims in the local Lebanese culture. Lebanon is a land of coexistence and interreligious dialogue, with its 18 religious communities, 'a country message' as defined by St John Paul II. Although coexistence between different religious communities over the course of history has been marked by repeatedly bloody conflicts, belief in the concept of 'living together' ('*Al aaych al-mushtarak*') has never been lost.

In May 1997, Pope John Paul II proclaimed Lebanon as a 'Holy Land' for its privileged place in the Bible, its martyrs and its sacred places. Following this statement, many civil society initiatives as well as the Ministry of Tourism aimed to place Lebanon as a pilgrimage destination. In 2007, the assembly of the Catholic patriarchs and bishops created an association for the development of pilgrimages in Lebanon (APL) and this association was organized in July 2010 during the first meeting around religious tourism in Lebanon. Following this conference, the Cultural and Religious Tourism project (CRT) was created by Prime Minister Saad El Hariri and the second religious tourism conference in Lebanon was organized in November 2010. Unfortunately, however, the situation in Lebanon and the Middle East, especially regarding political and security issues will dramatically stop the evolution of religious tourism (Farra Haddad, 2015).

On 29 May 2016, the Ministry of Tourism in collaboration with the United Nations World Tourism Organization (UNWTO) organized an event to introduce 'Our Lady of Mantara' (Maghdouche village, South Lebanon) as an international tourism destination, a symbol for religious tourism promotion in Lebanon. On 16 February 2018, the Lebanese Minister of Tourism and the Minister of Foreign Affairs, together, declared 2018 as the year for religious tourism during a press conference (Farra Haddad, 2018).

On 7 June 2018, 'Holy Lebanon', the first mobile phone app for interreligious tourism in Lebanon was launched officially, available for free on Google Play and Apple Store. The app introduces Lebanon as an attractive destination full of interesting religious sites to discover from different religious communities, as well as rituals to experience and events to share. Among others, the projects described in this chapter are listed in this app and pages related to these projects experience busy traffic. In July 2018, Lebanon was reintroduced in the Vatican annual pilgrimage list, the official Roman Catholic Church Pilgrimage destinations, after 12 years of absence. Knowing that thousands of faithful Christians from around the world refer to this list to organize their pilgrimages, a significant increase in traffic towards Lebanon is expected.

All these initiatives and conferences elaborate on the economic benefits of the development of religious tourism, and thus are believed to encourage the development of the monumental pilgrimage projects. These sites have the aim of becoming major religious landmarks on the touristic map of Lebanon.

Religious Destinations in Lebanon: Seven Major Christian Landmarks across Lebanon

The seven major Christian landmarks across Lebanon are:

- In the north-east, close to the Syrian border, Qobayat is a city hosting one of the biggest crosses in Lebanon.
- The project of the 'City of Peace' in Ijdabra village, rural Batroun, hosts the biggest inhabitable cross in the world.
- A unique religious landmark, a huge concrete rosary, is under construction in the Deir El Ahmar area, Bekaa valley.
- The world's and Lebanon's largest statue of St Charbel installed in Faraya on top of Jabal El Salib.
- The biggest illuminated cross in the world was inaugurated in Qanat Bakish, Baskinta mountainous area.
- The biggest rotatable statue of Jesus Christ in the Middle East (Ghosta-Mount Lebanon) is today accessible in Jabal El Rahme.
- The biggest tower with a Virgin Mary statue, Saydet Em El Nour Project, in Ain Ebel, South Lebanon is still under construction.

The piety of the Lebanese faithful, observable in the religious sites across the country, encourage religious authorities to develop such projects. The trend of implementing such landmarks is related to the importance of local pilgrimages but also to the phenomenon of the development of religious tourism in Lebanon and finally, these monuments are identity markers.

Territorial identity is a complex political, social and cultural concept using territorial markers. The study limitation was mainly because for the respondents it can be taboo to talk about religious fanaticism and identity. Most of the time there is a bias as respondents feel that they must glorify the 'living together' in Lebanon despite the tensions between the communities.

Qobayat is a northern capital. It is the biggest Christian settlement in the Akkar district with many traditional Lebanese houses and old churches located in a sensitive area threatened by political neighbourhood conflicts remarkably close to the Syrian border. The entire population of Qobayat is Maronite facing the majority Sunni Muslim of the residents in the district, home to the largest population of Alawites in Lebanon. On 26 October 2013, Qobayat inaugurated one of Lebanon's biggest crosses, 40 m long and 20 m wide, made in Romania. It was erected on Jabal Blat (Blat Mountain) and inaugurated in the presence of thousands of people with an important panel of Lebanese religious and political figures. The Cross overlooks all the area and is believed to be protecting the city and its inhabitants from all the problems that can occur. Locals are convinced that, since the time of the erection of the Cross, security issues have been minimized.

In Ijdabra, a village in rural Batroun, 65 km north-east of Beirut, the 'City of Peace' project was implemented in an area of 50,000 m^2 in 2015. The biggest inhabited cross in the world is being built to reach 55 m height and at 48 m and in the middle of the cross, a church inside the concrete structure can host up to 200 people (Fig. 9.1). A workshop and an exhibition for rural products were inaugurated in 2018 on site

Fig. 9.1. Ijdabra Cross (rural Batroun) 'City of Peace', the biggest inhabitable cross in the world. Photo taken by author.

and the rest of the land will host a prayer and meditation area, and a camping area; 25,000 m^2 will be protected as a forest for recreation with hiking trails. The project aims to preach a culture of peace and dialogue in Lebanon and across the world as Father Charbel Khachan, the Maronite parish priest of Ijdabra village, insists. Roads are being developed to simplify access to the City of Peace. The official inauguration of the site was supposed to take place in late 2019 but it was postponed because of the pandemic caused by coronavirus disease 2019 (COVID-19) and the economic crises in Lebanon.

Deir El Ahmar, a Maronite village of Baalbek (a district in the Bekaa valley with a majority of Shia Muslims), located 110 km from Beirut, has more than 14 religious sites and is the seat of the District Maronite Archdiocese. On the route between Deir El Ahmar and Beshouat (a village renowned for its miraculous site dedicated to the Virgin Mary), a unique religious project is under development: the 'Rosary of Lebanon' (Fig. 9.2). This project, inaugurated in October 2008, is the result of the commitment of a diligent prayer group, which later became the 'Association of the Rosary for Protection and Salvation' in collaboration with the Lebanese Maronite Order (OML). The rosary marks the landscape dramatically with concrete cells, 59 illuminated beads: the six 'Our Fathers' and the 53 'Holy Mary's'. Each bead is 4.90 m long and 3.5 m wide. The rosary, which is 600 m long, leads to the cross of the resurrection and to a large amphitheatre for prayer and outdoor celebrations, which was under construction at the same time as our fieldwork. A collaboration with the Mosaic school of Madaba (Jordan) aims to cover all the beads with mosaics.

The monastery of the Holy Savior (Kreim-Ghosta), surrounded by beautiful forests, is the Mother House where the Congregation of Maronite Lebanese Missionaries was founded in 1865. Father Jean Bou Khalifeh developed the Mount of Mercy (Jabal El Rahme)

Fig. 9.2. The 'Rosary of Lebanon', Deir El Ahmar, Bekaa valley. Photo Rosary of Lebanon (2024).

project, a few hundred metres from the monastery. The project includes the biggest statue of Jesus Christ on a rotatable base, which can turn 360° (Fig. 9.3), as well as a chapel to Saint Theresa of Calcutta and a museum dedicated to St John Paul II. The statue, which is 12 m high and weighs 8 t, was installed on its pedestal on 8 May 2017. Counting the subterranean chapel dedicated to St Faustina under the statue, the total height of the construction is 22 m. Surrounding the round base of the statue is an esplanade with 12 fountains representing the 12 disciples. Every 6 hours, the statue rotates, so that sometimes it is oriented towards the mountains and at other times it overlooks the coast. The project is not limited to the statue, the chapels and the museum, but also includes different pedestrian trails leading to the site, one from the old Kreim convent, as well as the hill area called 'Jabal El Zeitoun' (the mountain of the olive trees). These trails are meant to attract not only pilgrims but also ecotourism visitors.

Before August 2017, Mount Chabrouh was renowned for its cross on top of a high mountain, Jabal El Salib, only accessible by hiking or with a four-wheeled-drive vehicle. Chabrouh has an altitude of 1650 m and is located about 50 km from the capital, Beirut. In 2007, Chabrouh Dam was inaugurated, creating one of the highest reservoirs in the Middle East region, but as a result, more than three-quarters of the orchards were flooded. Ten years later, Lebanon's tallest statue of St Charbel was inaugurated overlooking the dam (Fig. 9.4). The statue was transported to the top of Jabal El Salib on 20 August 2017. On that day, the statue was transported with a police escort, and all along the way the inhabitants prepared rice and rose petals to honour the saint and celebrate this exceptional event. The 24 m long and 40 t statue was conceived by a Lebanese Maronite artist, Nayef Alwan, and contains a relic of the saint that is believed by some to produce miracles. The statue was placed on a pedestal by a

Fig. 9.3. The biggest rotatable statue of Jesus Christ in the Middle East (Ghosta-Mount Lebanon), Jabal El Rahme (Mount of Mercy). Photo taken by the author.

Fig. 9.4. The world's and Lebanon's largest statue of St Charbel, Mount Chabrouh, Faraya. Photo taken by the author.

huge metal cross already installed there in 1951. This statue was offered by the family of the miraculously healed Michel Salameh. Elie Salameh, the father of the teenager, is convinced that St Charbel intervened to heal his son and financing the statue project was a way of giving thanks for this miracle.

In the high plateau of Qanat Bakish, at 1800 m altitude, near to Baskinta village and close to the ski station and resort, stands the monumental iron cross of Baskinta. The project of the Baskinta cross began with the dream project of a jeweller, Souhel Abou Samra, who decided first to renovate an abandoned church dating back to 1898;

he called it the Unified Church of Saint Joseph for All Christians. Later, in 2008, the project of building the biggest cross in the area and the largest illuminated cross in the world, the 'Cross of All Nations', 73.8 m in height and 37 m wide, was developed (Fig. 9.5). The cross was inaugurated on 13 September 2010, on the Eve of the Feast of the Exaltation (which is celebrated on 14 September). The cross is massive: regardless of its 5 t steel base, the cross's body itself is made up of 170 t of steel and is lit by 1800 spotlights (Najib, 2010).

The project of the biggest statue of the Virgin Mary, elevated over a tower in Ain Ebel village, is the most recent project on the list. Ain Ebel is situated on the Lebanese side of the 'Holy Land', in the south of Lebanon close to the border, about 130 km from the capital Beirut (number 7 on the map shown in Fig. 9.6). Local traditions and some historians explain that Jesus Christ, after leaving Caphernaum, passed by Ain Ebel before reaching Cana and Tyre. Different sites and remains confirm these oral traditions as a spring carrying the name of Jesus Christ (Ain El Massih). Locals, authors and scholars are interested in relocating all this area to the map of the Holy Land (Roncaglia, 2004; Harb, 2008; Abou Nakoul, 2013). 'On the footsteps of Jesus Christ', a project retracing the routes that were followed by Christ include Ain Ebel in their trails[1].

The site of Saydet Em El Nour (Our Lady of the Light) sits on a hill open to an unobstructed 360° horizon. The tower consists of three columns that unite at 60.3 m above the ground. A 14 m-high statue of the Mother of Light carrying her baby Jesus will sit on top. This tower will also host an underground multipurpose hall and a second building will host a souvenir shop, a cafeteria, private offices and public restrooms. A third building was planned as well to accommodate a guesthouse and a meeting room.

The project idea was originally created in 2010 by Canadian emigrants of the village, and it took 6 years to collect the necessary funds, finalize the designs and buy the land. The project

Fig. 9.5. The 'Cross of All Nations', the biggest cross in the Middle East and the biggest illuminated one in the world, Qanat Bakish, Baskinta. Photo taken by G. Moukheiber.

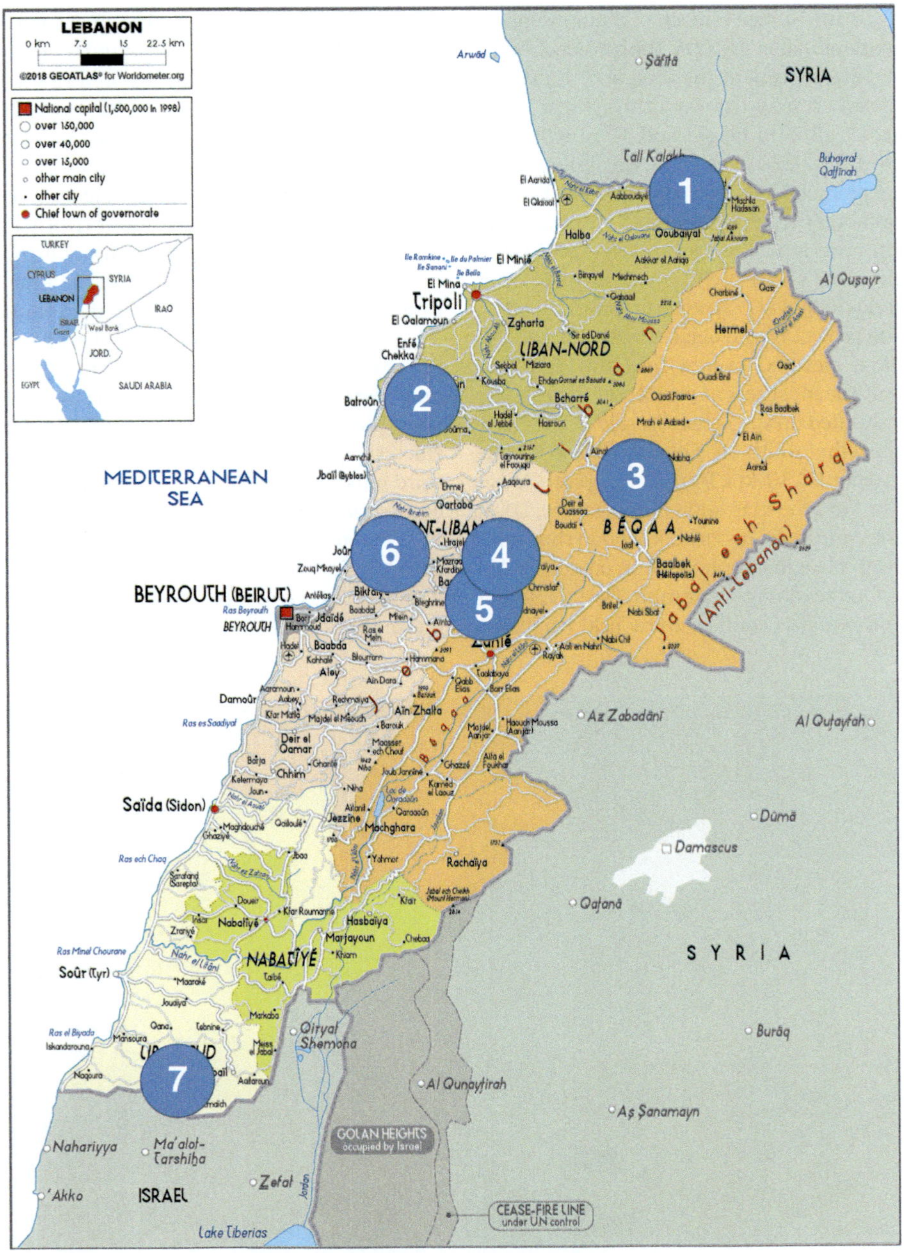

Fig. 9.6. Map of Lebanon showing location of the oversized pilgrimage projects developed by Christian host communities promoting Lebanon. 1, one of the biggest crosses in Lebanon, Qobayat Cross; 2, the project of the 'City of Peace' in Ijdabra village, rural Batroun, hosts the biggest inhabitable cross in the world (see Fig. 9.1); 3, a unique religious landmark is under construction in Deir El Ahmar area, Bekaa (Béqaa) valley: a huge concrete rosary, the 'Rosary of Lebanon' (see Fig. 9.2); 4, the world's and Lebanon's largest statue of St Charbel was installed in Faraya on top of Jabal El Salib (see Fig. 9.4); 5, the biggest illuminated cross in the world, the 'Cross of All Nations', Qanat Bakish, Baskinta (see Fig. 9.5); 6, Jabal El Rahme (the Mount of Mercy) with the biggest rotatable statue of Jesus Christ in the Middle East (Ghosta-Mount Lebanon) (see Fig. 9.3); 7, the biggest tower with a Virgin Mary statue in Lebanon, Saydet Em El Nour Project, Ain Ebel, South Lebanon.

was officially launched in the summer of 2016 and a crowdfunding campaign followed dating from June 2017 (Saydet Ainebel, 2021). Today, the concrete base can be clearly observed but construction works are getting slower because of insufficient funds, political tensions and security issues. Local communities state that it is a peaceful Christian act of resistance in South Lebanon facing the occupied territories of the Holy Land. They genuinely believe that the Virgin Mary will protect the border from potential new intrusions without the use of arms and violence.

Religious Identity, Territory Sacralization and Landscape Markers

Regardless of the economic value of these religious monuments supporting tourism development and their role in beautifying the environment, they have undoubtedly a cultural and political value/dimension. As Daba (2017, p. 58) stated 'statues are central to cultural identity and crucial for awareness and development of one's identity and transfer the cultural practices to the next generation.'

In general, such monuments can serve as global witnesses to sociopolitical divisions, conflicts, identities, dominance and patriotism. Reasane (2018) demonstrates how statues, symbols and signages in South Africa report socio-political divisions serving history and the memory. In India, Muslims are alarmed by the enormous statues of Hindu deities and leaders (Prasad, 2019). The meaning of 'Cristo Redeemer' in Rio de Janeiro, varies among Brazilians. Some view it as a broadside against secularism, while others regard it as a salute to Catholicism (Borstein, 2016). As a symbol of peace dedicated to the unification and peace of all people and nations in South Asia, the largest Marian statue in the world (nearly 100 m high) was constructed in the Philippines at the Montemaria pilgrimage site (Mauro, 2021).

The projects studied in Lebanon illustrate the ongoing negotiations of spatial and social identities that have persisted since the war period. In Lebanon, religious disputes for the territory and the skyline are recorded regularly. One example in this context is the famous 'Battle for the sky' between the height of minarets of Mohmad El-Amine Mosque and the bell tower of Saint George Maronite Cathedral in downtown Beirut, the Lebanese capital. Mermier (2015) develops this topic to explain the sensitive identity tensions reflected in the Beirut skyline. Mermier (2015) also explains how when in 1894 the Saint George Maronite Cathedral was built, the Orthodox community considered this as a provocation as well.

Another example in the Lebanese capital of territorial fights is the Saint George Bay issue. With the establishment and development of the 'Solidere' Project (the Lebanese company for the development and reconstruction of Beirut central district s.a.l. post-war rehabilitation plan), the famous Saint George Bay is going to be called 'Zaitunay Bay'. Once publicly accessible, St George Bay has been rebranded as the corporate-controlled space 'Zaitunay Bay' with its multi-million-dollar yachts and expensive restaurants by the sea. Saint George is the patron saint of Beirut city and many churches from different denominations in the city are dedicated to the saint and some authors and scholars locate the episode of the dragon killed by the saint in Beirut Bay. Saint George (*Mar Jiryes* or Al Khodr for Muslims) is universally renowned and venerated by Christians, Muslims and Druze alike. Marvellous legends, transported to Europe, tell of his heroic actions and his passion. When 'Solidere' decided to call Saint George Bay 'Zaitunay Bay' the Christian community was offended and the Metropolitan Greek Orthodox Bishop, Elias Audi, blamed the authorities for giving the name of an old red-light district 'Zeitoun' to a sacred area.

This brings us to the complexity of the Lebanese multi-confessional identity. Different conceptions of identity have resulted in painful confrontations and remain an unsolved issue. Each Lebanese religious community with its cultural heritage interpretates the war history differently. The Lebanese war (1975–1990) questions the 'living together' motto and the peaceful coexistence in Lebanon, as religious bias and tensions were exacerbated during this period. During the war and the post-war period, politicians, as well as civil initiatives, insisted on 'living together' and the duty of preserving real dialogue and coexistence between all the different religious communities in Lebanon. The evolution of religious demography in Lebanon is a sensitive political issue, because before the war period, a balance existed between Christian and Muslim communities, as the last national census in 1932 stated. In 1997, after the elections, the Ministry of Interior estimated that the Muslims represented 52% and the Christian 44.8% of the population (Assaf and Barakat, 2003). Little reliable information exists, as there is an absence of accurate data on the relative percentages of the population of the major religions and groups; for example Faour (2007) estimated that the Christians in Lebanon represent around 35% whereas Muslims 65%.

Considering these potential figures, some Lebanese Christians feel threatened by a growing Muslim population and a rise in fanaticism is evident. This perceived threat can also be interpreted as the subsequent development of symbolic religious identity markers, in a bigger complex of symbolic demarcation of physical as well as existential space. During and after the war period, the phenomenon of marking the top of mountains in Christian areas with metallic crosses of various sizes became quite common. Some of these crosses were installed by individuals, such as the metallic cross in the top of the Mzaar Mountain (2465 m high summit, Kfardebian), while others were erected by religious orders, parishes or municipalities.

The sites discussed in this chapter can also be considered as Christian markers, even if officially the objective of their development is to beautify the environment, boost the local tourism economy and attract the faithful. The fieldwork observations help define the motivations behind the projects and analyse them in the context of the religious landscape related to Christian sacralization of the land. 'Religious landscapes' have become a relevant topic for the last 20 years (Horster, 2010). Horster (2010) elaborates on the fact that the sacred land can be used for economic and non-economic activities. The economic activities of a sacred territory could serve its multifaceted functions, financing not only the cult itself but also the solidarity of the citizens, allowing them diverse types of economic use of the land such as agriculture and agropastoralism. The research around the Lebanese monumental sites highlights the economic benefits that such religious landmarks can generate for locals.

At the inauguration of the discussed projects, religious figures and politicians explained that these landscape markers would instill confidence in Lebanese Christians that Lebanon can remain their homeland. Speeches during the inaugurations also addressed the local communities and insisted that these initiatives will offer new job opportunities during the construction and development period, as well as after their inauguration by attracting many tourists. The economic prospects of sustainable development in most of the rural areas concerned was highlighted. In relation to this, the development cannot benefit a single community only, but the local communities all along the way to these sites and surrounding areas as well.

In most interviews, religious leaders, locals and Christian visitors were enthusiastic to develop tourism and encourage visits to these sites, being proud to attract tourists from different religious communities. However, at the same time, they stated that these territories are Christian and should be marked as such.

'*Lezem yefhamo enno hal ared la elna*' [They should understand that this is our land] – this same sentence was heard from three different Christian visitors (Rami, Josephine and Tamara) in three different sites. 'They' refers to Muslims – Christians wanted to show them that they are strongly rooted in their land and will not compromise it. From another side, we also heard twice '*Holeh el mazarat la kell el lebneniyeh ou qalbna maftouh lal Jami*' [These shrines are for all Lebanese and our hearts are open to all]. This dichotomy reminds us of the Lebanese paradox of the 'living together' belief, while feeling being 'threatened' by the 'Other'. This also reminds us of the debates around the political Lebanese system, the question of federalism in Lebanon raised by many politicians, especially Christian ones, and the question of secularism.

In the Taif Agreement in 1989 seeking to end the Lebanese war (1975–1990), officially known as the National Reconciliation Accord, federalism was rejected in favour of administrative decentralization. However, some politicians still see this as a solution for all the problems of the country (social, political and economic) (Aboultaif, 2021). On the other hand, some parties believe strongly that the process of deconfessionalism is a departure towards modernizing the state and the mentality of the citizens (Habib, 2009).

Regardless of the size, the developers of such statues or monuments are always trying to find a superlative qualification to describe their projects and to draw the attention of the tourist. For example: (i) the Ijdabra Cross is said to be the tallest cross 'inhabited' in the world, even if it is not the tallest in general; (ii) the 'Cross of All Nations' at Baskinta is said to be the biggest 'illuminated' cross in the world; and (iii) the statue of Jesus Christ in Ghosta is the biggest 'rotating' statue of Jesus Christ in the world (Najib, 2010; Grantham, 2015). Looking to score records with the size and the characteristics of the monuments concerned, developers are trying to attract the maximum visitors, hoping to collect from them donations and for their collaboration in local economies and tourism services.

There are no accurate statistics regarding the increasing number of visitors to these exceptional sites, but local observations show that most of these sites become more popular over time and keep attracting tourists. Excluding the Qobayat Cross, which is located in a distant and insecure area, and the uncompleted site construction of the tower of the Virgin Mary in Ain Ebel, the statue of St Charbel in Faraya may be considered the most visited pilgrimage site, followed by the rosary of Deir El Ahmar, followed equally by Jabal El Rahme (the Mount of Mercy) in Ghosta, the 'City of Peace' in Ijdbra and the 'Cross of All Nations' at Baskinta. According to our analysis, the site of Jabal El Rahme and the Ijdbra cross will quickly become immensely popular, and visitors will increase significantly to these sites for many reasons. These include that they are easily accessible during all seasons; this is in contrast to the statue of St Charbel in Faraya and the Baskinta cross which are unreachable during the winter season. They are also eco-friendly sites with the opportunity of enjoying nature around the site, walking to the site and extending a hike into the surroundings. Jabal El Rahme is only a few kilometres from Our Lady of Lebanon, the most visited site in Lebanon, allowing many tourists to extend their visit. When the project of the 'City of Peace' in Ijdabra is completed, it is expected to attract many Christians from Lebanon and Father Kachan, the project initiator, is willing to promote the site to all religious communities and develop an important ecotourism project around the cross itself. The promotion of these sites significantly supports the development of religious tourism on a local and international level. From another aspect, the implementation of such landmarks comfort Christians in their belonging to the land.

This topic and the research will not elaborate on persistent political dilemmas of a plural society such as Lebanon but it shows how the development of religious projects also reflect political and social considerations. Just considering the panel of politicians and religious figures from different religious communities at the launch of these projects and their speeches, we can understand that from one side Christian identity markers on the territories are considered to be a must. However, from another side there should be insistence on interreligious solidarity and living together, with an invitation to all communities to enjoy the attractions.

Summary

In general, the statues and religious landmarks around the globe have different values and dimensions: (i) cultural; (ii) aesthetic; (iii) political; and (iv) economic. They are built, most of the time, as sites to attract tourists thereby supporting economic development of the country. Daba (2017) demonstrates the importance of tourism around monumental statues:

> Among the tourism sites, statues play a vital role to the economy of one country due to such sites are the nodal points of tourist travel, both local and foreign. They are powerful sites that could depict the civilization of one society.
>
> (Daba, 2017, p. 57)

In Lebanon, every year, hundreds of religious sites are still built all over the country and, as elaborated in this chapter, monumental pilgrimage sites are added to the landscape marking territory. The study reported in this chapter questioned the motivations behind the construction of these sites but also the proliferation of such oversized Christian pilgrimage sites. The sacred geography in Lebanon is not static and keeps evolving with new religious sites. Examples are the big statue of St Charbel installed in the top of a church, serving as a podium (13 m in diameter and 7.5 m in height) in Hammana village (inaugurated in July 2018), and what is said to be the biggest statue in the world of St John the Baptist in Karkha (Shouf district) inaugurated in 2022. From the Muslim side there is a trend of building huge mosques in urban and rural areas, such as the mosque of Mohamad El Amine (Sunni) of Beirut or the Mosque-Maqam of El Sayyida Khawla (Chiite) in Baalbeck. The important question for the debate can be: When are the Lebanese to stop adding Christian or Muslim signs in Lebanese public spaces?

All the leaders, initiators and religious people behind the Lebanese landmarks introduced in this chapter insist on the objective of developing pilgrimages and religious tourism in order to attract visitors from all around Lebanon and abroad. They also insist on the fact that these initiatives will benefit local host communities by creating jobs and economic opportunities for service providers in the localities concerned. Finally, most of them insist on inviting all communities to share these sites by integrating an ecotourism dimension into the project even if they display Christian symbols. The major objective of the construction of these sites is first to develop sustainable religious tourism. The initiators of these projects intentionally tried to hit some records by using special themes and huge sizes of construction to gain maximum attention and attract the maximum number of visitors. Investments to establish these projects in Lebanon range between US$800,000 and US$1500,000 due to the high price of construction and materials used. In general, donations and cash are gathered over time through a variety of methods (on site, through the Internet, social media pages, crowdfunding, etc.). Most of these expensive projects open before they are finished to raise money for the site's adornment and development.

In the Western world, by contrast, a regression in constructing Christian religious sites is observed as well as the secularization of some churches, cathedrals and convents along with the reconsideration of all religious signs in the landscape. In 2014, the French magazine *Le Figaro* published an article: 'Secularism: are we going to remove all Christian signs from our public space?' (Chemain, 2014). The religious signs in the landscapes of some European countries are not ready to disappear; instead, they often remain as traces of the past or are only valorized for their cultural heritage and artistic interest. The time of major construction of Christian worship sites, which had characterized the period between World War I and World War II, is a bygone age. After World War II, the erasure of religious visibility often accompanied the sale or destruction of churches (Prudhomme, 2016). Prudhomme (2016) explains how this evolution varies according to diverse cultural areas, and how in Africa and Asia monumental religious shrines are still erected to mark the landscape. In all cases, the relationship to space reflects the various objectives of the actors. Within this complex relationship, religion is not always associated with a hegemonic will and the aim to control territory to expand its influence.

Note

[1] The chapter was written prior to the shelling of Ain Ebel, December 8th, 2023.

References

Aboultaif, E.W. (2021) The federal question in Lebanon: myths and illusions. *50 Shades of Federalism*. Available at: http://50shadesoffederalism.com/case-studies/the-federal-question-in-lebanon-myths-and-illusions (accessed 14 September 2022).

Abou Nakoul, N. (2013) *Haramoun Min Adam Ila El Massih [Haramoun from Adam to Jesus Christ]*. Daccache Printing, Beirut.

Assaf, R. and Barakat, L. (2003) *Atlas du Liban: Géographie, Histoire, Économie*. Saint Joseph University Press, Beirut.

Boorstein, M. (2016) The many meanings of Rio's massive Christ Statue. *The Washington Post*. Available at: www.washingtonpost.com (accessed 7 May 2023).

Catholic News Agency (2017) Croatia will soon hold one of the largest Marian statues in the world. Catholic News Agency, 7 March. Available at: www.catholicnewsagency.com/news/35571 (accessed 28 October 2022).

Chemain, J.F. (2014) Laïcité: va-t-on supprimer tous les signes chrétiens du paysage public? [Secularism: are we going to remove all Christian signs from the public landscape?] *Le Figaro*, 28 March. Available at: http://www.lefigaro.fr/vox/histoire/2014/03/28/31005-20140328ARTFIG00401-laicite-va-t-on-supprimer-tous-les-signes-chretiens-du-paysage-public.php (accessed 4 October 2022).

Daba, M. (2017) The historiography and values of statue construction: focus on global perspectives. *Humanities and Social Sciences* 5(2): 53–59. DOI: 10.11648/j.hss.20170502.11.

Faour, M. (2007) Religion, demography, and politics in Lebanon. *Middle Eastern Studies* 45(6), 909–921.

Farra Haddad, N. (2015) Planning for sustainable tourism development in a context of regional instability: the case of the Lebanon. In: Morpeth, D.N. and Yan, H. (eds) *Planning for Tourism: Towards a Sustainable Future*. CAB International, Wallingford, UK, pp. 186–202.

Farra Haddad, N. (2018) The sanctuary of Saydet El Mantara. *Lebanon Traveler*, 21 June. Available at: http://www.lebanontraveler.com/en/magazine/lebanon-traveler-the_sanctuary_of_saydet__el_mantara (accessed 17 September 2022).

Grantham J. (2015) Lebanese build biggest inhabitable cross in the Middle East. *Catholic Outlook*. Available at: https://catholicoutlook.org/lebanese-build-biggest-inhabitable-cross-middle-east/ (accessed 26 September 2022).

Habib, C. (2009) Lebanese politics and the tyranny of confessionalism. *Confluences Méditérarnée* 70, 63–70. Available at: www.cairn.info/revue-confluences-mediterranee-2009-3-page-63.htm (accessed 10 May 2022).

Harb, A. (2008) *The Roots of Christianity in Lebanon*. Lebanese Heritage Foundation Publications, Beirut.

Horster, M. (2010) Religious landscape and sacred ground: relationships between space and cult in the Greek World. *Revue de l'histoire des religions* 4. DOI: org/10.4000/rhr.7661.

Mauro, J.-P. (2021) Philippines finishes construction of largest Marian statue in the world. *Aleteia*, 6 May. Available at: www.aleteia.org/2021/05/06/philippines-finishes-construction-of-largest-marian-statue-in-the-world/ (accessed 10 May 2023).

Mermier, F. (2015) *Récits de villes : d'Aden à Beyrouth [Stories of Cities: From Aden to Beirut]*. Actes Sud/Sindbad, Paris.

Miratico (2016) World's top 19 largest crosses (Reach high for the sky!). Miratico, 3 April. Available at: http://miratico.com/worlds-largest-crosses-reach-high-for-the-sky. (accessed 15 October 2022).

Najib (2010) The biggest lighted cross in Bekish – Lebanon. Blog Baladi, 14 September. Available at: http://blogbaladi.com/the-biggest-lighted-cross-in-bekish-lebanon (accessed 7 November 2022).

Prasad, I. (2019) Giant statues of Hindu gods and leaders are making Muslims in India nervous. *The Print*, 28 April. Available at: http://www.theprint.in (accessed 19 November 2022).

Prudhomme, C. (2016) Quand la religion modèle l'espace [When religion shapes space]. *Histoire, Monde et Cultures Religieuses* 1(37), 3–6. DOI: 10.3917/hmc.037.0003.

Reasane, K. (2018) Statues, symbols and signages: monuments towards socio-political divisions, dominance and patriotism? *HTS Teologiese Studies* 74(4). DOI: 10.4102/hts.v74i4.4895.

Roncaglia, M. (2004) *In the Footsteps of Jesus, the Messiah, in Phoenicia*. Dar El Ahad, Beirut.

Rosary of Lebanon (2024) The Rosary of Lebanon. Available at: https://www.rosaryoflebanon.com (accessed 21 February 2024).

Saydet Ainebel (2021) Em El Nour the 'Mother of Light!' Available at: https://www.saydetainebel.com (accessed 27 September 2022).

10 In the Global Spotlight: Japanese Buddhist Destinations in Regional Areas

Ricardo Nicolas Progano*
Wakayama University, Wakayama, Japan

Abstract

This chapter explores the globalization of sacred sites in regional areas through the case study of Kōyasan, a Shingon Buddhism site located in Wakayama Prefecture, Japan. Traditionally, this site, situated in a mountainous area, has been a centre for pilgrimage and scholarly Buddhist studies. While its religious significance remains important, its development as a tourist destination has been on the rise since the 20th century, facilitated by modern transport infrastructure. Its designation as a UNESCO World Heritage Site in 2004 drew further attention from international visitors, particularly Western audiences interested in Buddhism. The Silk Road, a well-known trans-Eurasian commercial network, provides a clear example of pre-modern global networks of trade and traditions. It played a key role in the spread of major religions such as Islam and Buddhism as they moved beyond their original cultural environments (Foltz, 2010). Furthermore, this chapter also analyses an economic-centric interpretation of globalization. This understanding of globalization, often seen as the integration of international commodity markets, primarily focuses on the movement of capital, goods and technology.

Introduction

While religions have historically been 'on the move', globalization and access to fast and low-priced transport since the 20th century have brought different impacts to sacred sites, previously visited only by pilgrims from similar national or religious groups. In consequence, different regional sacred sites have been brought to the global spotlight, integrating themselves into the contemporary tourism industry. Local governments, to solve the socio-economic challenges of present rural areas, have favored this approach as a solution. This chapter explores the globalization of sacred sites in regional areas through the case study of Kōyasan, a Shingon Buddhism site located in Wakayama prefecture, Japan. This site, located in a mountainous area, was traditionally a site of pilgrimage and scholarly Buddhism studies. While its religious significance is still important, its tourism development has been increasing since the 20th century, facilitated by modern transport infrastructure. Its designation in 2004 as a United Nations Educational, Scientific and Cultural Organization (UNESCO) World Heritage Site brought further attention from international visitors, particularly Western audiences interested in Buddhism.

It can be argued that religions have been historically 'on the move' since the dawn of human civilization. Academics from different fields have long studied the origin of

*nprogano@gmail.com

DOI: 10.1079/9781800623675.0010

Proto-Indo-European cultures on the Pontic-Caspian steppes, estimated to be around 4000 BCE, and their subsequent migrations to the Altai and Carpathian Mountains (Anthony and Ringe, 2015), to later become the cultural, linguistic and mythological basis of numerous historical Indo-European civilizations such as the Greek, Indic, Iranian and Germanic (Forston, 2010). Moore and Lewis (2009), while recognizing the differences in scale, technology and complexity, mention that globalization had its precursors during ancient times in a proto-global economy, with examples from Egypt, India, China and Mesopotamia. The Silk Road, the well-known trans-Eurasian commercial network, provides a clear instance of pre-modern global webs of trade and traditions, having a key role in the formation of major religions such as Islam and Buddhism as they left their original cultural environments (Foltz, 2010). For example, the first recorded version of the Lotus Sutra, one of the most influential scriptures of Mahayana Buddhism, was brought to China from Central Asia by Buddhist monk and translator Kumarajiva (344–413 CE), born in the ancient Kucha kingdom, located in the Silk Road (Watson, 1993). In this sense, globalization cannot be defined as solely a modern phenomenon. An economic-centric interpretation of globalization, understood as the integration of international commodity markets, solely focuses on the movement of capital, goods and technology, bypassing the flow of ideas, culture and people (Herrington, 2013). Indeed, the idea that both economic and cultural interconnections began in modern history has been criticized as being Eurocentric and ignoring comprehensive insights into human history (Nederveen Pieterse, 2012).

Despite these criticisms, modern globalization is commonly understood to have started between the 1950s and 1980s, or the neoliberal globalization of 1980–2000, following studies from the economy, political science, media studies and international studies (Nederveen Pieterse, 2012). Friedman (2006), utilizing a Western-centric approach, describes three globalization eras. The first one, Globalization 1.0, is situated in the Age of Discovery when European nations expanded global integration from 1492 to 1800. Next, Globalization 2.0 is situated between 1800 and 2000 and was driven by American and European multinational companies which went global for new markets and labour forces and were aided by new transport and communication technologies. Finally, Globalization 3.0 started in 2000 and is driven by individuals directly accessing global markets to collaborate and compete, with global companies being empowered as well. In this last globalization era, non-Western nations such as China and India play a significant role. In its development, globalization influences various aspects of human life such as capital flows, environmental pollution and diseases (Hjalager, 2007). The role of information and communications technologies (ICT) has been key in modern globalization, particularly since the creation of the Internet allowed a means for worldwide fast and accessible exchange of data. These technologies have also facilitated access to global markets for a wider range of both consumers and suppliers (Friedman, 2006; Jyoti, 2021).

Tourism is observed as one of the many causes and results of globalization (Hjalager, 2007). As Sharif *et al.* (2021) explain, tourism development contributes to the globalization of the economy, society and policy, while also benefitting from globalization. Theuns (2008) notes that tourism presents an interesting case since tourism production is location specific, and thus cannot be outsourced or offshored like other economic activities. However, it can be affected by traditional takeovers, mergers and hedge fund activity. Hjalager (2007) describes the globalization of the tourism industry in four stages as a gradually evolving process in which companies, destinations and countries control costs, obtain market gains, and obtain access to resources. ICT and transport infrastructure further enhanced this process. Despite its gradual nature, Chiu *et al.* (2021) note that globalization has a non-linear influence on inbound tourism, as globalization increases international arrivals, but does not necessarily increase international tourism receipts and net tourism service exports. This may be explained by the increased and intense competition that destinations undergo in the tourism market. Similarly, Hjalager (2007) notes that globalization develops differently both internationally and intra-regionally due to factors such as energy prices. Other issues such as economic crises or terrorism can also negatively impact globalization processes (Friedman, 2006; Sharif *et al.*, 2021).

The case study of the Buddhist site of Kōyasan, a Shingon Buddhism site located in Wakayama prefecture, Japan, is examined. To further contextualize this case study from a cultural and historical perspective, modern Western discourses on Buddhism and the general situation of Japanese Buddhist temples will be introduced as key factors.

Case Study: Kōyasan Site

Kōyasan is located in a mountainous area, far away from the main urban centres (especially at the time when it was founded), it has always been interwoven in global networks of cultural, religious and commercial exchange. Kōyasan is a large Buddhist complex and the seat of Shingon Buddhism, established by the celebrated Japanese monk Kūkai (774–835 CE), also known posthumously as Kōbō Daishi (Nicoloff, 2008; Matsunaga, 2014). Shingon Buddhism is part of esoteric Buddhism (*mikkyō* – 密教 in Japanese) which is understood to be a major branch of Mahayana Buddhism that existed in India from the 6th century until the advent of Islam in the 13th century. It was transmitted to China in the 7th century from Central Asia through the Silk Road, although it disappeared in the late Tang dynasty (618–907 CE) (Matsunaga, 1991).

To bring the teachings of Shingon Buddhism to Japan, in 804 Kūkai travelled to Chang'an, the capital city of China during the Tang dynasty, along with a delegation including other monks, officials, scribes and translators. At the time, the Chinese capital was a centre of esoteric Buddhism learning, after being introduced in the early 8th century. Here, Kūkai went to be initiated into Shingon under master Hui-kuo. Having been recognized as the inheritor of Shingon teachings, he returned to Japan in 806 to spread this Buddhism branch (Nicoloff, 2008). After obtaining permission from Emperor Saga, Kōyasan was consecrated in 819 under the name of Kongōbu-ji, with Kūkai supervising its construction to make Kōyasan resemble both physically and spiritually a lotus flower (Nicoloff, 2008). This location was selected because Kūkai judged these remote mountains as an adequate place for esoteric training and meditation (Matsunaga, 2014). In 835, Kūkai passed away and was entombed in Kōyasan. According to tradition, however, he did not die and is currently in a state of deep meditation until the future Buddha Miroku (Maitreya in Sanskrit) appears in our world (Nicoloff, 2008).

As observed, the establishment of Shingon Buddhism and the construction of Kōyasan are deeply embedded in global networks of commercial and cultural exchange, which can be traced further beyond the Japanese embassies to imperial China, when Shingon Buddhism was transmitted from Central Asia through the Silk Road (Matsunaga, 1991). However, its modern growth into an international tourism destination cannot be solely explained by these historical reasons. The next section aims to explain the recent background of this development.

Examining Kōyasan as a global tourism destination

Modern Western discourses on Buddhism

An interesting point of discussion is why Western visitors have been attracted by Kōyasan or Buddhist sites in general as shown in previous research (Choe *et al.*, 2015). While there have been numerous contacts between Western cultures and Buddhism, this religion historically did not have adopters in Europe or the Americas, so what are the roots of this interest by Western travellers? Two main factors may help to explain this travel phenomenon: (i) the historical ties of Buddhism and the Western world, and its introduction to modern Western society; and (ii) the complex socio-economic situation of Japanese Buddhist temples and their associated communities since the post-war period. The first factor will be discussed here and the second factor in the next section 'The situation of Japanese Buddhist temples'.

First, it is crucial to discuss the historical ties of Buddhism and the Western world, as well as its introduction to modern Western society. The first contact of Buddhism with Western culture is represented in the art of Gandhara, a region located within the modern borders of Afghanistan and Pakistan that conserves an important range of historic cities. It developed from the 1st century BCE until the 6th century CE following the military campaigns of Alexander the

Great in the 4th century BCE that paved the way for the introduction of Greco-Roman artistic influence in the area (Falser, 2015). Gandhara art is by no means a small footnote in Buddhism's history, as it showcases the development of Mahayana through images of the future Buddha Maitreya and bodhisattvas, as well as the transformation of Buddha's images (Luczanits, 2008). In Japan's case, in the 16th century, European travellers encountered Buddhism, leading to different religious polemics. Certain Japanese thinkers not only criticized Christianity as demonic dharma (*mahō* – 魔法) but also as a vehicle of Western conquest and colonization, while the Tokugawa shogunate conducted a harsh suppression of Christianity (Inoue, 1990; Josephson, 2012). In turn, Western travellers condemned Buddhism as heresy and focused on Christian missionary work, although these contacts led to modern anthropological studies (Kitagawa, 1987; Josephson, 2012). This first contact was interrupted by the Tokugawa shogunate's edict of national exclusion, following the Christian uprising at Shimabara of 1637–1638 (Kitagawa, 1987).

While an in-depth exposition of the premodern interactions between Buddhism and Western cultures is outside this chapter's scope, these examples showcase that the West had contact with Buddhism throughout history. Still, Buddhism was not historically a widespread religion in the West, with the Kalmyks, an Oirat Mongolian ethnic group living in the Russian Caucasus, being the sole exception (Dondukov, 2015). The popularization of Buddhism in modern Western society can be traced back to the 19th century, with the World's Parliament of Religions, which took place in Chicago in 1893, being a key event in this regard (McMahan, 2004). During this event, educated Western audiences, disenchanted with Christianity while supporting liberal and positivistic ideas, were presented with rhetorical discourses by reformist Asian Buddhists who aimed at presenting a 'scientific' and 'rational' version of Buddhism, devoid of aspects perceived as superstitions, and seeking parallels with Western modern science. Certain reformists also claimed that Buddhism made discoveries only found by Western science millennia later. The role of the Theosophical Society, the well-known occult society founded in New York in 1875 and headed by Russian mystic Helena Petrovna Blavatsky (1831–1891), is particularly interesting here. The Theosophical Society had different contacts with Buddhist organizations in Asian countries such as Sri Lanka, India and Japan (Blavatsky, 1890), with Henry Steel Olcott (1832–1907), co-founder and first president of the Theosophical Society, doing a successful trip to Japan in 1889 with the support of Buddhist societies in his attempts to create a global Buddhism network (McMahan, 2004; Yoshinaga, 2009). The Theosophical Society also aimed at presenting a 'scientific' rationale for Buddhism, although based on the occult, while criticizing rituals and ceremonies as degeneration of ancient, 'pure' Buddhism. These ideas influenced certain reformist Buddhist figures, such as Sri Lankan Buddhist activist Anagarika Dharmapala (1864–1933), who introduced a 'Protestant' Buddhism to the West (McMahan, 2004). Many other Western figures presented their understanding of Japanese Buddhism to Western audiences from other perspectives, such as Ernest Fellonosa (1853–1908), an American art historian interested in the preservation of Japanese art (Yoshinaga, 2009).

Still, this popularization of Buddhism was not solely the work of non-Japanese organizations and figures. Indeed, Buddhism thinkers also played a role as well, to gain ground against Christianity, which was perceived as a religion of civilization due to its association with Western science and culture, through nationalistic rhetoric (McMahan, 2004; Snodgrass, 2012). For example, Buddhist reformist Inoue Enryō (1858–1919) aimed at showing that Buddhism contained all the wisdom of the West, and even surpassed it. He particularly emphasized the preeminence of Japanese Buddhism, where he claimed its higher meanings were better conserved (Snodgrass, 2012). These confrontative discourses were eventually toned down when the 1893 Parliament of Religions took place, with the role of young Buddhist reformers being of importance here, including well-known writer Suzuki Daisetsu (1870–1966). The interest of Westerners in Buddhism was also interpreted as proof of Buddhism's validity and by association, Asian cultures (Yoshinaga, 2009). The 'scientific' Buddhism discourses mentioned above can be understood in this light, as they challenged the perceived negative image of Buddhism of the time, conceptualized as a nihilistic and superstitious religion, as well as the general condescending conception towards Asia (McMahan, 2004).

These novel discourses on Buddhism, shaped by both Western figures and Asian reformers, introduced the West to an attractive doctrine that could be accommodated along with modern science and provide a viable alternative to Christianity. At the same time, it provided Asian reformists with a way to challenge the preeminence of Christianity as the religion of civilization and legitimize their cultures in the modern world. For example, Yamamoto (2014) notes that the works of American writer Ellery Sedgwick, invited by the Japanese government in 1930, heavily emphasize Buddhist temples in a positive light, portraying them as abodes of serenity and 'Eastern' tradition. During the 1960s and the 1970s, Buddhism was further popularized in the West by leaving its traditional hierarchical structure and attitudes towards women. This helped Buddhism to fit better into the social issues of the time and become more acceptable to Western individualism (Cusack, 2011). This period also coincided with the expansion of alternative spiritualities in the West, such as the New Age movement, a counterculture that emerged during the 1970s in the USA. This movement involves a wide range of eclectic elements including alternative medicine, reinterpreted Asian religions, neo-paganism and astrology, with no unifying dogmas or institutions (Hanegraaff 2000; Timothy and Conover, 2006; Progano and Kato, 2018). These discourses on Buddhism are still of relevance in contemporary society, as positivistic studies on meditative states and the psychologizing view of cosmological doctrines play a vital role in legitimizing and explaining Buddhism (McMahan, 2004; Yoshinaga, 2009; Cusack, 2011; Snodgrass, 2012). Its consequences can still be observed a century later, as it paved the way for the introduction and adoption of Tibetan Buddhism in the late 20th century (Snodgrass, 2012), which was also influenced by media, film and celebrities (Mullen, 1998; Cusack, 2011). At the same time, since the end of World War II, Japanology transformed from a small niche academic subject to a growing and attractive subject for researchers (Kitagawa, 1987), increasing published studies on Buddhism.

The situation of Japanese Buddhist temples

The second factor is the complex socio-economic situation of Japanese Buddhist temples and their associated communities since the post-war period, which has partially motivated their involvement in the global tourism industry. Kitagawa (1987) notes that the war itself led to numerous losses of Buddhist property and clergymen's lives. Additionally, Reader (2012) mentions that, since the end of World War II, there has been a steady decrease in faith among the Japanese, and the youth have abandoned their interest in Buddhism for communism, socialism or Japanese new religions (Kitagawa, 1987). This situation particularly affects temples in rural areas, which are already suffering from ageing and depopulation processes (Nelson, 2012). This disinterest in religion also negatively impacts the recruitment of future monks. Moreover, Kitagawa (1987) and Covell (2005) mention the financial challenges that Japanese Buddhism faces, including: (i) a decrease in donors; (ii) loss of political privileges; (iii) diminished confraternities; (iv) loss of traditional private estates (*shōen* – 荘園); and (v) rural depopulation. Also, since the post-war period, temples have extensively focused on providing funeral and memorial services to secure funding, as well as graveyard fees (Covell, 2005; Nelson, 2012). While this focus on mortuary services is criticized, they are not a simple solution to the temples' financial needs, as the loss of affiliated member households, due to ageing and rural migration, and the disinterest of Japanese society in its traditional religious institutions leads to fewer funeral services being performed (Covell, 2005; Nelson, 2012). Finally, it is important to note the impact on the public of the 1995 sarin attack on public opinion towards religion and its institutions. This negativity and suspicion towards religious institutions may prevent them from conducting proselytization activities (Nelson, 2012).

Facing this situation, Buddhism temples have attempted to tackle the new challenges of modern times. Borup (2015) and Dessi (2017) mention that the clergy attempted to create a positive image of Buddhism by increasing participation in social affairs and events, as well as engaging in activities including welfare, environmentalism and bioethics. Reader (2020) mentions that temples are increasingly engaging with public and private stakeholders to promote themselves as tourism destinations and receive more visitors. Yanata and Sharpley (2022) note that tourism gives an additional opportunity to spread Buddhist teachings. At the same time, temples are becoming a vital resource for destinations, as city planners, transport companies and businesses notice the consumer traffic they bring. Temples with designated cultural heritage can also apply for government funding (Covell, 2005). Partially due to the positive reception of Buddhism in Western countries, international visitors have become a recurring sight on temple grounds. This phenomenon is not limited to Japan, as other countries such as Thailand receive Western visitors interested in Buddhism (Schedneck, 2020). Buddhist cuisine, known as *shōjin ryōri* (精進料理), has been offered as another valuable experience for inbound tourism in Japan and South Korea (Son and Xu, 2013; Progano and Kato, 2021). Temple stays have also become a staple of Buddhism-related tourism experiences, with South Korea (Kaplan, 2010) and Japan (Matsunaga, 2014; Progano and Kato, 2021) being examples of this.

Kōyasan as a global tourism destination

Kōyasan, while still being an important centre of Buddhism training and scholarly, is in a comparable situation. Historically, Kōyasan was, like other religious institutions, under the patronage of the aristocracy and military clans, providing the temples with financial and political support. In turn, Kōyasan provided them with religious and accommodation services, known as *shukubō* (宿坊). However, this patronage relationship ended when the Meiji government (1868–1912) ordered its termination. Due to this loss of financial support, the temples decided to host guests from the public (Nicoloff, 2008; Matsunaga, 2014), commodifying *shukubō* stays into the popular tourism service of today and refurbishing them to hospitality standards. The *shukubō* are vitally important in Kōyasan's current tourism industry, as they provide all types of accommodation to visitors by counting for 6166 of 6240 available beds (Wakayama Tourism Agency, 2022). The degree of tourism commodification of each temple is diverse and depends on the views of the corresponding head monk (Yanata and Sharpley, 2022). The development of modern infrastructure was key in this development, as it facilitated the transportation of both domestic and international visitors. Different infrastructure developments such as the opening of railways from Osaka in 1925, the construction of a cable cart in 1930, and the construction of road links in the 1960s cemented Kōyasan's position as a popular destination and increased the number of day visitors (Yanata and Sharpley, 2022). Finally, the designation of Kōyasan in 2004 as part of the 'Sacred Sites and Pilgrimage Routes in the Kii Mountain Range' UNESCO World Heritage Site became a key event in the globalization of the site. While their main objective is heritage protection and preservation, World Heritage Sites often become international tourism destinations due to the wide recognition they enjoy across the globe (Gonzalez Santa-Cruz and Lopez-Guzman, 2017).

In this sense, while Kōyasan was already a well-known location in Japan, the 2004 UNESCO designation provided a tool to further develop it for inbound tourism. According to official statistics published by the Wakayama Tourism Agency (2022), in 2007 a total of 1,247,600 visitors went to Kōyasan, and in 2019, 1,481,788 visitors were registered. However, the impact of coronavirus disease 2019 (COVID-19) caused visitor numbers to plummet to 1,183,189 in 2021 (Wakayama Tourism Agency, 2022), with the overall tourism sector being negatively impacted (Progano, 2021). Regarding international visitors, pre-pandemic statistics showed that most of them were from Western nations, with a growing Asian market as well. In 2019, 108,993 international visitors (48.57% of all lodgers) stayed the night at Kōyasan; the most represented countries were France (22,834), the USA (12,174), Germany (8884), China (6975) and the UK (6923) (Wakayama Tourism Agency, 2019). These numbers, however, were negatively impacted by the COVID-19 travel restrictions conducted by the Japanese government, as only 398 foreign visitors were registered in 2021 (Wakayama Tourism Agency, 2022), showing the severe impact on the Japanese tourism industry (Muneta, 2020). Regarding their travel motivations, Western visitors are interested in: (i) the religious traditions of Kōyasan such as temples; (ii) Buddhist cuisine; and (iii) *shukubō* accommodation (Matsunaga, 2014; Jimura, 2016). Here, accommodation is facilitated by an online reservation system established by the local Shukubō Association, as some temples still rely on telephone reservations, showing the importance of ICT in facilitating global tourism as mentioned by Hjalager (2007) and Jyoti (2021). Kōyasan's local government has also incentivized inbound tourism through infrastructure development, provision of tourism information, creation of new tourism products, hospitality seminars for locals, and the establishment of a web page aimed at international visitors. It is currently preparing for the 1200th anniversary of Kūkai's passing in 2034 when large crowds are expected to visit (Kōya Town, 2019).

Apart from the efforts conducted by the local tourism stakeholders, the popularity of Kōyasan among international visitors, particularly from Western countries, can be explained by the historical processes described before. In contemporary Western societies, Westernized versions of Buddhism are popular among the general population, due to their general lack of doctrinal, hierarchical or ritual content which makes them attractive to larger population segments (Cusack, 2011). At the same time, Buddhist temples are inclined to participate in contemporary ways and secure funding (Covell, 2005; Borup, 2015; Reader, 2020). Likewise, local governments recognize the importance of their Buddhist heritage for tourism commodification and consumption as an alternative to vanishing traditional economies such as forestry and agriculture. This approach, which seeks to mobilize the local stakeholders and valorize local resources to engage outside markets (Woods and McDonagh, 2011), is certainly common in Japan (Kinoshita, 2021) and elsewhere (Woods and McDonagh, 2011), but it has been criticized. Woods and McDonagh (2011) mention that conceptualizing globalization as a 'saviour' for rural communities is a simplistic approach and that not all rural localities can compete equally in the global economy. In Japan, Kinoshita (2021) notes that inbound tourism is still only 17.3% of travel expenditure and calls for a re-evaluation of domestic tourism's importance in rural areas. He also notes the numerous challenges that are found in Japan's rural destinations such as inferior work conditions, low salaries, a focus on offering low prices over quality, the lack of qualified human resources, and the conservative mindsets of local stakeholders. In Kōyasan, the involvement of religious stakeholders, the *shukubō* temples, brings additional challenges to tourism development (Fig. 10.1). For example, the monks do not receive hospitality education during their religious training, and thus may face challenges when managing visitors who may not display the type of behaviour expected by the monks (Progano, 2021; Yanata and Sharpley, 2022). Similar observations were made by previous researchers (Shackley, 2001). The growing workload of monks, who must perform both religious and hospitality duties, is also of concern (Yanata and Sharpley, 2022).

Summary

This chapter examined the globalization of sacred sites, through the case study of Kōyasan and Western visitors. As observed, while globalization is not a modern phenomenon (Moore and Lewis, 2009; Foltz, 2010; Herrington, 2013), in our current age this process has accelerated with modern transport systems and ICT playing a leading role (Friedman, 2006; Jyoti, 2021). Here, tourism is observed as both a cause and a consequence of globalization (Hjalager, 2007).

While Kōyasan is located in a rural area, it enjoys a situation that is not as concerning as that of minor temples due to its recent tourism development and transport infrastructure. Still, its population is characterized by ageing and steady decline and its traditional economic industries continue to wane (Kōya Town, 2019), showing similar trends to other regional areas in Japan. In such a situation, sacred sites in regional areas may opt to jump into the 'global spotlight' and develop themselves for international tourism consumption, showing the association between rural areas and globalization (Woods and McDonagh, 2011). However, there are existing concerns regarding the commodification

Fig. 10.1. A *shukubō* temple in Kōyasan. Photo taken by author.

of temples (Yanata and Sharpley, 2022). On the other hand, commodification provides temples with opportunities for obtaining funding, spreading Buddhist teaching, creating a positive image, and playing a larger role in contemporary society (Covell, 2005; Borup, 2015; Reader, 2020; Progano, 2021; Yanata and Sharpley, 2022). While commodification may be dismissed by tourism stakeholders as a mere 'anxiety', it is important to note that religious institutions have different aims beyond visitor growth and the provision of goods and services (Shackley, 2001). This religion-consumerism dialectic is a key aspect of contemporary religious tourism. Further study of the globalization process of sacred sites in tourism contexts may shed further light on this chapter's subject. Dessi (2017) mentions that Japanese new religious movements are often the subject of academic studies on globalization and religion, while traditional religions such as Buddhism are perceived as being unaffected by globalization. Therefore, as religious destinations become more involved in global tourism networks, studies on the subject may be of interest.

Finally, it is important to note that Buddhism is not the only religion or cultural group trying to find its place within an increasingly uninterested society and reinterpreted through globalized narratives. Rots (2015) notes that the modern notion of Shinto as an apolitical, ancient religion centred on nature worship and ecological knowledge is a historically inaccurate depiction that can be traced back to the end of the 19th century. This depiction has served as the basis for modern discourses on environmentalism and sustainability related to Shinto. Japanese Buddhism adopted similar ecological discourses during the 20th century, reflecting global discourses on ecology and environmentalism (Dessi, 2013). Similarly, Nadasdy (2005) notes the Western reinterpretation of Indigenous traditions for environmentalism discourses, portraying Indigenous groups as living attuned to nature in a modern twist of the 'noble savage' colonial stereotype, thus denying their historical realities and cultural richness. At the same time, Indigenous groups may utilize this 'ecological noble' image as a political strategy due to its widespread appeal and rhetorical power among government bureaucrats. However, environmentalists and Indigenous groups have clashed over environmental policies due to their different stances and beliefs (Chapin, 2004; Hames, 2007). These 'ecological noble' conceptions also tend to diminish 'uncomfortable' topics, such as the importance of human sacrifice and ritual bloodletting in Mesoamerican and South American Aboriginal religions (Faux, 2012; Carrasco, 2014). While these modern reinterpretations are valuable as objects of anthropological study, researchers should be aware that they are not historically accurate.

References

Anthony, D.W. and Ringe, D. (2015) The Indo-European homeland from linguistic and archeological perspectives. *Annual Review of Linguistics* 1, 199–2019.

Blavatsky, H.P. (1890) Recent progress in theosophy. *The North American Review* 151(405), 173–186.

Borup, J. (2015) Easternization of the East? Zen and spirituality as distinct cultural narratives in Japan. *Journal of Global Buddhism* 16, 70–93.

Carrasco, D. (2014) *Religions of Mesoamerica*, 2nd edn. Waveland and Press, Long Grove, Illinois.

Chapin, M. (2004) A challenge to conservationists. *World Watch November/December*, 17–31.

Chiu, Y., Zhang, W. and Ding, K. (2021) Does globalization influence inbound tourism? Evidence from a dynamic panel threshold analysis. *Journal of Travel Research* 60(5), 1074–1084.

Choe, J., Blazey, M. and Mitas, O. (2015) Motivations of non-Buddhists visiting Buddhist temples. *Current Issues in Tourism* 18(1), 70–82.

Covell, S.G. (2005) *Japanese Temple Buddhism: Worldliness in a Religion of Renunciation*. University of Hawai'i Press, Honolulu.

Cusack, C.M. (2011) The Western reception of Buddhism: celebrity and popular cultural media as agents of familiarisation. *Australian Religion Studies Review* 24(3), 297–316.

Dessi, U. (2013) *Japanese Religions and Globalization*. Routledge, New York.

Dessi, U. (2017) *The Global Repositioning of Japanese Religions: an Integrated Approach*. Routledge, New York.

Dondukov, B. (2015) The global space of dialogue and conflict: the case of Russian Buddhists. *Kultura – Historia – Globalizacja* 18, 45–56.

Falser, M.S. (2015) The Graeco-Buddhist style of Gandhara – a 'storia ideologica', or: how a discourse makes a global history of art. *Journal of Art Historiography* 13, 1–53.

Faux, J.L. (2012) Hail the conquering gods: ritual sacrifice of children in Inca society. *Journal of Contemporary Anthropology* 3(2), 3–15.

Foltz, R. (2010) *Religions of the Silk Road: Premodern Patterns of Globalization*, 2nd edn. Palgrave Mcmillan, New York.

Forston, B.W. (2010) *Indo-European Language and Culture: an Introduction*, 2nd edn. Wiley-Blackwell, Oxford.

Friedman, T.L. (2006) *The World is Flat: a Brief History of the Twenty-first Century*. Farrar, Straus and Giroux, New York.

Gonzalez Santa-Cruz, F. and Lopez-Guzman, T. (2017) Culture, tourism and World Heritage Sites. *Tourism Management Perspectives* 24, 111–116.

Hames, R. (2007) The ecologically noble savage debate. *Annual Review of Anthropology* 36, 177–190.

Hanegraaff, W.J. (2000) New Age religion and secularization. *Numen* 47(3), 288–312. https://doi.org/10.1163/156852700511568

Herrington, L.M. (2013) Globalization and religion in historical perspective: a paradoxical relationship. *Religions* 4, 145–165.

Hjalager, A. (2007) Stages in the economic globalization of tourism. *Annals of Tourism Research* 34(2), 437–457.

Inoue, N. (1990) Globalization and modern Japanese religion within the context of Sect Shinto's policy towards Christianity. *Senri Ethnological Studies* 29, 21–35.

Jimura, T. (2016) World Heritage Site management: a case study of sacred sites and pilgrimage routes in the Kii mountain range, Japan. *Journal of Heritage Tourism* 11(4), 1–13.

Josephson, J.A. (2012) *The Invention of Religion in Japan*. The University of Chicago Press, Chicago, Illinois.

Jyoti, D. (2021) Information and communication technology (ICT) and globalization. *International Journal of Innovative Science and Research Technology* 6(5), 224–226.

Kaplan, U. (2010) Images of monasticism: the Temple Stay Program and the re-branding of Korean Buddhist temples. *Korean Studies* 34, 127–146.

Kinoshita, H. (2021) *The Illusion of Community Development*. SB Shinsho, Tokyo. (in Japanese)

Kitagawa, J.M. (1987) *On Understanding Japanese Religion*. Princeton University Press, Princeton, New Jersey.

Kōya Town (2019) *Kōya Town Fourth Long-term General Plan*. Kōya Town, Japan. (in Japanese)

Luczanits, C. (2008) Buddhism in a cosmopolitan environment: the art of Gandhara. *Orientations* 39(7), 44–50.

Matsunaga, Y. (1991) *Mikkyō*. Iwanami Shoten, Tokyo. (in Japanese)

Matsunaga, Y. (2014) *Kōyasan*. Iwanami Shoten, Tokyo. (in Japanese)

McMahan, D.L. (2004) Modernity and the early discourse of scientific Buddhism. *Journal of the American Academy of Religion* 72(4), 897–933.

Moore, K. and Lewis, D. (2009) *The Origins of Globalization*. Routledge, New York.

Mullen, E. (1998) Orientalist commercializations: Tibetan Buddhism in American popular film. *Journal of Religion and Film* 2(2), n.p.

Muneta, Y. (2020) *The Regeneration of Inbound Tourism*. Gakugei Shuppansha, Kyoto, Japan. (in Japanese)

Nadasdy, P. (2005) Transcending the debate over the ecologically noble Indian: indigenous peoples and environmentalism. *Ethnohistory* 52, 291–331.

Nederveen Pieterse, J. (2012) Periodizing globalization: histories of globalization. *New Global Studies* 6(2), n.p.

Nelson, J.K. (2012) Japanese secularities and the decline of temple Buddhism. *Journal of Religion in Japan* 1, 37–60.

Nicoloff, P.L. (2008) *Sacred Kōyasan: a Pilgrimage to the Mountain Temple of Saint Kōbō Daishi and the Great Sun Buddha*. State University of New York Press, New York.

Progano, R.N. (2021) The impact of COVID-19 on temple stays: a case study from Koyasan, Japan. *International Journal of Religious Tourism and Pilgrimage* 9(5), 116–127.

Progano, R.N. and Kato, K. (2018) Spirituality and tourism in Japanese pilgrimage sites: exploring the intersection through the case of Kumano Kodo. *Fieldwork in Religion* 1(1), 22–43.

Progano, R.N. and Kato, K. (2021) Pilgrimage and tourism: shift in values and motivation. In: Liutikas, D. (ed.) *Pilgrims: Values and Identities*. CAB International, Wallingford, UK, pp. 117–126.

Reader, I. (2012) Secularisation, R.I.P.? Nonsense! The 'Rush Hour Away from the Gods' and the decline of religion in contemporary Japan. *Journal of Religion in Japan* 1, 7–36.

Reader, I. (2020) Turning to tourism in a time of crisis? Buddhist temples and pilgrimage promotion in secular(ized) Japan. In: Bruntz, C. and Schedneck, B. (eds) *Buddhist Tourism in Asia*. University of Hawai'i Press, Honolulu, pp. 161–179.

Rots, A.P. (2015) Sacred forests, sacred nation: the Shinto environmentalist paradigm and the rediscovery of *chinju no mori*. *Japanese Journal of Religious Studies* 42(2), 205–233.

Schedneck, B. (2020) Loss and promise: the Buddhist temples as tourist space in Thailand. In: Bruntz, C. and Schedneck, B. (eds) *Buddhist Tourism in Asia*. University of Hawai'i Press, Honolulu, pp. 66–83.

Shackley, M. (2001) *Managing Sacred Sites*. Continuum, London.

Sharif, A., Ullah, S., Shahbaz, M. and Mahalik, M.K. (2021) Sustainable tourism development and globalization: recent insights from the United States. *Sustainable Development* 29, 957–973.

Snodgrass, J.M. (2012) Japan's contribution to modern global Buddhism: the World Parliament of Religions revisited. *The Eastern Buddhist* 43(1–2), 81–102.

Son, Z. and Xu, H. (2013) Religious food as a tourism attraction: the roles of Buddhist temple food in Western tourist experience. *Journal of Heritage Tourism* 8(2–3), 248–258.

Theuns, H.L. (2008) Globalization and tourism: pros and cons. *Tourism Recreation Research* 31(1), 99–105.

Timothy, D.J. and Conover, P.J. (2006) Nature religion, self-spirituality and New Age tourism. In: Olsen, D.H. and Timothy, D.J. (eds) *Tourism, Religion and Spiritual Journeys*. Routledge, London, pp. 139–155.

Wakayama Tourism Agency (2019) *Reiwa 1 Tourist Motivation Research Report*. Wakayama Tourism Agency, Wakayama, Japan. (in Japanese)

Wakayama Tourism Agency (2022) *Reiwa 3 Tourist Motivation Research Report*. Wakayama Tourism Agency, Wakayama, Japan. (in Japanese)

Watson, B. (1993) *The Lotus Sutra*. Columbia University Press, New York.

Woods, M. and McDonagh, J. (2011) Rural Europe and the world: globalization and rural development (editorial). *European Countryside* 3, 153–163.

Yamamoto, K. (2014) Representation of Japanese tourist sites and inbound tourism from the United States in pre-war Japan: Ellery Sedgwick's travel writing in 1930. *Kankō Hyōron* 2(1), 55–66.

Yanata, K. and Sharpley, R. (2022) Coexistence between tourists and monks: managing temple-stay tourism at Koyasan, Japan. In: Olsen, D.H. and Timothy, D.J. (eds) *The Routledge Handbook of Religious and Spiritual Tourism*. Routledge, London, pp. 398–410.

Yoshinaga, S. (2009) Theosophy and Buddhist reformers in the middle of the Meiji period. *Japanese Religions* 34(2), 119–131.

11 Pilgrims or Tourists? Visitors Seeking Out Experiences on Mount Athos

Polyxeni Moira,* Spyridon Parthenis and Dimitrios Mylonopoulos
University of West Attica, Egaleo, Greece

Abstract

This chapter investigates and discusses the region of Mount Athos which holds a unique distinction as the sole self-governing monastic community within the Greek state, serving as a spiritual centre of the Eastern Orthodox Christian world since AD 1054. Its cultural and natural significance led to its inclusion on the UNESCO World Heritage List in 1988. Mount Athos is governed by 20 holy monasteries, including 17 Greek, one Russian, one Serbian and one Bulgarian. Mount Athos is home to approximately 2000 permanent monks. Access to the area is regulated, with a strict prohibition, known as 'Avaton' in Greek, excluding women and female animals. Furthermore, the daily entry limit allows 100 Orthodox and ten non-Orthodox pilgrims to visit the Holy Mountain, contingent upon obtaining an entrance permit. This chapter explores diverse motives for visitors travelling to the Holy Mountain and examines the specific experiences they seek during their journey. These vary from spiritual guidance and healing to the pursuit of tranquillity, encounters with nature and a desire for escapism. This chapter endeavours to provide insights into the complex blend of motivations and aspirations that shape the visitor landscape on Mount Athos.

Introduction

The district of Mount Athos has been the sole self-governing monastic community of the Greek state and a spiritual centre of the Eastern Orthodox Christian world since AD 1054. In 1988 it was inscribed on the United Nations Educational, Scientific and Cultural Organization (UNESCO) World Heritage List under Cultural and Natural Criteria. Mount Athos is governed by its 20 holy monasteries (17 Greek, one Russian, one Serbian and one Bulgarian). About 2000 monks reside there permanently. Entry into the area is strictly controlled. Women and female animals are banned from Mount Athos under a prohibition called 'Avaton' in Greek. In addition to this, there is a daily quota of 100 Orthodox and ten non-Orthodox pilgrims that can enter the Holy Mountain after having obtained an entrance permit. This chapter examines visitors' potential motives for travelling to the Holy Mountain and the type of experience they seek out during their stay there (e.g. spiritual guidance or healing, peace and relaxation, contact with nature, escapism).

Since antiquity and throughout history, religious feelings have motivated people to travel to experience communion with the divine in sacred places. Almost all major religions have urged believers to visit a holy place in order to find relief

*polmoira@uniwa.gr

 Religious Tourism and Globalization. D. Liutikus and R. Raj (eds)
DOI: 10.1079/9781800623675.0011

from all the difficulties of life, spiritual or non-spiritual. The visit to a sacred place that meant purification, redemption of sins, fulfilment of a vow, healing or satisfaction of a request was called pilgrimage.

Movements for religious reasons have been recorded since prehistoric times. Menhirs, megalithic formations and cave paintings had the same function as cathedrals. In ancient times, Celts from all parts of Europe gathered in sacred places (e.g. forests, caves, burial grounds, springs and waterfalls) where they performed rituals, offerings and sacrifices. Herodotus described in his work *Histories*, written around 430 BC, the movement of thousands of Egyptians who sailed on the Nile in their feluccas, traditional wooden sailing boats, to go to the temple of Memphis to perform their religious duties (Μοίρα, 2009, 2019). In ancient Egypt there were many sanctuaries such as Abydos, Heliopolis, Luxor, Thebes, Karnak. The Assyrians worshipped their gods in Aleppo and Hierapolis (Syria), while the Babylonians gathered in Nippur, about 160 km southeast of Baghdad, praying for peace or asking the gods for longevity (Rinschede, 1992, p. 54). The journeys of groups of believers to sacred pilgrimages are known as 'sacred expeditions' (Μοίρα, 2009, 2019).

Numerous pilgrimage sites were located in ancient Greece. Pilgrimages were divided into two categories: (i) those that attracted individual pilgrims (e.g. the oracle shrine at Dodona, the oracle of Delphi) and the sanctuaries of Asclepius (god of medicine and son of the god Apollo) that believers travelled to in the hope of being healed; and (ii) those that attracted groups of pilgrims, such as the sites where large religious sporting events were held (e.g. Olympia (Olympian Games), Delphi (Pythian Games), Isthmus (Isthmian Games) and Nemea (Nemean Games)) from the 6th century BC to the 4th century AD. The word 'sanctuary' (hieron in Ancient and contemporary Greek meaning 'holy place') was used in the sense of the holy or sacred, meaning an area between the earthly and the sacred celestial world, where man could communicate with God. Thus, the ancient Greeks visited the sanctuary of Zeus at Dodona and the sanctuary of Apollo at Delphi to seek an oracle.

Travelling for religious purposes satisfied a spiritual need and doing so was considered a redemptive act. But the reasons for people going on pilgrimages were often twofold, satisfying both their spiritual and their material needs. Said (1979, p. 168) writing about the 19th-century French and British pilgrims to the Orient and the Holy Land: 'Every pilgrim sees things his own way, but there are limits to what a pilgrimage can be for, to what shape and form it can take, to what truths it reveals'. The passage above gives the outline of an activity that, however much it has changed over time, still retains today many of its ancient characteristics. Indeed, it is difficult to find another human habit so widespread in space and so resistant to time as travel to holy places for the purpose of pilgrimage. From antiquity to the present day, it has been firmly held that prayer or the performance of religious duties is most effective when it takes place in specific places. These may include the places where gods and saints were born, died, performed miracles, were martyred or simply appeared, places where statues, temples, churches, relics of saints or miraculous icons were found, places where the most significant events in the history of each religion took place (Moira and Mylonopoulos, 2017).

Pilgrimage, in the sense of human movement in space in search of the divine, along with the manifestation of faith, is an important psychological need for every human being regardless of race, gender, nationality or religion. In contrast, tourists travel for relaxation and leisure, so the motivations for travelling are varied. Smith (1992, p. 1) claims that 'the contemporary use of the terms, identifying the pilgrim as a religious traveler and the tourist as a vacationer, is a culturally constructed polarity' that conceals the motives of the travellers' search. In any case, it is extremely difficult to investigate the visitors' incentives and what motivations prevail, whether the reasons for travelling are purely pilgrimage related or cultural or whether they are both present.

To date, there has been a lot of research on the experiences of pilgrimage to holy places, such as the studies of Gross (1971) on the pilgrimage centre of Bom Jesus de Lapa, in Bahia, Brazil; Dubisch (1995, 2000) on the pilgrimage to the Greek island of Tinos; Abbate and Di Nuovo (2013) on the pilgrimage in Medjugorje, Bosnia and Herzegovina; Lopez (2013), Oviedo *et al.* (2014) and Lois-González and Santos (2015) on the pilgrimage to Santiago de Compostela; Shackley (1998) on Saint Catherine's Monastery on Mount Sinai, Egypt; etc. Other research focuses on the study of religious tourism and pilgrimages and on issues of pilgrimage authenticity (Belhassen *et al.*, 2008). Research often concerns specific sacred sites of cultural interest where visitors are not only pilgrims but also tourists. For this reason, it is difficult to examine visitors' motivations and experiences to these sites.

In addition, different typologies have been proposed on the distinction between pilgrims and religious tourists in relation to the experiences they have (MacCannell, 1973; Cohen, 1979, 1992a, b; Smith, 1989, 1992). Since the 1990s the analysis of this relationship has focused on the similarities and differences between the pilgrim and the tourist (MacCannell, 1973; Turner and Turner, 1978; Smith, 1992; Vukonic, 1996; Cohen, 1998; Frey, 1998; Ebron, 1999; Collins-Kreiner and Kliot, 2000; Fleisher, 2000; Digance, 2003; Timothy and Olsen, 2006; Weidenfeld, 2006; Μοίρα, 2009, 2019; Moira *et al.*, 2009; Bozonelos and Moira, 2023).

Taking into consideration the tourist–pilgrim continuum (Collins-Kreiner, 2010; Olsen, 2010; Damari and Mansfeld, 2016) it is difficult to distinguish the varying levels of visitors' motivation (Liutikas, 2021). What are the common elements and what are the differences? What are the needs and desires of each group and ultimately, what is their behaviour? Defining the dominant motive for visiting a sacred place can largely determine the conduct of the individual. In the case of Mount Athos, a sacred site with specific access restrictions in terms of both gender and daily quota of visitors, should we refer to tourists or only pilgrims? What are the visitors' possible motives for travelling to Mount Athos? How has globalization (commodification, materialism, consumerism, secularization, individualism) affected visitors in the quest of spirituality?

Mount Athos

The area of the Holy Mount Athos on the Athos peninsula, the easternmost of the three peninsulas of Chalcidice (Greece) is the oldest surviving monastic community in the world (Fig. 11.1). It is one of the most significant centres of Eastern Orthodox Christianity with about 2000 monks (Findler and Fidler, 2017). It was inhabited in the 9th century AD by monks while the first monasteries were founded in the 10th century AD. The region is a self-governing part of the Greek state and it is governed by a holy community, ecclesiastically under the direct jurisdiction of the Ecumenical Patriarch of Constantinople, based in the capital of Mount Athos, Karyes. Monastic foundations are divided into six classes: (i) 20 holy monasteries; (ii) 12 sketes; (iii) huts; (iv) cells; (v) seats; and (vi) hermitages.

Only men are allowed to enter Mount Athos while women are not allowed access ('Avaton'). The principle of the prohibition of women's entry to Mount Athos goes back to Canon I of St Antony's address to young hermit monks: 'Do not allow a woman to approach you nor an anemone to set foot in your dwelling for wrath comes from behind' (Πλατής, 2000; Ρομφαία, 2009).

Moreover, the Charter of Megisti Lavra (AD 970) states: 'You will not possess animals of the female gender to do the work you need because you have absolutely renounced every female being' (*iellada*, 2018). Today the prohibition of women's entry to Mount Athos is part of the living intangible cultural tradition and heritage. The ban works for both sides. By virtue of Canon 47 of the Quinisext Council, held in AD 692 at Constantinople, it was forbidden for a woman to stay overnight in a male monastery and a man in a female one. A century later, by virtue of Canon 18 of the Seventh Ecumenical Council, held in Nicaea in AD 787, women staying in dioceses and monasteries was prohibited. Therefore, the 'Avaton' for women in male monasteries and for men in female monasteries was a general rule (with very few exceptions, under restrictive conditions), provided for by all monastic rituals, since ascetic life meant leaving the worldly life (Moira and Mylonopoulos, 2017).

Mount Athos and its monasteries are unique for their history, architecture, art and Byzantine chant – which was inscribed on the UNESCO Representative List of the Intangible Cultural Heritage of Humanity in 2019 (ICH-UNESCO, 2019) – but also for their unrivalled natural environment (rich flora and fauna, astonishing natural beauty), isolation, tranquillity and mysticism (Fig. 11.2). The uniqueness of the natural and cultural environment of Mount Athos led to its inscription, based on six criteria, on the UNESCO World Heritage List in 1988 as a mixed property. In addition to its rich material heritage,

Fig. 11.1. Map of Mount Athos. Courtesy of Claude Zygiel from Wikipedia. Available at: https://commons.wikimedia.org/wiki/File:Mount_Athos_community_(complete_data).jpg (CC-BY-sa 4.0 DEED) (accessed 28 January 2023).

Mount Athos is a repository of preserving traditions of intangible cultural heritage as the brotherhoods of Mount Athos keep alive a unique Orthodox tradition and way of life (Kadas, 2002). It is a religious spatial ensemble in which 'the natural and man-made environment interact in harmony, summarizing a rich, century-long spiritual and cultural tradition and delimiting a sacred, inaccessible [to women and children] space where the God-bearing ideal of monasticism flourishes and grows' (Δωρής, 1994, p. 21). This is the key to the interpretation of the term 'treasures of Mount Athos', which refers to both tangible and intangible cultural heritage, in the sense of spiritual heritage (Alexopoulos and Fouseki, 2016).

Entry to and Hospitality on Mount Athos

Entry to Mount Athos

The ways visitors come to Mount Athos varies. Many visitors arrive individually, which means

Fig. 11.2. Mount Athos, Moni Simonopetra © GNTO/H. Kakarouhas with permission. From Greek National Tourism Organization_8625_HKakarouhas.

that they are not members of a group of travellers or a family. This choice, conscious or not, is recommended since individual tours allows pilgrims to devote themselves fully to the pilgrimage as well as concentrate themselves spiritually and mentally (Μιχελουδάκης, 2013). Alternatively, members of the same family, for example fathers with their son(s), often visit Mount Athos, which strengthens family relationships and bonds between fathers and sons.

Visits in the form of 'companionship' (people accompanying pilgrims) are also recorded. The phenomenon of organized group visits by pilgrims is also frequent. They include members of groups of the church, education, the army and various associations and collectives. These groups can be classified into university, ecclesiastical-educational, school, teaching, catechetical, Christian, ecclesiastical, military, mountaineering, hiking, touring, scouting, local, conference and other groups (Μιχελουδάκης, 2013, p. 43).

The special entrance permit (diamonitirion)

In order for a visitor to enter the Athos peninsula, he should be provided with a special entrance permit called a 'diamonitirion' (διαμονητηριον in Greek). There are two types: (i) the general diamonitirion, which allows a visitor to stay overnight in any of the monasteries but only for 4 days (3 nights); and (ii) the special diamonitirion, which allows a visitor to visit only one monastery or skete,[1] but he can stay as many days as he has agreed with the monks. If a visitor wishes to extend his stay, he should apply to the Holy Community in Karyes. Entrance to Mount Athos is allowed daily to 100 Orthodox and ten non-Orthodox pilgrims, provided that they respect the site, observing the rules in force. Clerics and those who have a special invitation from the monasteries are excluded from this daily quota. The quota is managed by the Mount Athos Pilgrims' Bureau in Thessaloniki to whom visitors should apply for entry. Non-Orthodox pilgrims should have obtained a letter of recommendation from their country's embassy and a special document from the Greek Ministry of Foreign Affairs. A fee, payable in cash only, for the issue of diamonitirion is required at the time of collection. It is €35 for non-Orthodox pilgrims, €25 for Orthodox pilgrims and €18 for students while it is free for large families with four or more children. Hospitality in the monasteries is provided free of charge. Diamonitirion contains advice on how visitors should behave and dress. Smoking and audiovisual devices (e.g. video cameras, television, radio) are prohibited (Averitt, 2006). Photography is permitted outside the monasteries but if visitors are in doubt, they should ask for the monk's 'evlogia' (blessing). Mobile phones are allowed but the reception is uneven and it is often difficult to charge the mobile-phone battery either because the monasteries have no electricity or because of low-voltage electricity. It is argued that the complexity of the issue of diamonitirion combined with the daily quota of visitors is an attempt to protect the seclusion of Mount Athos, which is a key factor of its very existence. The aim is to protect the holy place from the influences of the modern world and significantly enhance the whole 'experience' characterized by asceticism, abstinence and humility.

Visitors must present their diamonitirion to gain entry to Athos. They will be refused entry if they do not do so. Minors under 18 years of age may enter Mount Athos, provided that they have their parents' consent. Intending visitors should first telephone or e-mail the Mount Athos Pilgrims' Bureau to ask if there is a place in the quota for the specific date on which they wish to enter. Reservations can be made any time up to 6 months before the planned departure. Visitors are expected to receive a letter of confirmation within 2 weeks with further instructions. Once they receive it, they should contact the monasteries to arrange their accommodation. Ferry boats and speedboats depart daily in the morning either from Ouranoupolis (the main point of departure, on the west side of the peninsula), calling at six monastery jetties ('arsanas'), or from Ierissos (on the east side), calling at eight monastery jetties. Tickets should be bought on the spot on the day before departure. The monasteries can also be viewed by men and women from the sea on tour boats out of Ouranoupolis (Averitt, 2006). In special cases, when the weather conditions have been extreme for a significant period of time and a ferry connection is not possible, only after instructions from the police authorities and permission from the Holy Community have been granted, are pilgrims and necessary products allowed to enter the Holy Mountain from the land border. On Mount Athos there are no retail stores, souvenir shops or other tourist facilities. Visitors can buy some religious items (e.g. icons, rosaries, crosses, amulets, books), herbal infusions, wine and monastery-made raki (a Greek traditional strong grape-distilled spirit) from each monastery or at Karyes, where there are two taverns, a bakery and a few shops. The major church at Karyes is called Protaton (the 'Protos' is the oldest surviving church), where the renowned Axion Estin icon of Theotokos is kept.

Archontaris

Monasteries are famed for their hospitality ('philoxenia'), also known in Ancient Greece (Μοίρα, 2022, pp. 12, 22). Pilgrims to each monastery on Mount Athos are welcomed by the guest master ('archontaris') at the guest house ('archontariki'), a special reception area. The institution of archontaris dates back to Byzantium and marks the provision of hospitality by churches and monasteries in special guesthouses (Σιμόπουλος, 1994, p. 35), the operation of which is entrusted to competent and trustworthy monks who hold the title of Archontaris (Θεοχαρίδης και Πολυβίου, n.d., p. 28). Monks welcome pilgrims by offering them a traditional loukoumi (Turkish delight), served with water, Greek coffee or raki (alcoholic drink made of twice-distilled grapes). Accommodation in a shared dormitory and meals are also offered to guests. Visitors have the opportunity to actively participate in the monks' ministries if they wish to do so. For example, they can help in the refectory or in the gardens, or they can make handicrafts such as incense and soap. Pilgrims usually discuss with the monks spiritual, theological and ecclesiastical issues and traditions that stimulate them spiritually. The long-lasting church services (usually lasting 5–6 hours), the communal worship, the lighting of candles, monks praying at the night vigils, the fragrance of the incense and the Divine Liturgy hymns sung by the fathers fascinate every visitor.

In contrast to a pilgrim, a simple tourist can stay on Mount Athos without being obliged to join any work or provide assistance to the monks, nor is it necessary for him to attend to the liturgical-worship needs. The only thing that everyone participates in is the common 'trapeza', the table for food, as every day the Archontaris writes down the total number of visitors and informs the trapezaris (monk responsible for the refectory) so that he can arrange the rations to be served. Therefore, as there is complete freedom of movement for everyone, many visitors go to Mount Athos to enjoy the uniqueness of the Athonite State.

A visitor to Mount Athos explains:

> on Mount Athos no one will 'force' you to follow a specific programme. However, because it is part of the experience, even those who do not visit the area for the purpose of pilgrimage (but out of curiosity or to see the monuments and nature) generally end up following the monastery's programme ...
>
> (Λιάλιος, 2020)

Motivations for Visiting Mount Athos

As visitors are not always coreligionists and often not even believers, legitimate questions arise about their motivations for visiting the holy monasteries and the needs they seek to satisfy. Nolan and Nolan (1992) claim that visitors to sacred sites have multiple motivations, interests and activities, some of which have nothing to do with religion but are associated with holidays or trips undertaken for social and cultural reasons. In the case of Mount Athos, the motivations are varied, as many seek out peace and tranquility, an escape from the stress of the modern city, others seek out communion with God, while others have personal quests. Some visitors wish to get to know the unique treasures and relics kept in the monasteries. The motivation for visiting Mount Athos is often combined with the challenge of climbing or hiking the sacred mountain and enjoying the beautiful landscape. Mountaineering clubs in Greece organize pilgrimages to the Holy Mountain, in combination with mountaineering. Therefore, it is appropriate to distinguish between the believer-pilgrim, the religious-cultural tourist, the hiker seeking out physical and mental well-being and exercise and the researcher. Visitors often identify themselves as tourists rather than pilgrims (Della Dora, 2012, p. 967). When a pilgrim enters Mount Athos, he is possessed by feelings of reverence, prayer and humility, unlike the religious tourist who comes for the purposes of simple knowledge and exploration (Μοίρα, 2019, pp. 43–44). The pilgrim actively engages in the rituals of the monasteries while the religious tourists carry out other activities.

The main problem that emerges is that both these groups of visitors, distinct in terms of motivations and needs, claim the same space, the former for reasons of religious and spiritual uplift, the latter for cultural, educational and recreational reasons as well as physical exercise. Similar findings have been made by Gatrell and Collins-Kreiner (2006, p. 2) who examined the visitors' motivations to the Bá'hai Gardens in Haifa, Israel. As visitors have different motivations for travelling to Mount Athos, they often disturb the natural-religious-cultural environment of the holy places they visit.

Visitors' Experiences on Mount Athos

Mount Athos, as a UNESCO World Heritage Site, offers diverse and authentic experiences to pilgrims or tourists and, besides, the boundaries of motivations and experiences between different visitors are often blurred and indistinct (Fig. 11.3). As Turner and Turner (1978), Cohen (1979) and Dicks (2003) have pointed out, both pilgrims and religious tourists seek out extraordinary experiences. In sacred places, people can experience unique culture, seek out spiritual support, and generate life meaning (Fletcher, 2004; Mazumdar and Mazumdar, 2004); they can enjoy the landscape, the natural environment and the architecture; they can initiate themselves into tradition and get acquainted with the tangible and intangible/living monastic heritage as well as rid their bodies from standard food, the stress of everyday life, their lifestyle and enjoy the quietness.

Cohen (1979) argues that the category 'existential experiences' mainly characterizes those who travel to sacred places in search of spirituality,

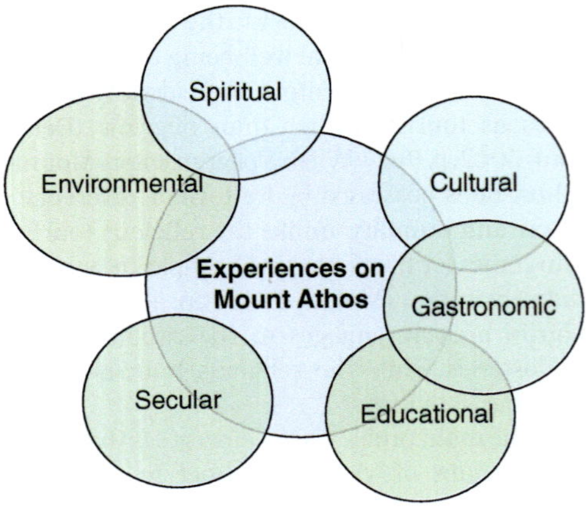

Fig. 11.3. Experiences on Mount Athos. Polyxeni Moira's own figure.

the divine. MacCannell (1973) has stressed that natural and cultural heritage have a significant impact on the tourist experience. These experiences are interrelated and often one enhances the other. In addition, Stokols (1990) has pointed out that Mount Athos, as an environment with symbolic qualities, generates strong emotional responses that contribute to group or individual identity building (Mazumdar and Mazumdar, 2004).

Spiritual experiences

Mount Athos is a unique place where the visitor experiences what Preston (1992) described as 'spiritual magnetism'. The spirituality of the place impacts directly on visitors, whether they are pilgrims in the traditional religious sense of the word, or simply tourists. Staying in the monasteries allows pilgrims to step away from the worries, constraints and problems of everyday life, seeking out spiritual guidance from the monks.

Pilgrims undertake simultaneously a physical journey – although not in the sense of pilgrimage in the Catholic Church (Sherrard, 1977) – and a spiritual journey which dominates the former. For Catholics, the concept of pilgrimage involves a journey on foot, under adverse conditions, towards a holy place/destination, during which walking and hardship and sharing with fellow travellers is a necessary element for the approach to the Divine, for the realization of inner change (Sherrard, 1977, p. 102). The driving force that draws visitors to monasteries is the strengthening of faith, the approach to God, inner contemplation and prayer. In this case, pilgrims follow the ritual of each monastery and do not show interest in worldly tourist activities. Still, the magnetism of the place often affects non-pilgrims, seekers of the spiritual element (Bowers and Cheer, 2017) as heterodox pilgrims often refer to the mysticism, the magic of the place, enhanced by the Byzantine chant and hymnography. It is worth notice that the Byzantine chant was inscribed on the UNESCO Representative List of the Intangible Cultural Heritage of Humanity in 2019 and 'it is inextricably linked with spiritual life and religious worship'. (ICH-UNESCO, 2019).

In May 2014 two Muslims from Turkey, the journalist Ertuğrul Özkök and the architect Ali Esad Göksel, spent 3 days at Saint Dionysius Monastery on the Holy Mountain, after they had been granted a special diamonitirion by the Ecumenical Patriarch of Constantinople, Bartholomew I. Özkök (2014) wrote a newspaper article, describing their stay. The scenery reminded them of the *Game of Thrones* TV series. Their first impression was the following:

> Those scary, black-bearded priests we've been taught from childhood to be horrible people are not horrible people after all. On the contrary, they are all very close to us. We even get the feeling that we are receiving special attention because we are Turks.
>
> (Özkök, 2014)

While they were astounded by the ambiance during the early morning Mass:

> The chant came once from the right, once from the left and turned into an enthusiastic torrent flowing down the aisle. An enchanting vocal trumpeting. Can art be metaphysical? That night we believed it could. We watched, speechless and enchanted ... The trembling of the candles and the light of the candles, the colorful worlds of icons and paintings, the golds, the shadows ... .
>
> (Τζίμας, 2015)

It is also worth noting that the meaning of religious places can be generated through interpersonal interaction (Trauer and Ryan, 2005). On Mount Athos, as in any stay in monasteries, participation in the daily life of the monks, in the works and the rituals and sharing food with the monks allows visitors to live an authentic experience. In addition, an important element that allows visitors to deepen their spiritual experience and gain a deeper understanding of the place is communication and dialogue with the monks. According to Trauer and Ryan (2005), communication with the monks can deepen visitors' understanding of a place and such close encounters can create positive, even sacred, memories.

A similar spiritual and cultural experience is also sought through the Temple Stay program, which consists of visitors (Buddhist and non-Buddhist) staying at Buddhist temples in South Korea (OECD, 2009) and participating in traditional ceremonies (e.g chanting, meditation). On Mount Athos, the possibility of staying in the monasteries allows visitors to actively participate and have a genuine, authentic, 'Byzantine' experience, to experience the 'behind the scenes', therefore, avoiding what MacCannell (1973) has described as 'staged authenticity'.

In the following extract from the pilgrims' book of the Monastery of Saint Gregory the experiences of Greek high-school students who visited Mount Athos in 1950 are recorded:

> The unquenchable breath of the sea, which only in the quiet night under our feet can be heard, as if caressing the curbs of the holy Monastery that rises imposingly on the sea rock ... the solemn liturgy in the fragrant and beauteous Church of the Monastery, the Byzantine chant, the communal dining, everything motivates you to contemplation, to self-reflection and self-control, to spiritual uplifting ... . The Fifth Grade of the Third Gymnasium of Thessaloniki (Βιβλίο Εντυπώσεων [Guest Book] 1923–1955, p. 328, 16 July 1950, cited in Μιχελουδάκης, 2013, p. 53)

Finally, it should be noted that the difference in the calculation of the date and time also contributes to the whole experience, as on Mount Athos the Julian Calendar remains in use, which is 13 days behind the Gregorian Calendar. All the establishments at Athos operate according to the Byzantine 24-hour system. When the sun sets, the time in Athos is 12:00 during all seasons with the exception of the Iviron monastery where the time is 12:00 when the sun rises. In addition, the monks have their own concept of time. Each hour has its own quality, its own content within the alternation of prayer and work. Sunrise and sunset, noon and midnight, light and darkness are sacred hours, separate but also connected to each other in the daily cycle of divine worship. The Monk Patapios Kafsokalivitis (Μοναχός Πατάπιος Καυσοκαλιβίτης, 2021) describes the sense of time in the following way: 'On Mount Athos one moves towards the limits of time. Here the flame that burns the fuse of time is very small. If one pulls the fuse a little more, one will be released from time'.

Cultural experiences

Religious tourism is considered by many scholars (Rinschede, 1992, p. 52; Μοίρα, 2009, 2019) a subcategory of cultural tourism. Of course, in the case of a religiously active monument, such as Mont-Saint-Michel in France, St Peter's Cathedral in the Vatican City or the Monasteries of Meteora in Greece, it is difficult to distinguish a visitor seeking out prayer and spiritual peace from someone admiring the work of great artists or simply motivated by curiosity (Μοίρα, 2009, 2019, pp. 69–71). The monasteries of Mount Athos, as already mentioned, are a living museum of culture, a museum for the preservation of both tangible and intangible cultural heritage. The architecture, the frescoes, the ecclesiastical art, the old icons, the ancient manuscripts, but also the preservation of the traditions of the intangible heritage, the Byzantine chant and Athonite hymnography, make Mount Athos a unique cultural destination. The treasures of the holy monasteries are exhibited in the monastery museums and are opened to visitors on special occasions. The monasteries also have important libraries with rare books, manuscripts and scrolls. The acquaintance with culture and scientific research is the reason why many visitors (Orthodox and non-Orthodox) who wish to get to know the unique treasures of the monasteries are researchers, architects, historians or theologians.

Educational experiences

A visit to Mount Athos is an important educational experience for many visitors. It is common for secondary school students accompanied by their teachers to visit it for educational purposes, to get acquainted with the monastic Athonite tradition and the treasures of Mount Athos (Anatolia High School, 2018; Ταχυδρόμος Μαγνησίας, 2022) as well as for pilgrimage reasons (2nd Vocational Senior High School of Karditsa, 2018).

Gastronomic experiences

Visitors' participation in the Common Refectory is also part of the overall cultural experience. The diet of the monks has become known worldwide after the publication and release of the cookbook by the Monk Epiphanios of Mylopotamos (2010) entitled *The Cuisine of the Holy Mountain Athos*, originally written in Greek in 2008, which contains 126 recipes. The Elder Epiphanios lived in the Skete of St Eustathius, a dependency of the Monastery of Great Lavra located in Mylopotamos, where he built and maintained a vineyard and a winery. The book and the recipes of the monastic cuisine of Mount Athos, which to date has been translated into eight languages, have attracted visits from celebrities from all over the world, such as the English chef and cookbook author Jamie Oliver and the famous Italian novelist Umberto Eco.

Mount Athos' monastic diet is dictated by the religious context and the prohibition of eating certain foods, such as meat. In essence, it is a special, frugal and healthy diet, characterized by culinary art which is based on fasting and often dry eating, meaning that they consume only raw food without using olive oil. Besides, it is extremely tasty. Cooking for the monks is considered an 'act of love' towards their neighbour. Epiphanios himself wrote: 'I hope that the book you hold in your hands will help you to reach even closer to those you love by cooking. Open your home, invite your friends, cook together, set the table, drink wine, and rejoice together' (Monk Epiphanios of Mylopotamos, 2010). So, guests during their stay on Mount Athos have the opportunity to have a different tasty and healthy eating experience, which they are likely to adopt in their daily lives.

Secular/social experiences

The community, monastic traditions and rituals, and participation in daily activities encourage and enhance social interaction (Trauer and Ryan, 2005; Singh, 2006, p. 377). Visitors usually become more sociable, chatting with strangers, sharing experiences more easily than in their daily lives. The special nature of the sacred space often allows participants to express their thoughts and feelings to family members, friends, relatives and others with whom they share similar interests and religious beliefs (Wang, 1999; Yagi and Pearce, 2007; Belhassen *et al.*, 2008). Of course, it should also be noted that a visitor to Mount Athos is respected and loved by his social environment (Madrigal and Kahle, 1994; Fletcher, 2004; Μοίρα, 2019).

Environmental experiences

On Mount Athos isolation and contact with nature contribute to spirituality. Apart from the mystical setting created by the monasteries and the life of the monks themselves, the place preserves a rich flora and fauna thanks to the careful management of the forests and the application of traditional agricultural practices. Due to its rich natural landscape, the Athos peninsula is a member of the European NATURA 2000 network. The isolation in the wild natural environment, the possibility of hiking in pristine places and enjoyment of tranquility are part of the attraction and experience of Mount Athos. Many visitors follow the designated hiking trails. The hiking trails/routes are grouped into two categories: (i) in the central part of Mount Athos with a tour of large monasteries; and (ii) in the southern part of the peninsula, visiting sketes, cells and climbing Athos on more specialized routes. Mount Athos is a unique paradise for hikers, as the magnificent nature combined with the monasteries, the sketes and the cells create a feeling of well-being and are mentally uplifting during the hike. The whole natural environment contributes to the spiritual experience, often in an inextricable way. According to a visitor, 'Sitting alone in nature, gazing at the sea, the mountain peaks, being "elevated" spiritually, is a unique experience. I feel like I'm catching a glimpse of paradise ...' (Μιχελουδάκης, 2013: 256). In addition:

> the Mount Athos environment consists of 'magnificent and majestic nature', which is 'connected in a harmonious union with the

> holiness of the space'. The 'beauty' ... 'the majesty of nature is unmatched', while the natural environment 'of the sea and the green steep mountains', the combined 'beauty of the landscape and the serenity of nature' carry the soul away: 'it leaves you stunned with its splendor'. (Excerpts from visitors' impressions in Μιχελουδάκης, 2013, p. 138) (Fig. 11.4)

Summary

According to the findings outlined in this chapter, the motives of the visitors to Mount Athos are diverse. Likewise, their experiences vary (spiritual, cultural, educational, gastronomic, etc.) depending on their origin, religion, needs, life experiences, and so on, as well as the inherent spirituality of the place. Undoubtedly, the unique environment of Mount Athos (natural, human, spiritual), the monasteries and the monks have the power to transform every visitor into a 'pilgrim'.

An American scholar-priest, who visited Mount Athos, describes his departure from the holy place as follows:

> But here comes the moment of leaving the mountain. One is initially happy to go home to the family and share the experiences, the beautiful places, the chanting, the relics, the conversations. But as the boat takes the pilgrim away from the mountain and the mountain fades in the haze of the horizon there is an unexplainable longing that starts settling in his soul. As he gets closer to the world there is something that calls him back and that call, he does not realize it now, will be with him forever. This is the gift of the mountain. The greatest shock, however, is when he is back into the world and meets the first 'man from the city' (Luke, 8: 27) as Jesus met the demoniac in Gadara. The colors of the city hurt the pilgrim's eyes now, the loud and rhythmic music inflicts pain in his ears, everything disturbs the inner peace he was able to briefly experience on the mountain.
>
> (Vasile, 2010)

Fig. 11.4. Mount Athos. Moni Agiou Grigoriou © GNTO/H. Kakarouhas with permission. From Greek National Tourism Organization_8822_HKakarouhas.

Note

[1] The skete represents an alternative for a monk of Mount Athos who does not wish to live in the strict form of monastic seclusion. In sketes, small monastic communities are formed, characterized by relative isolation, and are governed by elders-advisors. In addition to manual labour, such as handicrafts, agriculture and animal husbandry, the monks also engage in iconography, woodcarving and music courses.

References

2nd Vocational Senior High School of Karditsa, Greece (2018) *Προσκυνηματική επίσκεψη στο Άγιο Όρος* [*Pilgrimage to Mount Athos*]. Available at: https://blogs.sch.gr/2epalkardits/archives/1450 (accessed 12 January 2023). (in Greek)

Abbate, C.S. and Di Nuovo, S. (2013) Motivation and personality traits for choosing religious tourism: a research on the case of Medjugorje. *Current Issues in Tourism* 16(5), 501–506. https://doi.org/10.1080/13683500.2012.749844

Alexopoulos, G. and Fouseki, K. (2016) Gender exclusion and local values versus universal cultural heritage significance: the Avaton debate on the monastic community of Mount Athos. *European Journal of Post-Classical Archaeologies* 6, 235–254.

Anatolia High School (2018) *Επίσκεψη του Welfare Club στο Άγιο Όρος* [*Visit of Welfare Club to Mount Athos*], 7 May. Available at: https://anatolia.edu.gr/el/high-school/nea/3781-episkepsh-toy-welfare-club-sto-agio-oros (accessed 25 January 2023). (in Greek)

Averitt, N. (2006) Mt. Athos, Greece: of monks and men. *The Washington Post*, 6 August. Available at: https://www.washingtonpost.com/archive/lifestyle/travel/2006/08/06/mt-athos-greece-of-monks-and-men/397afc86-956e-44ed-85e7-cfdba2891554/ (accessed 23 November 2023).

Belhassen, Y., Caton, K. and Stewart, W.P. (2008) The search for authenticity in the pilgrim experience. *Annals of Tourism Research* 35(3), 668–689. https://doi.org/10.1016/j.annals.2008.03.007

Bowers, H. and Cheer, J.M. (2017) Yoga tourism: commodification and Western embracement of Eastern spiritual practice. *Tourism Management Perspectives* 24, 208–216. https://doi.org/10.1016/j.tmp.2017.07.013

Bozonelos, D. and Moira, P. (2023) Introduction to the politics of religious tourism. In: Bozonelos, D. and Moira, P. (eds) *The Politics of Religious Tourism*. CAB International, Wallingford, UK, pp. 1–12.

Cohen, E. (1979) A phenomenology of tourist experiences. *Sociology* 13(2), 179–201. https://doi.org/10.1177/003803857901300203

Cohen, E. (1992a) Pilgrimage centres: concentric and excentric. *Annals of Tourism Research* 19(1), 33–50. https://doi.org/10.1016/0160-7383(92)90105-X

Cohen, E. (1992b) Pilgrimage and tourism: convergence and divergence. In: Morinis, A. (ed.) *Sacred Journeys: the Anthropology of Pilgrimage*. Greenwood Press, New York, pp. 47–61.

Cohen, E. (1998) Tourism and religion: a comparative perspective. *Pacific Tourism Review* 2(1), 1–10.

Collins-Kreiner, N. (2010) Researching pilgrimage: continuity and transformations. *Annals of Tourism Research* 37(2), 440–456. https://doi.org/10.1016/j.annals.2009.10.016

Collins-Kreiner, N. and Kliot, N. (2000) Pilgrimage tourism in the Holy Land: the behavioural characteristics of Christian pilgrims. *GeoJournal* 50, 55–67. https://doi.org/10.1023/A:1007154929681

Damari, C. and Mansfeld, Y. (2016) Reflections on pilgrims' identity, role and interplay with the pilgrimage environment. *Current Issues in Tourism* 19(3), 199–222. https://doi.org/10.1080/13683500.2014.957660

Della Dora, V. (2012) Setting and blurring boundaries: pilgrims, tourists and landscape in Mount Athos and Meteora. *Annals of Tourism Research* 39(2), 951–974. https://doi.org/10.1016/j.annals.2011.11.013

Dicks, B. (2003) *Culture on Display: the Production of Contemporary Visitability*. Open University Press, Maidenhead, UK.

Digance, J. (2003) Pilgrimage at contested sites. *Annals of Tourism Research* 30(1), 143–159. https://doi.org/10.1016/S0160-7383(02)00028-2

Dubisch, J. (1995) *In a Different Place: Pilgrimage, Gender and Politics of a Greek Island Shrine.* Princeton University Press, Princeton, New Jersey.

Dubisch, J. (2000) *Το θρησκευτικό προσκύνημα στη σύγχρονη Ελλάδα. Μια εθνογραφική προσέγγιση* [*Religious Pilgrimage in Contemporary Greece: an Ethnographic Approach*]. Αλεξάνδρεια, Αθήνα. (in Greek)

Ebron, P.A. (1999) Tourists as pilgrims: commercial fashioning of transatlantic politics. *American Ethnologist* 26(4), 910–932 https://doi.org/10.1525/ae.1999.26.4.910

Findler, R. and Fidler, M. (2017) The Holy Mountain: monks of Mount Athos – photo essay. *The Guardian*, 5 January. Available at: https://www.theguardian.com/artanddesign/2017/jan/05/the-holy-mountain-monks-of-mount-athos-photo-essay (accessed 19 March 2023).

Fleisher, A. (2000) The tourist behind the pilgrim in the Holy Land. *International Journal of Hospitality Management* 19(3), 311–326. https://doi.org/10.1016/S0278-4319(00)00026-8

Fletcher, S.K. (2004) Religion and life meaning: differentiating between religious beliefs and religious community in constructing life meaning. *Journal of Aging Studies* 18(2), 171–185. https://doi.org/10.1016/j.jaging.2004.01.005

Frey, N.L. (1998) *Pilgrim Stories: On and Off the Road to Santiago*. University of California Press, Berkeley, California.

Gatrell, J.D. and Collins-Kreiner, N. (2006) Negotiated space: tourists, pilgrims and the Bahá'i terraced gardens in Haifa. *Geoforum* 37(5), 765–778. DOI: 10.1016/j.geoforum.2006.01.002.

Gross, D.R. (1971) Ritual and conformity: a religious pilgrimage to Northeastern Brazil. *Ethnology* 10(2), 129–148.

iellada (2018) Πώς δημιουργήθηκε το Άβατο στο Άγιο Όρος - Η μοναδική ιστορία [How the Avaton was created on Mount Athos: the unique story]. *iellada*, 10 May. Available at: https://www.iellada.gr/thriskeia/pos-dimioyrgithike-avato-sto-agio-oros-i-monadiki-istoria-deite-ta-vinteo (accessed 28 January 2023). (in Greek)

Intangible Cultural Heritage (ICH) UNESCO (2019) Byzantine chant. Intangible Cultural Heritage UNESCO. Available at: https://ich.unesco.org/en/RL/byzantine-chant-01508 (accessed 28 January 2023).

Kadas, S. (2002) *Mount Athos: an Illustrated Guide to the Monasteries and their History*. Ekdotike Athenon, Athens.

Liutikas, D. (2021) Travel motivations of pilgrims, religious tourists, and spirituality seekers. In: Olsen, D. and Timothy, D. (eds) *The Routledge Handbook of Religious and Spiritual Tourism*. Routledge, London, pp. 225–242.

Lois-González, R.C. and Santos, X.M. (2015) Tourists and pilgrims on their way to Santiago: motives, Caminos and final destinations. *Journal of Tourism and Cultural Change* 13(20), 149–164. https://doi.org.10.1080/14766825.2014.918985

Lopez, L. (2013) How long does the pilgrimage tourism experience to Santiago de Compostela last? *International Journal of Religious Tourism and Pilgrimage* 1(1), article 2. https://doi.org/10.21427/D7C133

MacCannell, D. (1973) Staged authenticity: arrangements of social space in tourist settings. *American Journal of Sociology* 79(3), 589–603. https://doi.org/10.1086/225585

Madrigal, R. and Kahle, L.R. (1994) Predicting vacation activity preferences on the basis of value-system segmentation. *Journal of Travel Research* 32(3), 22–28. https://doi.org/10.1177/004728759403200304

Mazumdar, S. and Mazumdar, S. (2004) Religion and place attachment: a study of sacred places. *Journal of Environmental Psychology* 24(3), 385–397.

Moira, P. and Mylonopoulos, D. (2017) Marian pilgrimages in Greece and their management. In: Calò Mariani, M.S. and Trono, A. (eds) *Le Vie Della Misericordia – the Ways of Mercy*. Mario Congedo Editore, Lecce, Italy, pp. 465–480.

Moira, P., Parthenis, S., Kontoudaki, A. and Katsoula, O. (2009) Religious tourism in Greece: the necessity to classify religious resources for their rational valorization. In: Trono, A. (ed.) *Tourism, Religion and Culture. Regional Tourism Development through Meaningful Experiences. Proceedings of the International Conference, Lecce, Tricase, Poggiardo 27–29 October*. University of Salento/Mario Congedo Publishers, Salento, Italy, pp. 465–480.

Monk Epiphanios of Mylopotamos (2010) *The Cuisine of the Holy Mountain Athos*. Synchroni Orizontes, Athens.

Nolan, M.L. and Nolan, S. (1992) Religious sites as tourism attractions in Europe. *Annals of Tourism Research* 19(1), 68–78. DOI: https://doi.org/10.1016/0160-7383(92)90107-Z

Olsen, D.H. (2010) Pilgrims, tourists and Max Weber's ideal types. *Annals of Tourism Research* 37(3), 848–851. http://dx.doi.org/10.1016/j.annals.2010.02.002

Organisation for Economic Co-operation and Development (OECD) (2009) *The Impact of Culture on Tourism*. OECD Publishing, Paris.

Oviedo, L., de Courcier, S. and Farias, M. (2014) Rise of pilgrims on the Camino to Santiago: sign of change or religious revival? *Review of Religious Research* 56(3), 433–442. https://doi.org/10.1007/s13644-013-0131-4

Özkök, E. (2014) 'Bakire Bahçesi'nde 3 gün [3 days in the 'Garden of Virgin']. *Hürriyet*, 8 June. Available at: https://www.hurriyet.com.tr/yazarlar/ertugrul-ozkok/bakire-bahcesi-nde-3-gun-26570747 (accessed 19 March 2023). (in Greek)

Preston, J. (1992) Spiritual magnetism: an organizing principle for the study of pilgrimage. In: Clifford, J. and Marcus, G.E. (eds) *The Poetics and Politics of Ethnography*. University of California Press, Berkeley, California, pp. 27–50.
Rinschede, G. (1992) Forms of religious tourism. *Annals of Tourism Research* 19(1), 50–67.
Said, E.W. (1979) *Orientalism*. Vintage, New York.
Shackley, M. (1998) A golden calf in sacred space?: the future of St Katherine's Monastery, Mount Sinai (Egypt). *International Journal of Heritage Studies* 4(3–4), 124–134. https://doi.org/10.1080/13527259808722230
Sherrard, P. (1977) The paths of Athos. *Eastern Churches Review* 9(1–2), 100–107.
Singh, S. (2006) Tourism in the sacred Indian Himalayas: an incipient theology of tourism? *Asia Pacific Journal of Tourism Research* 11(4), 375–389. https://doi.org/10.1080/10941660600931226
Smith, V.L. (1989) *Hosts and Guests: the Anthropology of Tourism*. University of Pennsylvania Press, Philadelphia, Pennsylvania.
Smith, V.L. (1992) Introduction: the quest in guest. *Annals of Tourism Research* 19(1), 1–17. DOI: 10.1016/0160-7383(92)90103-V.
Stokols, D. (1990) Instrumental and spiritual views of people–environment relations. *American Psychologist* 45(5), 641–646.
Timothy, D.J. and Olsen, D.H. (eds) (2006) *Tourism, Religion and Spiritual Journeys*. Routledge, London.
Trauer, B. and Ryan, C. (2005) Destination image, romance and place experience: an application of intimacy theory in tourism. *Tourism Management* 26(4), 481–491. DOI: 10.1016/j.tourman.2004.02.014.
Turner, V. and Turner, E. (1978) *Image and Pilgrimage in Christian Culture: Anthropological Perspectives*. Columbia University Press, New York.
Vasile, F. (2010) The way of the pilgrim. *Gladsome Light Dialogues – An Orthodox Blog*, 25 October, Available at: http://dialogues.stjohndfw.info/2010/10/the-way-of-the-pilgrim/ (accessed 28 January 2023).
Vukonic, B. (1996) *Tourism and Religion*. Elsevier Science, London.
Wang, N. (1999) Rethinking authenticity in tourism experience. *Annals of Tourism Research* 26(2), 349–370. DOI: 10.1016/S0160-7383(98)00103-0.
Weidenfeld, A. (2006) Religious needs in the hospitality industry. *Tourism and Hospitality Research* 6(2), 143–159. https://doi.org/10.1057/palgrave.thr.6040052
Yagi, C. and Pearce, P.L. (2007) The influence of appearance and the number of people viewed on tourists' preferences for seeing other tourists. *Journal of Sustainable Tourism* 15(1), 28–43. DOI: 10.2167/jost528.0.
Δωρής, Ε. (1994) *Το Δίκαιον του Αγίου Όρους* [*The Law of Mount Athos*]. *τ. Β', Πηγαί του Αγιορείτικου Δικαίου. Αντ. Σάκκουλας, Αθήνα-Κομοτηνή*. (in Greek)
Θεοχαρίδης, Π. and Πολυβίου, Μ. (n.d.) Το Αρχονταρίκι και η Φιλοξενία στα Μοναστήρια του Αγίου Όρους [Arhontariki and Hospitality in the Monasteries of Mount Athos]. *Αρχαιολογία* 50, 27–37. (in Greek)
Λιάλιος, Γ. (2020) Άγιον Όρος: Ταξίδι Μύησης [Mount Athos: Initiation Voyage], *Η Καθημερινή*, 27 January. Available at: https://www.kathimerini.gr/k/travel/1061716/agion-oros-taxidi-myisis/ (accessed 28 January 2023). (in Greek)
Μιχελουδάκης, Σ.Ι. (2013) *Προσκυνηματικές Περιηγήσεις στο Άγιον Όρος μέσα από τα βιβλία εντυπώσεων της Ιεράς Μονής Οσίου Γρηγορίου Αγίου Όρους* (1923–1973) [Pilgrimage to Mount Athos through the Impression Books of Saint Gregorios (Grigoriou) Monastery of Mount Athos(1923–1973)]. Διδακτορική Διατριβή, Αριστοτέλειο Πανεπιστήμιο Θεσσαλονίκης, Θεολογική Σχολή, Τμήμα Ποιμαντικής και Κοινωνικής Θεολογίας. Available at: https://ikee.lib.auth.gr/record/289170?ln=el (accessed 19 March 2023). (in Greek)
Μοίρα, Π. (2009) *Θρησκευτικός Τουρισμός [Religious Tourism]*. Interbooks, Αθήνα. (in Greek)
Μοίρα, Π. (2019) *Θρησκευτικός Τουρισμός και προσκύνημα. Πολιτικές-διαχείριση-αειφορία [Religious Tourism and Pilgrimage. Policies-Management-Sustainability]*. Φαίδιμος, Αθήνα. (in Greek)
Μοίρα, Π. (2022) *Τουρισμός. Ιστορία-εξέλιξη-προοπτικές [Tourism. History-Evolution-Prospects]*. Τζιόλα, Θεσσαλονίκη. (in Greek)
Μοναχός Πατάπιος Καυσοκαλιβίτης [Monk Patapios Kafsokalivitis] (2021) *Ο Χρόνος στη ζωή του Αγιορείτη μοναχού* [*Time in the Life of the Mount Athos Monk*]. *Πεμπτουσία*, 15 January. Available at: https://www.pemptousia.gr/2021/01/o-chronos-sti-zoi-tou-agioriti-monachou/ (accessed 28 January 2023). (in Greek)
Πλατής, Ν. (2000) *Αθωνικό Λεξικό [Athonic Dictionary]*. Άγκυρα, Αθήνα. (in Greek)
Ρομφαία (2009) Άγιον Όρος: Ο πολυσυζητημένος θεσμός του άβατου και πότε παραβιάστηκε [Mount Athos]. Ρομφαία, 2 September. Available at: https://www.romfea.gr/afieromata/16726-agion-oros-o-polusuzitimenos-thesmos-tou-abatou-kai-pote-parabiastike (accessed 28 January 2023). (in Greek)

Σιμόπουλος, Κ. (1994) *Ξένοι ταξιδιώτες στην Ελλάδα [Foreign Travellers in Greece]. 333 μ.Χ.-1700*, 8η έκδοση. Πιρόγα, Αθήνα. (in Greek)

Ταχυδρόμος Μαγνησίας (2022) Επίσκεψη μαθητών στο Άγιον Όρος [Students' Visit to Mount Athos]. 2 February. Available at: https://www.taxydromos.gr/magnisia/volos/227149/episkepsi-mathiton-sto-agio-oros/ (accessed 28 January 2023). (in Greek)

Τζίμας, Σ. (2015) Μουσουλμάνοι στον Άθω [Muslims in Athos]. *Η Καθημερινή*, 21 June. Available at: http://www.kathimerini.gr/820321/article/epikairothta/ellada/moysoylmanoi-ston-a8w (accessed 28 January 2023). (in Greek)

12 Experiencing the Passion Play: Case Studies from Germany, Slovenia and Greece

Dimitrios Mylonopoulos*, Polyxeni Moira and Spyridon Parthenis
University of West Attica, Egaleo, Greece

Abstract

This chapter critically evaluates the commonalities and distinctions between the religious dramatic presentations of Passion Plays in three European countries, each representing different Christian denominations: (i) Paros, Greece, symbolizing Eastern Orthodoxy; (ii) Škofja Loka, Slovenia, representing Roman Catholicism; and (iii) Oberammergau Passion Play, Germany, reflecting Protestantism. Religions worldwide incorporate a rich tapestry of religious events that carry deep spiritual significance for local communities. However, due to their unique characteristics, originality and authenticity, these events also pique the interest of tourists. Religious festivals, ceremonies and rituals often extend to encompass various activities such as fairs, gastronomic events and exhibitions. It describes how these events are organized and operate as religious or tourist acts and will investigate visitors' potential motives (e.g. religious instruction, spiritual exhilaration, enjoyment of the simple country life). Passion Plays hold a distinctive place among religious events. Passion Plays are a matching blend of tradition and contemporary relevance, fostered through intergenerational dialogue and religious devotion. Through these case studies, this chapter aims to shed light on the multifaceted nature of Passion Plays and their role as bridges between religious traditions and the world of tourism.

Introduction

All religions encompass various religious events which used to have, or still have, a religious significance for local populations. Nevertheless, because of their peculiarities, originality or authenticity they also capture tourists' attention. Religious festivals, ceremonies and rituals are accompanied by different types of activities (e.g. fairs, gastronomic events, exhibitions). Passion Plays are particular types of religious events which consist of dramatic presentations depicting scenes of the Passion of Jesus Christ, following the biblical accounts (e.g. teachings, trial, suffering, crucifixion, death on the cross, resurrection). Passion Plays combine dedication to tradition and openness for the present that is facilitated by the dialogue between the generations and religious beliefs.

Typical examples are *La Pasione of Sordevolo* in Italy, performed every 5 years, which dates back to 1850 (Piva and Cerutti, 2015, p. 7), the *Passion Play of Kalwaria Zebrzydowska* near Krakow in Poland (Kalwaria Zebrzydowska, 2023), *De Passiespelen* in the Netherlands (Openlucht theater hertme, 2023), the *Škofja Loka Passion Play* in Slovenia, the *York Mystery Plays* and the *Coventry Corpus Christi Plays* (England). Similar performances take place

*dimilon@uniwa.gr

DOI: 10.1079/9781800623675.0012

every 4 years in the market town of Leominster in Herefordshire in the UK as well as at the Spanish villages of Esparreguera, Olesa de Montserrat and Cervera of Catalonia every Easter, which dates back to 1538. Beyond Europe, similar re-enactments of the Passion of Christ take place in the Americas, Asia and Oceania (Mylonopoulos *et al.*, 2023). They are a continuation of historical re-enactments (e.g. Oberammergau, Sordevolo) or were created subsequently for tourism purposes (e.g. the Nova Jerusalém Paixão de Cristo in the state of Pernambuco in Brazil, which began in 1951) (Landy, 2013), for worship or educational purposes (e.g. the re-enactment of the Passion of Jesus Christ on the island of Paros in Greece, which started in 1924).

This chapter aims to investigate the similarities and differences between religious re-enactments of Passion Plays in three Christian countries with different denominations:

- Oberammergau in Germany (Protestantism and Catholicism);
- Škofja Loka in Slovenia (Catholicism); and
- Paros Island in Greece (Eastern Orthodoxy).

It describes how these events are organized and operate as religious or tourist acts and will investigate visitors' potential motives (e.g. religious instruction, spiritual exhilaration, enjoyment of the simple country life). Analysis will be sought regarding the organization and function of these activities, whether they are religious worship, tourist attractions, or a combination of both, as well as the motivations of the visitors (e.g. religious edification, spiritual lift, experiencing simple country life, alternative to bourgeois art theatre) (Mohr and Stenzel, 2022). Findings are based on a review of the literature and interviews with: (i) Maria Frantzi-Trivyza, former President of the Community of Marpissa, and former President of the Tourism Committee of the Municipality of Paros, Greece (interview 27 September 2022); (ii) Spyridoula Kavali, former member of the cultural association 'Skopas o Parios', Paros, Greece (interview 27 September 2022); and (iii) Elli Terzoglou of the University of West Attica (interview 20 September 2022).

Theoretical Framework

Since ancient times, all peoples and religions have recorded various events with religious content. Since Greek antiquity, groups of citizens from neighbouring city states have gathered over fixed periods of time to honour a deity. The most important festivals or competitions were the Olympic, the Isthmian, the Pythian and the Nemean. The celebrations were an often inseparable network of religious, sporting and cultural events. The most important one was the Olympic Games, held in Olympia every 4 years in honour of Zeus. Thousands of people flocked to these events either as spectators or as competitors. These celebrations had all the features of modern mass tourist trips to watch major sporting events or religious pilgrimages (Μοίρα, 2022).

Many of these events are being revived today. Events with religious content are particularly important as they capture tourists' interest. These events have had, or still have, a religious meaning for the local population. Still, because of their special character and the element of sensationalism they draw tourists' attention. Religious ceremonies are often accompanied by various activities such as music, gastronomy, exhibitions and other cultural events. Due to globalization, many representations have acquired an international character (e.g. Oberammergau, Škofja Loka), but at the same time, their international character raises questions of authenticity and commercialization (Mylonopoulos *et al.*, 2023).

The term 'events' has started to be used by both governments and private bodies in recent years. 'An event is any gathering of people for a specific purpose' (Carter, 2007, p. 3). *Cambridge Dictionary* (2022) defines 'event' as 'an activity that is planned for a special purpose and usually involves a lot of people, for example, a meeting, party, trade show, or conference'.

According to Arcodia and Barker (2003), events are grouped under three broad types – sporting, cultural and professional. In contrast, Bowdin *et al.* (2011) classify events by their form or content: (i) business events (e.g. meetings, conferences, exhibitions); (ii) cultural events; (iii) local or community events; (iv) outdoor events (e.g. concerts, theatre shows); (v) sporting events; (vi) charity events; and (vii) social events associated with the human life cycle (e.g. weddings, christenings). Getz (2005, 2008, p. 404) and Μοίρα (2009, 2019) refer to religious events which they rightly include in cultural events. Raj *et al.* (2013) consider religious events a distinct category from cultural events.

Religious-cultural events can be classified in subcategories depending on the way they are organized as days of celebration of a deity or a saint, religious festivals, re-enactment of the Divine Drama and the Crucifixion, religious fairs and dances (Moira *et al.*, 2009; Μοίρα, 2019, pp. 84–93), and other religious events, for example indulgence feasts (Liutikas, 2015, p. 155). An event can often include all subcategories at the same time. These events, regardless of their categorization, whether they are organized by the public or private sector, whether they are commercial or charitable, festive or commemorative, religious or sporting, attract large numbers of visitors, create jobs (Kose *et al.*, 2011) and make a major contribution to urban regeneration (Chalkley and Essex, 1999, p. 370) and economic growth (Ferdinand and Kitchin, 2012). They also bring people together and create unique experiences. In the case of religious celebrations, the events that accompany them are an attraction for both believers (pilgrims, seekers of the spiritual element) and religious-cultural tourists (Μοίρα, 2009, 2019), for whom the events are a way of getting to know the local culture. In modern times, a multitude of pilgrims and tourists alike travel to various destinations to participate in or attend religious events that take place on important dates and anniversaries (Vukonic, 1996; Blackwell, 2007, p. 45; Μοίρα, 2009, 2019).

The incentives to participate in religious events (e.g. festivals, ceremonies, events, re-enactments, fairs) are varied, complex, multilevel and vary depending on whether the visitor is a pilgrim or a religious tourist. Thus, the motives for visiting the Wailing Wall in Jerusalem are different for a Jewish pilgrim and different for an Orthodox one. In the former case, the visit to the Wailing Wall by a Jew is part of a spiritual-emotional experience as it is part of his/her faith while in the latter case visiting is part of a religious-cultural tourist visit (Poria *et al.*, 2003). Similarly, participating in a ritual at Saint Peter's Basilica in Vatican City has a different meaning for a Catholic believer and a totally different one for an Orthodox tourist who seeks the religious-cultural element of the place (Μοίρα, 2009, 2019). In any case, exploring travellers' motivations should be a key objective for tourism policy makers as they determine travellers' needs, their consumption in the destination and their corresponding behaviours (Μοίρα, 2022). At the same time, the particularities of religious events make it necessary to adapt their management, taking into consideration the specific characteristics of the religious tourism experience. It is also imperative to avoid 'commercialization' (MacCannell, 1984) and 'staged authenticity' (Greenwood, 1989) of religious-cultural events.

Historical Overview of the Re-enactment of the Divine Drama (Passion Plays)

Cultural heritage is a co-creator of the modern tourism offer. The visibility of cultural heritage is increased when it is designed as a United Nations Educational, Scientific and Cultural Organization (UNESCO) World Heritage Site. The inclusion of a site on the UNESCO World Heritage List strengthens national identity and at the same time makes it internationally recognizable. Intangible cultural heritage, as defined in the 2003 UNESCO Convention for the Safeguarding of the Intangible Cultural Heritage, encompasses the following fields: (i) oral traditions and expressions, including language as a vehicle of the intangible cultural heritage; (ii) performing arts; (iii) social practices, rituals and festive events (including religious rituals, pilgrimages, religious dances, music, burial ceremonies and other events); (iv) knowledge and practices concerning nature and the universe; and (v) traditional craftsmanship.

A particular form of event with religious content among the three most widespread Christian European or Slavic denominations (Catholicism, Protestantism and Eastern Orthodoxy) are Passion Plays (Križnar, 2018). A Passion Play is a dramatic depiction of the Passion of Jesus Christ, during which scenes of his journey are re-enacted, including his preaching, his arrest, his trial, his crucifixion and resurrection. Re-enactments are often enriched by scenes from the Old and New Testaments.

Historians have described the re-enactment of the Passion of Jesus Christ 'as an evolutionary throwback or curious hybrid of ancient Greek and medieval theatre' (Groeneveld, 2021, p. 67). The distinctive characteristics of the Passion Plays are 'the combination of loyalty to tradition and openness for the present day that is made possible by the dialogue between the generations, religious persuasions, representatives of orthodoxy and the modern' (UNESCO, 2014). The history of these re-enactments is long and complex. It is characterized by faith, spirituality and devotion to the Lord. Nevertheless, it has been accused of harbouring political agendas and anti-Semitism and it has been subjected to both political and religious censorship.

Ohm (2010, p. 5) argues that 'religious drama in the Catholic Church developed out of role-playing religious ritual, particularly during church festivals, for the benefit of illiterate and non-Latin-speaking worshippers'. According to Grafenauer (2015, p. 80): 'Like all Baroque processions, the ... Passion Play is a moving penitence sermon in live images and spoken word with the aim to shake the sinners and persuade them to convert'.

The term '*tableaux vivants*' is French and it means 'living pictures', which are also called 'dynamic pictures' or 'frozen pictures'. The technique of *tableau vivant* is the re-enactment of a historical scene or a well-known work of art of living fixed faces so that they give the impression of a painting. Participants are asked to create a 'photo' with their bodies. It is 'a hybrid genre having an ancestry part theatrical, part social, part religious' (Chapman, 1992, p. 22). It is the representation by a cast of characters of scenes from literature, art, history (Elbert, 2002, p. 237) or daily life. The characters represented a scene by remaining silent and frozen for about 30 seconds while staging, pose, costume, make-up, lighting and the facial expression of the characters became prominent. 'Sometimes a poem or music accompanied the scene, and often a large wooden frame outlined the perimeter of the stage, so as to reference the frame of a painted canvas' (Murphy, 2012).

According to Schenz (2022, pp. 15–16), the cause of the re-enactments was the bloody conflict between Protestants and Catholics over matters of faith, but also over territorial claims and power, which became known as the Thirty Years' War (1618–1648). The conflict devastated large parts of Europe, because of warfare, famine and plague. The resulting devastation brought about the inevitable end of existing political and social structures. In this context, the church tried in every way to defend the faith by promoting the re-enactment of the Passion of Jesus Christ.

Over time, these dramatized representations left the confines of the church (as a venue) and began to be performed in the streets and squares of Bavarian villages and towns as well as in cemeteries (Schenz, 2022, p. 16). Representations of the Passion of Jesus Christ were also recorded in York, Coventry and Chester in Great Britain from the 14th century to mid-16th century, but also in Italy, and other cities on the European continent. As these performances were often long and therefore tedious for the audience, the organizers began to intersperse them with comic elements. Thus, gradually, the re-enactments discarded their purely religious character and became more festive, developing secular elements. Then, new figures were added, the performances left many hints, such as the role of Mary Magdalene, and the figures of the devil evolved into comic figures that provoked laughter rather than fear from the audience. But this caused the wrath of the church, which considered that the 'farcical' elements that were part of the Divine Drama were offensive to the faith and distorted and devalued the religious message of the performance. As early as 1471 the Bishop of Havelberg ordered his clergy to ban the re-enactments in his parishes because he considered that they had lost their character and had developed into shameful and irrelevant farces. Similar bans were imposed in Great Britain resulting in the banning of the *York Mystery Plays* (York Mystery Plays, 2022) and the *Coventry Corpus Christi Plays* (Davidson, 2016).

In 1770 and under the weight of the Enlightenment, the Catholic Church banned the practice of re-enacting the Passion of Jesus Christ. Professor and historian James Shapiro (2000, cited in Ohm, 2010) argues that the ban resulted in the discontinuation of re-enactments throughout Bavaria and the Austrian Tyrol. The only re-enactment of the Passion that was preserved was that at Oberammergau which followed a process of requests to reconsider the ban and revisions that were eventually adopted.

The Modern 'Experience' of Re-enactment

Tourism activity is constantly changing and evolving, adapting to the new needs and new market trends. The modern tourist does not seem to be content with simply seeing places and events, as in the case of mass tourism, but seeks experiences (e.g. through acquaintance with the local population, culture, lifestyle, food, entertainment). In the context of these alternative activities, the modern tourist seeks experiences through active participation and experiential practices (Cohen, 1979; Prentice *et al.*, 1998; Pine and Gilmore, 1999; Richards, 2001; Swarbrooke, 2003). In this case, the various acts, in the form of events, festivals and re-enactments allow spectators to engage in them and gain meaningful experiences. Re-enactments also allow spectators to enhance and augment their experience as they generate emotional connections (Dolcos and Cabeza, 2002), evoke strong emotional arousal and reinforce memory (Pine and Gilmore, 1999; Gilmore and Pine, 2002), and stimulate a religious/spiritual feeling. These re-enactments act as a bonding agent for the inhabitants of local communities, strengthening their sense of identity and generating new ties.

These are the reasons why re-enactments are part of the tourism policy of many countries, which seek, through them, to create new or enrich their tourism product. Today the development of tourism has stimulated the revival of re-enactments of the Passion in various European cities. Some of the re-enactments have been inscribed on the UNESCO Representative List of Intangible Cultural Heritage of Humanity (e.g. the Škofja Loka Passion Play in 2016) (ICH-UNESCO, 2016). Some others have been inscribed on the National Inventories of Intangible Cultural Heritage by the National Commissions for UNESCO such as the Passion Play Erl, which was inscribed in 2013 on the Austrian Inventory of Intangible Cultural Heritage (UNESCO, 2013).

The Re-enactment of the Divine Drama and Crucifixion in Oberammergau (Germany)

Every 10 years, since 1634, a re-enactment of the Divine Drama and the Crucifixion has been held in the village of Oberammergau in the Bavarian Alps. The dramatic representation has its origins in the Thirty Years' War and the plague outbreak that followed. The villagers gathered in the church and vowed to the crucified Christ that if God would protect them from the illness, they would re-enact the Christ's Passion every decade (Harpur and Westwood, 2005, p. 206; Mork, 2005, p. 85). Historical records show that no deaths from the plague were recorded thereafter, so the villagers gratefully began performing the Passion of Jesus the following year. The first performance took place in the village cemetery next to the parish church, near the graves of the plague victims, on a simple wooden structure (Passion Play, 2022).

Over the centuries the Oberammergau Passion Play has managed to survive despite significant changes and revisions. Evil, which originally appeared in the form of the devil, was replaced by Jewish priests and merchants, portrayed as avaricious persecutors of Jesus and the Christians. The costumes and colours of the performance reflect this perception. This change, however, led to accusations of anti-Semitism being attributed to the representation. In 1771 with the advent of the Enlightenment, the Catholic Church banned the re-enactments of the Passion of Jesus, which resulted in the cessation of re-enactments in Bavaria (Schenz, 2022, p. 21). In 1780, after a special privilege was granted, re-enactments resumed for the following decades. In 1810 they were banned again. The monk Othmar Weis, from the Benedictine monastery in the neighbouring village of Ettal, wrote a new version of the re-enactment, based on the Gospels. This version set reconciliation as the basic idea of the re-enactment. Despite the changes in the structure of the representation, Othmar Weis retained the *tableaux vivants* (living images) that had been added to the work in the late 18th century. The performance is exclusively in German, but instruction books are available in several languages with a translation or explanation of the scenes.

In 1850, in order to prevent a new ban, the village tried to secularize the performance by entrusting the organization to the community and separating it from the Catholic Church (Schenz, 2022, p. 24). Thus, it established a committee responsible for organizing the performance, which was now housed in the community theatre and they started charging an entrance fee. In 1890 the stage of the theatre was furbished, and part of the seats were put under a canopy. In

1900 an iron structure was built, which was extended in 1930, forming the basis of the present theatre. The stage was also reconstructed taking a monumental shape that is still preserved today. The theatre where performances take place has been declared a listed monument (Passion Play, 2022). In the 1990s and 2000s the costumes and the way the scenes were performed were altered so that accusations of anti-Semitism were eliminated.

The performance has always been the focus of interest of the European aristocracy, kings, princes and other dignitaries, motivated not necessarily by piety or education, but also by pleasure (Schenz, 2022, p. 172). We can safely assume that watching the performance, due to the interest of contemporary aristocracy and contemporary media, was a fashion and a sign of social status. After World War II, the composition of the audience changed; but again, visitors still were members of government, businessmen and art enthusiasts. Later, the composition of the audience changed again as tourism started to develop, as families with children began to be attracted but this occurred without limiting the number of prominent visitors (Table 12.1).

The Passionsspiele Oberammergau were inscribed in 2014 on Germany's National Inventory of Intangible Cultural Heritage. For the German Commission for UNESCO (2014) the continuous work being carried out 'on the text, stage, props, music and acting' of the Passion Plays preserves its tradition over the centuries and encourages dialogue between the generations and religious doctrines. Nevertheless, there are obstacles as the younger people wish to modify and present a new work while both the older people and tour operators wish to retain the old elements, arguing that the audience wants to enjoy the traditional performance (Studarus, 2022).

In fact, the performance is subject to constant adaptations/changes, which in practice alter the work. According to Schenz (2022), the latest re-enactment in 2022 involved professional musicians in leading roles and new female characters were added. Today more than 2000 villagers participate in the re-enactment as actors, singers and musicians while all sets and costumes are made by them (Groeneveld, 2021, p. 68). The re-enactments take 7 hours and are interspersed, between acts, with scenes that resemble living paintings inspired by the works of great painters. Around 500,000 visitors from all over the world came to watch the performance (Fig. 12.1). The performances take place in a 4500-seat open-air theatre with a unique mountain landscape as a backdrop. The 42nd edition of the play should have taken place in 2020, but due to the coronavirus disease 2019 (COVID-19) outbreak it was held from 14 May to 2 October 2022, with 102 performances scheduled (Passion Play, 2022).

Škofja Loka (Slovenia)

Škofja Loka is a well-preserved medieval town in Slovenia which has been listed as a cultural monument (Fig. 12.2). The re-enactment of the Christ's Passion (Škofjeloški Pasijon) is held every 6 years during Lent and Easter on the streets of the old town centre. Records indicate that the first re-enactment took place in 1721. It started and ended at the 18th-century Capuchin monastery. It was repeated in the years 1727, 1728, 1745 and 1751. The procession was attended by people from the town and the surrounding villages of the wider area. The event became very popular and as a result, local people and foreigners came to watch it, considering it pious and unique (Grafenauer, 2015, p. 80).

The Škofja Loka Passion Play, based on the work of the Capuchin monk Lovrenc Marušič, presents 20 scenes of the Stations of the Cross. It includes a penitential passion procession which shows moving scenes from the Old and New Testaments with various allegorical motifs in a dramatized form, which are transported through the streets and squares of the medieval town centre. More than 900 inhabitants participate in the events, while more than 400 volunteers are involved in the production of the Passion Play, which is written and uttered in the old Škofja Loka dialect (Topler and Kužnik, 2019). The Passion Play was inscribed in 2016 on the UNESCO List of Intangible Cultural Heritage of Humanity (ICH-UNESCO, 2016), as it 'contributes to social cohesion giving residents involved an opportunity to connect with one another and feel like they are contributing to their community'.

What is interesting is that this re-enactment was inactive for a long time, from 1936 to 1999,

Table 12.1. Evolution of the Passion Play of Oberammergau (1760–2022). Adapted from Waybackmachine, 2022.

Year	Number of spectators	Occasions	Edition of the play
1760	14,000		14
1770			15
1771		Ban of performances by Catholic Church	
1780		Launch of Oberammergau privilege[a]	16
1790	11,000		
1800	3,000	Napoleonic Wars	17
1801		To generate revenues	18
1810		End of Oberammergau privilege	
1811		Banning was lifted due to changes made by Othmar Weis	19
1815		Special performance on the occasion of the end of the Napoleonic Wars	20
1830	13,000		22
1840	35,000		23
1850			24
1860			25
1870	40,000		26
1880	100,000	Easier access to Oberammergau due to railway and Thomas Cook tourism	27
1890	124,000		28
1900	174,000		29
1910	223,548		30
1922	311,127	Rescheduled 1920 performance which was postponed due to war – the Silesian Uprisings[b]	31
1930	420,000		32
1934	400,000	Special performances on the occasion of its 300th anniversary, attended by Hitler	33
1950	480,000	87 performances	34
1960			35
1970			36
1980			37
1984		Special performances on the occasion of its 350th anniversary in which the entire town collaborated	38
1990	480,000		39
2000	520,000		40
2010	500,000		41
2022	450,000	Rescheduled 2020 performance which was postponed due the COVID-19 pandemic[c]	42

[a]Performances were banned by the Catholic Church in 1771 with the advent of the Enlightenment. However, the Oberammergau privilege was granted in 1780 allowing enactments to resume.
[b]The Silesian Uprisings were a series of three uprisings from 1919 to 1921 in Upper Silesia, which, at the time, was part of the Weimar Republic.
[c]COVID-19, coronavirus disease 2019.

first because of the outbreak of World War II and later because of the political situation prevailing in the country. Its 'visibility' was gradually restored after the fall of Communism in 1989 and the breakup of the Socialist Federal Republic of Yugoslavia, when space was reassessed, and new identities and images were recreated. The re-establishment of the Škofja Loka Passion Play was achieved through a targeted strategy and planning, thanks to the support of the state, which provided the necessary institutional framework and funding, combined with the efforts of the local population and associations (Rátz *et al.*, 2008). In 1999, when the re-enactment was resumed, more than 77,000 spectators were recorded (Grafenauer, 2015,

Fig. 12.1. The entrance of the Passion Play Theatre at Oberammergau. Photo courtesy of E. Terzoglou, 2022, with permission.

p. 83). Emphasis was placed on enhancing the diffusion and promotion of the event and educating the young people. Another breakthrough was that Processio Locopolitana was inscribed in 2008 onto the Slovenian Register of Living Cultural Heritage of Slovenia kept by the Ministry of Culture and it has been part of the Strategy of the Sustainable Growth of Slovenian Tourism (2017–2021) (Topler and Kužnik, 2019). Further, a number of measures have been taken to prevent the effects of over tourism and over commercialization (Μοίρα, 2022) of the events although such a risk does not seem likely due to spatial constraints. Through this re-enactment,

Fig. 12.2. Škofja Loka. Image by M W from Pixabay (CC0).

which is integrated into the curricula of local schools, knowledge and skills linked with the practice of the play are conveyed from older to younger generations by families and craftspeople (ICH-UNESCO, 2016).

Easter Events on Paros Island, Greece

Paros is a Greek island in the central Aegean Sea with a population of about 14,520 people (ΕΛΣΤΑΤ, 2023, p. 52). In various villages of the island on the evening of Good Friday, a re-enactment of the Passion of Jesus Christ takes place. The custom was started by a female primary school teacher, Vassilia Kafourou-Assonitis, who proposed and succeeded in 1924 in re-enacting the Passion in the village of Marpissa, built on a hill at the east side of the island, with about 500 inhabitants. She was the first woman to be appointed as a teacher at the primary school of the village. She remained in the same position until 1951 when she retired. The original re-enactment was adopted by the residents who added new themes and events, culminating on Easter Sunday. Originally the re-enactments on the island were four in total and only involved girls. Today re-enactments take place in numerous villages throughout the island and men and women of all ages can participate in them. The number of participants in the re-enactment in Marpissa is 150 people.

According to testimonies 'all generations of the residents of Marpissa from 1922 to the present day are associated with the custom of Re-enactments. ... Everyone has taken part, either playing various roles or as group leaders' (Φραντζή-Τριβυζά, 2022, cited in Ιορδάνογλου, 2022). The representation is in the form of a *tableau vivant* and there is no speech or movement. Melodies and hymns are only sung during the Epitaph Procession (Figs 12.3 and 12.4). A distinguishing feature about this re-enactment is that there is no official body that organizes it; it is organized by associations of people who live in or come from the villages of Paros and live in Athens but every year they go back to their villages with their families to

Fig. 12.3. Re-enactment of the Passion of Jesus Christ on Paros Island, Greece: the Ascent to Calvary. From P. Moira's personal archive.

celebrate Easter. These associations cooperate with local schools, cultural associations and the local church so that the re-enactments are spontaneous, non-commercialized and more authentic.

On Good Friday, the Epitaph, which symbolizes the funeral cortege of Jesus Christ, is brought out of the church to lead a candle-lit procession in the alleys in the villages of Marpissa, Marmara, Prodromos, Lefkes and Aspro Horio. During the procession, the Epitaph makes 15 stops (stations) at various points, where scenes from the Passion of Jesus Christ are re-enacted by the villagers. Children and young people from the villages of the island and visitors from Athens who have family members in the island, dress up in period clothes as Roman soldiers or as disciples of Jesus and re-enact scenes from His Life and sufferings (e.g. Jesus entering Jerusalem on a donkey, the Resurrection of Lazarus, the Last Supper, Jesus praying on the Mount of Olives, the Ascent to Calvary, the Crucifixion). Each time a scene is re-enacted, a part of the mountain facing the village is illuminated. On Holy Saturday evening, the re-enactment of the Resurrection takes place. The events end on Easter Sunday with a traditional feast and the custom of the spit-roasted lamb over charcoal. Visitors and tourists are also offered free food and wine, accompanied by traditional music throughout the island. Easter events terminate with the ritual of the burning of Judas, where an effigy of Judas Iscariot (one of the original 12 disciples of Jesus) is burned and the Vespers of Love ('Agape') held in the afternoon of Easter Sunday, which portrays the first appearance of the risen Jesus to his disciples.

Organization and Symbolism of the Re-enactments

All three re-enactments primarily aim at preserving the religious-cultural heritage of the respective regions, but also strengthening local tourism development. They can also enhance respect for re-enactments of the Passion of Jesus at local, national and international levels and

Fig. 12.4. Re-enactment of the Passion of Jesus Christ on Paros Island, Greece: the burial of Jesus. From P. Moira's personal archive.

promote dialogue, cultural diversity and human creativity. This is evident from the Oberammergau Passionsspiele, which managed to transcend social, religious and political boundaries and attracts Protestants, Orthodox Jews as well as believers of other religions (Priest, 2018). In fact, UNESCO argues that the inscription of re-enactments on the UNESCO List of Intangible Cultural Heritage of Humanity will enhance awareness and visibility of European cultural heritage (ICH-UNESCO, 2016). Although the three case studies presented in this chapter have a common background in that the Passion of Jesus is represented in the form of *tableaux vivants*, they have significant differences (Table 12.2). In terms of organization, the Passion Play of Oberammergau is performed every 10 years and it is essentially a well-organized tourist show with thousands of tourists and visitors as spectators. However, the content of the re-enactments is filtered through political correctness, religious censorship and economic considerations. Moreover, the re-enactment is subject to constant adaptations/changes, altering the original Passion Play. For example, in the performance which was held in 2022 (due to the COVID-19 pandemic outbreak in 2020) professional actors were cast in the main roles while some roles were even assigned to refugees' children and Muslim actors. The role of women was also enhanced. In the same performance, Jesus was 'more political, more angry, someone looking for social justice' (Connolly, 2022). The producers argued that these alterations are necessary as modern audiences are less familiar with the Bible and may be less interested in faith and church but seek spectacle and entertainment (Schenz, 2022, p. 168). Undoubtedly, however, for some of the audience the performance is also a 'thrilling spectacle that evokes strong emotions even in non-believers' (Τερζόγλου, 2022, personal communication). Groeneveld, (2017, p. 131) claims that the spectacle of the Oberammergau Passion Play 'represented Christ's Passion more accurately and powerfully than any spoken sermon or written gospel could, establishing Oberammergau as an important religious as well as aesthetic pilgrimage site'.

Table 12.2. Comparative data of re-enactments.

Representation	Oberammergau	Škofja Loka	Paros island
Periodicity	Every 10 years, in years ending on a '0'	Every 6 years	Annually
Venue	Theatre stage	Open-air venues, streets Moving *tableaux vivants* Portable moving stages	Open-air venues at every village σε ανοικτούς. Fixed *tableaux vivants*
Year of first performance	1634	1721 (interrupted between 1936 and 1999)	1924
Number of participants	> 2000	900 + 400 volunteers	150
Number of spectators	450,000–500,000	77,000 (1999)	Not available
Participants	Local people of all ages	Local people of all ages	Local people of all ages
Organizing body		Municipality of Škofja Loka, which coordinates the other bodies (Processio Locopolitana, 2022)	Associations in Paros and Athens, schools, communities
Number of visitors	500,000	Not available	Not available
Number of scenes		20 UNESCO/13 WEBpages (Slovenia info, 2022)	15
Other events	No	Yes	Yes
Museum	Oberammergau Museum (cultural history museum)	The Škofja Loka Passion Play Collection	
Monument inscribed on the UNESCO List of Intangible Cultural Heritage of Humanity	German Commission for UNESCO	UNESCO Representative List of Intangible Cultural Heritage of Humanity	No
Religion	Protestants and Catholics	Roman Catholic Church	Eastern Orthodox Church

The Škofja Loka Passion Play, organized every 6 years, has also been a successful revival, partly for tourist purposes, of medieval re-enactments. Both believers and tourists engage in it. As this re-enactment has been revived relatively recently and is in the early stage of its development, it is premature to speak of authenticity or commercialization.

By contrast, in Paros, the re-enactments of Christ's Passion are organized and performed every year on Good Friday as they are part of the religious celebration of Easter and children and young people participate in them. Spectators consist of believers and tourists who flock to the island in large numbers to spend their Easter holidays. The re-enactments of the Christ's Passion have not been utilized in terms of tourism promotion because they are more of an emotional event, an expression of common identity and are therefore non-commercial and more authentic.

Spyridoula Kavali, from the village of Prodromos, says that 'everyone feels a great devoutness ... lives the moment ... listening to the hymns sung by the children ...' (S. Kavali, interview 27 September 2022). Maria Frantzi-Trivyza explains vividly the impact that re-enactments have on people who engage in them:

> my 13-year-old grandson's dilemma is whether to participate in the re-enactment of Jesus' Prayer in the Mount of Olives or watch all the re-enactments as a spectator. After his participation in the re-enactment, he described the intense emotions he developed from this experience.
>
> (M. Frantzi-Trivyza, interview 27 September 2022)

Summary

The Passion of Christ, with the intervention of national and international bodies and organizations, are converted into new events or are revived with the aim of preserving cultural heritage and attracting tourists. However, careful handling is required so that these events do not fall into the realm of mere 'spectacle', losing their meaning and symbolism, and being transformed into commodities.

According to Butler's (1980) Tourism Area Life Cycle (TALC) model, Oberammergau seems to be at the fourth stage (consolidation), Škofja Loka is at the second stage (involvement) while Paros' Passion Play is at the first stage (exploration). In addition, re-enactments of the Oberammergau Passion Play raise questions of staging and authenticity, mainly due to the constant interventions and modifications that the performance undergoes. Therefore, special attention is needed in the planning and management of tourism so that the negative elements do not prevail. For instance, re-enactments of the Passion Play in Oberammergau necessitate continuous redesign and renewal so that they are not led to the sixth stage of TALC (decline).

In contrast, in Škofja Loka the local community presently seems to be trying to decide on the form of tourism it wants and the extent to which it will be developed. The custom is being fully revived and it is being promoted in the national and international media. Furthermore, the construction of infrastructure has already started, and a suitable legislative framework has been introduced (Čok, 2012). At this point, the original rules for the sustainable development of tourism need to be set.

On the island of Paros, the local communities do not seem to have planned the commercial exploitation of the celebration, there are not many facilities, and the re-enactments are mainly the result of an unplanned cooperation and communication between locals and visitors. Over time the tradition has been embraced by the local people and continues to evolve. Maria Frantzi-Trivyza (interview 27 September 2022) and Spyridoula Kavali (interview 27 September 2022) claim that over the last few years some infrastructure has begun to be built and new restaurants and accommodation facilities have been added so that tourists can stay over in the region. Residents stress that care is needed so that the tradition is not commodified and that it retains its authenticity.

Due to the lack of data, a targeted survey addressed to the visitors/tourists to the aforementioned regions could be conducted so that their motivations and experiences from their participation in the re-enactments could be explored.

Acknowledgements

The authors would like to thank Ms Maria Frantzi-Trivyza (former President of the Community of Marpissa, and former President of the Tourism Committee of the Municipality of Paros, Greece) and Ms Spyridoula Kavali (former member of the cultural association 'Skopas o Parios', Paros, Greece) for their interviews and for sharing information on the re-enactments of the Christ's Passion on Paros Island. Also, we would like to thank Ms Elli Terzoglou (University of West Attica) for her interview and for sharing information on the re-enactments of the Christ's Passion in Oberammergau.

References

Arcodia, C. and Barker, T. (2003) The employability prospects of graduates in event management: using data from job advertisements. In: Braithwaite, R.W. and Braithwaite, R.L. (eds) *Riding the Wave of Tourism and Hospitality Research. Proceedings of the Council for Australian University Tourism and Hospitality Education (CAUTHE) Conference, Coffs Harbour, New South Wales, Australia, 5–8 February 2003*. Southern Cross University, Lismore, New South Wales, Australia, pp. 19–34.

Blackwell, R. (2007) Motivations for religious tourism, pilgrimage, festivals and events. In: Raj, R. and Morpeth, N.D. (eds) *Religious Tourism and Pilgrimage Management: an International Perspective*. CAB International, Wallingford, UK, pp. 35–47.

Bowdin, G., Allen, J., O'Toole, W., Harris, R. and McDonnell, I. (2011) *Events Management*, 3rd edn. Routledge, London.

Butler, R.W. (1980) The concept of a tourist area cycle of evolution: implications for management of resources. *Canadian Geographer* XXIV(1), 5–12. http://dx.doi.org/10.1111/j.1541-0064.1980.tb00970.x

Cambridge Dictionary (2022) Event. Available at: https://dictionary.cambridge.org/dictionary/english/event (accessed 14 August 2022).

Carter, L. (2007) *Event Planning*. AuthorHouse, Bloomington, Indiana.

Chalkley, B. and Essex, S. (1999) Urban development through hosting international events: a history of the Olympic Games, *Planning Perspectives* 14(4), 369–394. https://doi.org/10.1080/026654399364184

Chapman, M.M. (1992) 'Living pictures': women and *tableaux vivants* in 19th century American culture. PhD dissertation, Cornell University, Ithaca, New York. Available at: https://www.academia.edu/32303549/Living_Pictures_Women_and_Tableaux_Vivants_in_19th_century_American_Culture (accessed 8 March 2022).

Cohen, E. (1979) A phenomenology of tourist experiences. *Sociology* 13(2), 179–201. https://doi.org/10.1177/003803857901300203

Čok, G. (2012) Urbanistična delavnica Škofja Loka Kamnitnik [Škofja Loka Kamnitnik, Urban Design Workshop]. *AR Arhitektura, raziskave* 2–3, 114–115. (in Slovenian)

Connolly, K. (2022) 'We kept our beards': Oberammergau's passion play emerges from pandemic. *The Guardian*, 14 May. Available at: https://www.theguardian.com/world/2022/may/14/we-kept-our-beards-oberammergaus-passion-play-emerges-from-pandemic (accessed 8 March 2023).

Davidson, C. (2016) The Coventry mysteries and Shakespeare's histories. *Early Drama, Art and Music*, 6. Available at: https://scholarworks.wmich.edu/early_drama/6 (accessed 8 March 2022).

Dolcos, F. and Cabeza, R. (2002) Event-related potentials of emotional memory: encoding pleasant, unpleasant, and neutral pictures. *Cognitive, Affective & Behavioral Neuroscience* 2(3), 252–263. https://doi.org/10.3758/CABN.2.3.252

Elbert, M. (2002) Striking a historical pose: antebellum tableaux vivants, 'Godey's' illustrations, and Margaret Fuller's heroines. *The New England Quarterly* 75(2), 235–275. https://doi.org/10.2307/1559765

Ferdinand, N. and Kitchin, P. (2012) *Events Management: an International Approach*. Sage, London.

Getz, D. (2005) *Event Management and Event Tourism*, 2nd edn. Cognizant Communication Corporation, Putnam Valley, New York.

Getz, D. (2008) Event tourism: definition, evolution, and research. *Tourism Management* 29(3), 403–428. https://doi.org/10.1016/j.tourman.2007.07.017

Gilmore, H.J. and Pine B.J. II (2002) Differentiating hospitality operations via experiences: why selling services is not enough. *Cornell Hotel and Restaurant Administration Quarterly* 43(3), 87–96. https://doi.org/10.1177/0010880402433009

Grafenauer, B. (2015) Potential of heritage as the basis for the development of town tourism – the case of Škofja Loka. *Broj Turističko poslovanje* 15, 75–85. https://doi.org/10.5937/TurPos1515075G

Greenwood, D.J. (1989) Culture by the pound: an anthropological perspective on tourism as cultural commoditization. In: Smith, V.L. (ed.) *Hosts and Guests: the Anthropology of Tourism*, 2nd edn. University of Pennsylvania Press, Philadelphia, Pennsylvania, pp. 171–186.

Groeneveld, L. (2017) 'I felt as never before, under any sermon that I ever heard preached': word, image, and the Oberammergau Passion Play, 1840–1900. *Nineteenth Century Theatre and Film* 43(2), 131–159. https://doi.org/10.1177/1748372716687316

Groeneveld, L. (2021) Nineteenth- and early twentieth-century representations of the Oberammergau Passion Play: heredity, eugenic theatre, and 'epic selection'. *Theatre Survey* 62(1), 67–88. https://doi.org/10.1017/S0040557420000484

Harpur, J. and Westwood, J. (2005) *Τόποι των θρύλων [Places of Legends]*. Κοχλίας, Αθήνα. (in Greek)

ICH-UNESCO (2016) Škofja Loka Passion Play. Slovenia Inscribed in 2016 (11.COM) on the Representative List of the Intangible Cultural Heritage of Humanity. Intangible Cultural Heritage (ICH) United Nations Educational, Scientific and Cultural Organization (UNESCO). Available at: https://ich.unesco.org/en/RL/kofja-loka-passion-play-01203 (accessed 7 March 2023).

Kalwaria Zebrzydowska (2023) *Sanktuarium*. Available at: https://kalwaria.eu/strona/english (accessed 8 March 2023).

Kose, H., Argan, M.T. and Argan, M. (2011) Special event management and event marketing: a case study of TKBL All Star 2011 in Turkey. *Journal of Management and Marketing Research* 8, 1–11.

Križnar, F. (2018) Music and sustainability: Škofja Loka, a Passion Play in an interreligious context. *Facta Universitatis, Series: Visual Arts and Music*, 4(1), 1–10. https://doi.org/10.22190/FUVAM1801001K

Landy, T.M. (2013) Paixão de Cristo: Brazil stages star-studded passion play. *Catholics & Cultures*. Available at: https://www.catholicsandcultures.org/brazil/passion-christ (accessed 3 March 2022).

Liutikas, D. (2015) Indulgence feasts: manifestation of religious and communal identity. In: Jepson, A. and Clarke, A. (eds) *Managing and Developing Communities, Festivals and Events*. Palgrave Macmillan, London, pp. 148–164.

MacCannell, D. (1984) Reconstructed ethnicity: tourism and cultural identity in Third World communities. *Annals of Tourism Research* 11(3), 375–391. https://doi.org/10.1016/0160-7383(84)90028-8

Mohr, J. and Stenzel, J. (2022) The ways of things: mobilizing charismatic objects in Oberammergau and its Passion Play. *Religions* 13(1): 71. https://doi.org/10.3390/rel13010071

Moira, P., Parthenis, S., Kontoudaki, A. and Katsoula, O. (2009) Religious tourism in Greece: the necessity to classify religious resources for their rational valorization. In: Trono, A. (ed.) *Tourism, Religion and Culture. Regional Tourism Development through Meaningful Experiences. Proceedings of the International Conference, Lecce, Tricase, Poggiardo, 27–29 October*. University of Salento/Mario Congedo Publishers, Salento, Italy, pp. 465–480.

Mork, G.R. (2005) Dramatizing the Passion: from Oberammergau to Gibson. *Shofar* 23(3), 85–92.

Murphy, S. (2012) Tableaux vivant: history and practice. *Art Museum Teaching*, 6 December. Available at: https://artmuseumteaching.com/2012/12/06/tableaux-vivant-history-and-practice/ (accessed 9 March 2023).

Mylonopoulos, D., Moira, P. and Terzoglou, E. (2023) Passion Plays and their tourism utilization: authenticity or commercialization? In: *14th Annual International Religious Tourism and Pilgrimage (IRTP) Conference, 21–24 June 2023, Bosnia-Herzegovina, Medugorje, Mostar, Book of Abstracts*, p. 31.

Ohm, A.L. (2010) Oberammergau: Germany's 376-year-old Passion Play before and after the Holocaust, Vatican II, and ongoing research into early Christianity. *Headwaters* 27(2), 4–15.

Openlucht theater hertme (2023) Bellef het in Hertme! Available at: https://openluchttheaterhertme.nl/ (accessed 3 March 2023). (in Dutch)

Passion Play (2022) *Passion Play Theatre*. Available at: https://www.passionsspiele-oberammergau.de/en/visit/passion-play-theatre (accessed 2 March 2023).

Pine, B.J. II and Gilmore, H.J. (1999) *The Experience Economy: Work is Theatre & Every Business a Stage.* Harvard Business School Press, Boston, Massachusetts.

Piva, E. and Cerutti, S. (2015) Religious events and event management: an opportunity for local tourism development. *International Journal of Religious Tourism and Pilgrimage* 3(1), 55–65. DOI: 10.21427/D76F0J.

Poria, Y., Butler, R. and Airey, D. (2003) Tourism, religion and religiosity: a holy mess. *Current Issues in Tourism* 6(4), 340–363. https://doi.org/10.1080/13683500308667960

Prentice, R.C., Witt, S.F. and Hamer, C. (1998) Tourism as experience: the case of heritage parks. *Annals of Tourism Research* 25(1), 1–24. https://doi.org/10.1016/S0160-7383(98)00084-X

Priest, R.D. (2018) A Rabbi's impressions of the Oberammergau Passion Play: Joseph Krauskopf, antisemitism, and the limits of the transnational Jewish public sphere. *Jewish Social Studies* 24(1), 100–126. https://doi.org/10.2979/jewisocistud.24.1.04

Processio Locopolitana (2022) *Škofjeloški Pasijon*. Available at: https://www.pasijon.si/predstavitev/skofjeloski-pasijon/ (accessed 8 March 2023).

Raj, R., Walters, P. and Rashid, T. (2013) *Events Management: Principles and Practice*. Sage, London.

Rátz, T., Smith, M. and Michalkó, G. (2008) New places in old spaces: mapping tourism and regeneration in Budapest. *Tourism Geographies* 10(4), 429–451. https://doi.org/10.1080/14616680802434064

Richards, G. (2001) The experience industry and the creation of attractions. In: Richards, G. (ed.) *Cultural Attractions and European Tourism*. CAB International, Wallingford, UK, pp. 55–69.

Schenz, V. (2022) *The Story of the Oberammergau Passion Play. How a Village Captivates the World.* Volk Verlag Museum, Ljubljana.

Slovenia info (2022) The Škofja Loka Passion Play – a masterpiece on the UNESCO List of Intangible Cultural Heritage. Available at: https://www.slovenia.info/en/stories/the-skofja-loka-passion-play (accessed 8 March 2023).

Studarus, L. (2022) The 388-year-old play that's still going. *BBC*, 10 August. Available at: https://www.bbc.com/culture/article/20220809-oberammergau-passion-388-year-old-play-thats-still-going (accessed 8 March 2023).

Swarbrooke, J. (2003) *Adventure Tourism: the New Frontier*. Butterworth-Heinemann, Oxford.

Topler, J.P. and Kužnik, L. (2019) The role of UNESCO heritage in the boutique tourism offer in Slovenia. In: Bevanda, V. and Štetić, S. (eds) *4th International Thematic Monograph: Modern Management Tools and Economy of Tourism Sector in Present Era.* Association of Economists and Managers of the Balkans in cooperation with the Faculty of Tourism and Hospitality, Belgrade, pp. 629–638.

United Nations Educational, Scientific and Cultural Organization (UNESCO) (2013) *Intangible Cultural Heritage*. Passion Play Erl. UNESCO Austrian National Commission. Available at: https://www.unesco.at/en/culture/intangible-cultural-heritage/national-inventory/news-1/article/the-erl-passion-play (accessed 5 March 2023).

United Nations Educational, Scientific and Cultural Organization (UNESCO) (2014) Nationwide Inventory of Intangible Cultural Heritage. The Passion Play of Oberammergau. German Commission for UNESCO. Available at: https://www.unesco.de/en/passion-play-oberammergau (accessed 8 March 2023).

Vukonic, B. (1996) *Tourism and Religion*. Pergamon, Oxford.

Waybackmachine (2022) Passionsspiele2010. Available at: https://web.archive.org/web/20110626135128/http://passionsspiele2010.de/index.php?id=51 (accessed 8 March 2023). (in German)

York Mystery Plays (2022) The Stories. Available at: https://www.yorkmysteryplays.co.uk/the-stories/ (accessed 31 July 2023).

ΕΛΣΤΑΤ (2023) Greece Census results, 2021. Available at: https://elstat-outsourcers.statistics.gr/Census2022_GR.pdf (accessed 8 March 2023).

Ιορδάνογλου, Α. (2022) Τα πάθη του Χριστού ζωντανεύουν [*The Passions of Christ Come to Life*]. *Η Καθημερινή*, 19 April. Available at: https://www.kathimerini.gr/culture/561815755/ta-pathi-toy-christoy-zontaneyoyn/ (accessed 8 March 2023). (in Greek)

Μοίρα, Π. (2009) *Θρησκευτικός Τουρισμός [Religious Tourism]*. Interbooks, Αθήνα. (in Greek)

Μοίρα, Π. (2019) *Θρησκευτικός Τουρισμός και Προσκύνημα [Religious Tourism and Pilgrimage]*. Φαίδιμος, Αθήνα. (in Greek)

Μοίρα, Π. (2022) *Τουρισμός. Ιστορία-εξέλιξη-προοπτικές [Tourism. History-Evolution-Prospects]*. εκδ. Τζιόλα, Θεσσαλονίκη. (in Greek)

13 Seeking Identity and True Transformation: the Hajj Experience

Jaffer Idris*
Sheffield Hallam University, Sheffield, UK

Abstract
This chapter critically evaluates and discusses the practice of pilgrimage, which has been observed across various cultures and religions for centuries. The Hajj, an essential pillar of Islam, is not only a religious obligation but a potent force for fostering unity. It underscores the equality of all before Allah, transcending social status, wealth and pride. This chapter places its focus on religious pilgrims and their quest for identity and personal transformation before, during and after these profound journeys. The chapter also investigates the concept of pilgrimage as a means of seeking identity and transformation, using the Hajj as a case study to contextualize these ideas. Success in the Hajj journey extends beyond the completion of rituals; it resides in the profound transformations occurring within individuals and their redefined outlook on life. This transformation results in a deepened understanding of one's faith, conduct, and role within the family, community and society at large. The chapter ventures into the heart of the Hajj experience, unveiling the profound journey towards identity and true transformation that unfolds within the souls of pilgrims.

Introduction

As humans, we are drawn to what makes us feel full and in harmony with not only ourselves but also with our surroundings. Seeking this happiness and fulfilment frequently leads to a greater respect for our presence and coexistence on our planet. Pursuit of this greater purpose, whatever it may be, frequently results in transformation, which often results in change. This metamorphosis alters not just our perception of how we perceive the world, but also how we communicate, engage and deal with people.

Pilgrimages are a feature of all major world religions as well as spiritual and secular movements (Digance, 2003; Margry, 2008; Raj and Griffin, 2015). Timothy (2018) emphasizes that almost every religion in the world includes elements of route-based pilgrimage. Pilgrimage is a practice that has been observed across various cultures and religions for centuries and it is a journey that people undertake for a variety of reasons, such as to deepen their faith, connect with their heritage, seek healing or simply to experience something new. The number of pilgrims along the pilgrimage routes is constantly increasing (Oviedo *et al.*, 2014; Lois-González and Santos, 2015; Lois-González *et al.*, 2018). In modern pilgrimage, not only religiously motivated pilgrims participate, but also cultural tourists, heritage

*j.idris@shu.ac.uk

DOI: 10.1079/9781800623675.0013

tourists and spiritual tourists (Olsen and Trono, 2018). There are many different types of pilgrimages, and the purpose of each journey is unique to the individual pilgrim. Some pilgrims may travel to a sacred site to pray or meditate, others may simply want to find themselves, while some may travel to a sacred destination as a temporary escape from the general harsh existence in an agrarian-based society (Digance, 2006, p. 36).

Pilgrimages serve a variety of purposes depending on the individual and their aims. Many pilgrims undertake these journeys to develop their faith and demonstrate their dedication to their religion. According to Liutikas (2021), pilgrims travel to sacred or valuable places to manifest their identities and values, to gain spiritual or physical benefits, or to simply express gratitude. Muslims, for example, may embark on the Hajj pilgrimage to Makkah (Mecca) to complete one of Islam's five pillars, while Christians may visit the Holy Land to retrace the steps of Jesus. Some pilgrimages are culturally or historically significant, and pilgrims may be motivated to visit these sites to learn more about their heritage or the history of their faith, such as Jews visiting the Western Wall in Jerusalem to connect with their heritage and history. Some pilgrims may undertake these journeys to seek healing or blessings from holy sites or religious figures, such as Hindus seeking blessings from the goddess Vaishno Devi in India. However, one thing is for sure and that is that all pilgrimages bring people together.

This chapter will concentrate mainly on pilgrims travelling for religious purposes and the identity that they seek as well as transformation, if any, that comes about before, during or after the event. The chapter will investigate the concept of pilgrimage as a means of seeking identity and transformation and will use the Hajj as a case study to contextualize these ideas.

Pilgrimage: Seeking Identity, Transformation and Other Benefits

Before exploring the concept of identity, transformation and the other benefits that are so often linked with a pilgrimage, it is important to shed some light on what a pilgrimage is and why so many embark on such a journey. The main motivation for many pilgrims is often in search of the truth, inspiration and confirmation of the pilgrim's commitment to their faith. However, in doing so, the pilgrim generally has to sacrifice other components of daily life such as comfort, security, finance and family, to name a few. It is widely accepted that pilgrimage is a sort of journey taken for diverse religious and spiritual purposes (Morinis, 1984; Barber, 1993; Coleman and Elsner, 1995; Cousineau, 1998; Westwood, 2003; del Alisal *et al.*, 2007). It is a significant aspect of many religions and has been practised for thousands of years. Pilgrimage, one of the religious and cultural phenomena best known to human society, is an important feature of the world's major religions: Buddhism, Hinduism, Islam, Judaism and Christianity. Many pilgrims embark on pilgrimages as a form of spiritual practice to connect with a higher power, seek guidance and experience a sense of transcendence.

Pilgrimages are spiritual journeys performed by persons to a sacred or holy area for religious or spiritual purposes. Individuals may benefit from it by reflecting on their ideas and values, gaining new insights and views, and developing a stronger sense of purpose and direction. Such journeys involve visiting physical spaces that have religious importance (e.g. the Holy Land, Makkah or the Vatican), or locations with personal meaning (e.g. the birthplace of a religious figure or the location of a life-changing event) which are typically connected with significant historical events or individuals. According to Barber (1993, p. 1), a pilgrim's journey results from religious causes, externally to a holy site, and internally for spiritual purposes and internal understanding. Vukonić (1996, p. 80) argues that traditionally and historically, pilgrimage has been defined as a physical journey in search of truth, in search of what is sacred or holy. Leaving the comfort of your surroundings in search of a higher purpose or connection, in this case pilgrimage, has often been acknowledged as a transformative experience that leads to enhanced spiritual well-being, increased empathy, and decreased stress and anxiety levels. According to Turner and Turner (1978, pp. 3, 34), pilgrimages have the following attributes of liminality in their transition rites: (i) temporary release from the mundane structure that normally binds us; (ii) release from the burdens of stress, anxiety and guilt; (iii) movement from the mundane centre to a sacred periphery; (iv) homogenization of status; (v) simplicity of dress and behaviour; (vi) reflection on the meaning of basic religious and cultural values; (vii) the ritualized enactment of correspondence between religious paradigms and shared human experience; and (viii) experience of human brother and sisterhood.

Others are often seeking healing, closure or inspiration. Whatever the cause, pilgrimage creates a sense of community and shared experience among fellow travellers, which can foster a sense of belonging and connection as well as a sense of cultural identity and pride, enhance historical understanding, and promote intercultural dialogue and understanding. Some pilgrims embark on pilgrimages to connect with their cultural or historical roots as well as seeking connections with others on the same path. A study by Timothy and Olsen (2006) examined the transformative effects of the Camino de Santiago pilgrimage in Spain. The study found that the pilgrimage experience led to changes in the pilgrims' attitudes, beliefs and values.

Timothy and Olsen (2006) also noted that the pilgrimage facilitated the development of a sense of community among the pilgrims, and many of them reported increased spirituality and a greater appreciation of life. It can be a time of communal prayer, celebration and sharing of stories and experiences. The concept of a pilgrim is often associated with religious or spiritual journeys, but it can also refer to anyone on a journey of self-discovery, personal growth, 'meaning of life and to develop personal values' (Hassan *et al.*, 2023, p. 3).

Another explanation and justification for why so many people embark on pilgrimages is the search for identity. What is identity, and why is it significant in a pilgrimage context? How might a pilgrimage help someone discover who they are? Can one's identity be shaped by things such as surroundings, physical stamina, patience, knowledge and understanding, among others? It is significant to remark that because we play so many distinct roles in our everyday lives, each needing a different approach, it is fairly challenging to describe and, to some extent, define what we mean by identity. According to Abrams and Hogg (1988, p. 2), identity is 'people's concepts of who they are, of what sort of people they are, and how they relate to others'.

Deng (1995, p. 1) further adds that 'Identity is the way individuals and groups define themselves and are defined by others on the basis of race, ethnicity, religion, language, and culture'. To formulate what we mean by identity, this chapter aims to look at this concept from the lenses of a pilgrim and why identity is important to them when going on a pilgrimage. Identity plays a central role for pilgrims' motivations and behaviour (Liutikas, 2021). Wendt (1992, p. 397) states that there are 'relatively stable, role-specific understandings and expectations about self'. This insight fits nicely in the sphere of helping a pilgrim understand not only themselves but their purpose in the context of their faith, duties and responsibilities. Holy places, as well as the sacred route to those places, often allows pilgrims to gain an understanding of their religious identity and image, from both spiritual and profane elements of the place and the journey to it (Eliade, 1959; Holm and Bowker, 1994; Raj, 2007; Augé, 2008; Belhassen *et al.*, 2008). By contrast, Carmichael *et al.* (1994) and Inge (2003) argue that a dialectical relationship between the pilgrims' identity and place is embedded in the dimensions of everyday life where the experience takes place, and where the separation between the sacred and the profane realm is highly pursued.

The strengthening process of religious identity building stems largely from a dynamic interplay between this identity and the sacred places as well as the itinerary of the route leading to them (Damari and Mansfeld, 2016). As Bremer (2003, p. 73) interestingly claimed: 'the making of place, sacred or otherwise, always involves the making of identities; conversely, the construction of identity always involves the construction of place' (Cazaux, 2011 p. 73). Pilgrimage is a powerful tool for shaping religious beliefs and practices, as well as personal and communal identity and personal growth (Reader, 2005).

Pilgrimage can also be a transformative experience that can lead to a deeper understanding of oneself, others, and the world around us. Returning pilgrims will often speak about feeling a sense of peace and tranquillity during their journeys, and they may also find new insights into their lives and beliefs as well as a sense of inner transformation. According to Gothóni (1993), pilgrimage is a transformational journey as it includes components of departure, the journey and the return. Osterrieth (1989, p. 154) goes on to state that 'a great pilgrimage is a struggle against space, a sheer effort to overcome distance, and it is this struggle which has transformed the man'. Pilgrimage is a journey, travelling to a sacred place or shrine, often for spiritual or religious reasons. The act of pilgrimage can involve physical exertion, sacrifice and devotion, which can lead to a deeper connection with oneself, with others, and with a higher power but at the same time providing an opportunity for reflection, introspection and self-discovery. Pilgrimage is also a time for individuals to disconnect from the distractions of everyday life and focus on spiritual matters. This concept of breaking away from the norms and the everyday activities (i.e. components of life that we are comfortable with) allows us the space and time to learn something new. Badone and Roseman (2004) examine the transformative power of pilgrimage through the lens of anthropology and argue that pilgrimage can be a transformative experience because it involves a break from the routine of daily life, a sense of community and a heightened sense of awareness.

According to Gothóni (1993), every pilgrimage begins in a familiar place (i.e. at the pilgrim's station in life). Although the motives for going on a pilgrimage vary, as does the time of departure regarding the pilgrim's stage of adult life, the underlying sentiment for departure seems always to be that of perceived deficiency. This whole concept of the departure is often full of the sadness of separation but is also strangely exciting at the same time because of the anticipation of the adventures to come. The forthcoming scenario is unknown, despite that in current times the pilgrim is aware of the route, and has easy access to and from the sacred sites with the help of modern transportation.

Despite the ease of access, planning and modern commodities, there is always a sense of risk and things not going to plan. Pilgrimage, in all its forms, will often expose the pilgrim to a variety of elements and challenges such as 'transitions, setbacks and triumphs but the return, marked by final transformation, fulfilment and completion' (Gothóni 1993 p. 108). This transition is often associated with change and the ability of the pilgrim to transform themselves, both their inner and outer being, but the impact of the result depends on what the pilgrim considered the initial goal of their making the journey would be. Pilgrimage, therefore, serves as a multifaceted phenomenon that assists various purposes for different individuals and communities. Some pilgrims experience an ecstatic transformation, whereas others experience only a moderate transformation.

The notion is that those who consider themselves to be heavy sinners feel much greater relief than those who consider themselves to be merely ordinary sinners. The greater the contrast, the more miraculous the feeling (Gothóni 1993). The ideas of rebirth and purification are central to the spiritual transformation which the returning pilgrim hopes to undergo and the social prestige and legitimacy that comes with it (Eickelman and Piscatori, 1990).

The Hajj

For Muslims, Makkah is the most sacred city on Earth, and it has a powerful presence in their everyday life. The pilgrimage to Makkah is one of the most important religious events in the world due to the gathering of millions of people from all over the world in the holy city of Makkah in Saudi Arabia during a specific time of the year (the last month 'Dhul hajj' of the lunar calendar) to perform the Hajj (Raj, 2007; Wu and Mursid, 2019; Alasmari *et al.*, 2020; Alkhawaldeh, 2022). The Muslim pilgrimage to Makkah (Hajj) is one of the five pillars of Islam and a duty which adult Muslims must perform once in a lifetime if they are physically and financially able to do so (Peters, 1994; Aziz, 2001; Bianchi, 2004). The importance of pilgrimage (Hajj) is given in the Quran, in *Surah Al-Imran*, verses 96–97 (Quran Explorer, 2021):

> Verily, the first House (of worship) appointed for mankind was that at Bakkah (Makkah), full of blessing, and guidance for Al-'Alamin (the mankind and jinns). In it are manifest signs (for example), the Maqam (place) of Ibrahim (Abraham); whosoever enters it, he attains security.
>
> (*Surah Al-Imran*, verse 96)

as well as:

> And Hajj (pilgrimage to Makkah) to the House (Kaaba) is a duty that mankind owes to Allah, those who can afford the expenses (for one's

> conveyance, provision, and residence); and whoever disbelieves [i.e. denies pilgrimage to Makkah, then he is a disbeliever of Allah], then Allah stands not in need of any of the 'Alamin (mankind and jinns).
>
> (*Surah Al-Imran*, verse 97)

and

> Perform the pilgrimage and the visit (to Makka) for Allah.
>
> (*Surah Al-Baqarah*, verse 196)

The Hajj pilgrimage is an intensive spiritual journey that requires a great deal of preparation and commitment. During the days of the Hajj, pilgrims perform a series of religious and symbolic rites, following the footsteps of the prophets Abraham and Muhammad (Mawdudi, 1982; Peters, 1994; Wolfe, 1997; Raj, 2007). Pilgrims must be in a state of purity, known as Ihram, and must perform a series of rituals, including circling the Kaaba seven times (Fig. 13.1), travelling between the hills of Safa and Marwa seven times, and standing in prayer at the plain of Arafat. As stated in the Quran from *Surah Al-Baqarah* (Life of Muslim, 2022):

> And [mention] when We made the House a place of return for the people and [a place of] security. And take, [O believers], from the standing place of Abraham a place of prayer. And We charged Abraham and Ishmael, [saying], 'Purify My House for those who perform Tawaf and those who are staying [there] for worship and those who bow and prostrate [in prayer].
>
> (*Surah Al-Baqarah*, verse 125)

and

> Indeed, as-Safa and al-Marwah are among the symbols of Allah. So whoever makes Hajj to the House or performs 'umrah' – there is no blame upon him for walking between them. And whoever volunteers good – then indeed, Allah is appreciative and Knowing.
>
> (*Surah Al-Baqarah*, verse 158)

Fig 13.1. The Kaaba at the centre of the Hajj. Author's photo.

The journey can be physically and emotionally demanding, and it can take several days to complete. Therefore, by its design, the Hajj often represents the culmination of years of preparation and planning, both spiritual and logistical (Gatrad and Sheikh, 2005, p. 133). Through its intensive spiritual and physical experience, rituals and invocations, the Hajj exemplifies Schatzki's (1996) interpretation of embodied know-how: understandable by those who perform it (a form of experiential learning) and who espouse the same belief framework (Caidi, 2023). Many Muslims consider the pilgrimage to Makkah to be the ultimate realization of one's religious development and a way of achieving moral fulfilment or becoming a good, or better, Muslim. Being and becoming a good pilgrim for many people includes the perfection of virtuous behaviour, morality and is a demonstration of piety (Mahmood, 2005). According to Luz (2020), the primary incentive for Muslims travelling to Makkah and Madinah in Saudi Arabia for the purposes of pilgrimage is the benefit of receiving forgiveness of sins and to pray in Masjid al-Haram (the Great Mosque of Makkah) and Masjid-e-Nabawi in Madinah. Performing the Hajj is an important act of worship that purifies the soul, forgives sins and brings one closer to Allah. Abu Hurairah narrated that the Prophet Muhammad (peace be upon him (pbuh)) said:

> He who performs Hajj for Allah's pleasure and avoids all lewdness and sins will return after Hajj free from all sins as he was the day his mother gave birth to him.
>
> (Bukhari and Muslim)

as well as:

> Whoever performs Hajj for the sake of Allah and does not utter any obscene speech or commit any sin, will return (from Hajj) as pure and sinless as a newborn child.
> (Sahih al-Bukhari 1521, Book 25, Hadith 9; Sunnah.com, nd)

and:

> O people! Allah (swt) had ordained fardh the worship of Hajj. Hurry to perform the Hajj.
>
> (Muslim)

Hadrat Abu Huraira (ra) reported that Allah's Messenger (pbuh) said:

> (The performance of) 'Umra' is expiation for the sins committed (between it and the previous one). And the reward of Hajj Mabrur (the one accepted by Allah) is nothing except Paradise.
>
> (Bukhari; Umra; 1) (Muslim, Tirmidhi, Nasai, Ibn Maja)

Caidi (2019) states that during the Hajj, millions of Muslim men and women will congregate in Makkah to carry out specified rituals at a series of locations, with the aim of absolving themselves of sins. It also corresponds to a profound and unifying set of rituals performed in the Muslim world (Clingingsmith *et al.*, 2009, p. 1134), as well as an embodiment of the trans-historical and transnational Muslim community (ummah). The idea of millions of individuals, all assembled in the same locations, at specified dates, performing the same set of rituals that are steeped in history and tradition is the single most mesmerizing event in the whole world. The Hajj is an inherently communal experience, in a religion that gives particular importance to communal rituals (McCleary, 2007). Each ritual component of the Hajj is performed simultaneously with over two million participants (Fig. 13.2) and each participant's performance is believed to reinforce the others' performances, providing a shared aspect to individual worship (Clingingsmith *et al.*, 2009).

During the Hajj, the behaviour of pilgrims should be dominated by piety and morality, by abstention from all temptations, by tolerance when dealing with others and by the avoidance of disputes (Bianchi, 2004). The Hajj engenders substantial mixing across national, sectarian and gender lines in an atmosphere that emphasizes equality and unity (Clingingsmith *et al.*, 2009). Furthermore, the Hajj provides an opportunity for pilgrims to learn about the history and traditions of Islam. According to Hervieu-Leger (2001), the Hajj maintains a valuable chain of tradition as it is not only an obligation and a duty, but also a recall of history, the past and tradition. This also allows Muslim pilgrims to maintain and reinforce this knowledge of orthodoxy and the authenticity of their belief as they visit places and perform the rituals of the Hajj (Toguslu, 2017). Thus, the knowledge and experience received through the Hajj will deepen the pilgrims' understanding of their faith and provide them with a greater sense of purpose and direction in life.

Fig. 13.2. Millions of pilgrims, performing the same rituals during the same time during the Hajj. Author's photo.

Seeking Identity and Transformation Through the Hajj

For many Muslims, the Hajj pilgrimage is a life-changing experience that brings them closer to their faith and to Allah. It is an opportunity to reflect on one's life, seek forgiveness for past sins, and make a fresh start. The Hajj pilgrimage is also a unifying experience, as pilgrims from all over the world come together in a spirit of brotherhood and sisterhood, united by their faith. This is particularly significant given the diversity of the Muslim ummah, which spans different ethnicities, languages and cultures. Research has shown that this sense of unity and brother/sisterhood has a profound impact on pilgrims, as they feel a greater sense of connection to other Muslims and a stronger commitment to the principles of Islam. The identity of a pilgrim after the Hajj can be seen as a combination of religious, cultural and social factors, all of which contribute to the pilgrim's sense of self and place in society.

The identity of a pilgrim after the Hajj can be understood in various ways, depending on the perspective and the context. However, some common themes can be identified that reflect the spiritual, social and cultural significance of the Hajj pilgrimage in Islam. The identity of a pilgrim after Hajj is considered to be one of purity, piety and devotion to Allah. This identity is reflected in the pilgrim's behaviour, attitude and outlook towards life. As if to mark its personal and social significance, once they have completed the Hajj, pilgrims are given the honorific title al-hajj (or Hajji) for males, or al-hajja (or Hajjah) for females, and the legacy of the Hajj manifests itself in their everyday lives (Al-Ajarma, 2021).

According to Kenny (2007, p. 364), it is also a mark of religious prestige, a rebirth achieved through a new experience. In addition to the religious significance of Hajj, there are also cultural and social aspects associated with the identity of an al-hajj/al-hajja. In many Muslim societies, performing the Hajj is seen as a symbol of social status, respect and honour. Therefore, the identity of an al-hajj/al-hajja is often associated with prestige and recognition (Clingingsmith *et al.*, 2009).

One of the most fundamental aspects of the identity of a pilgrim after the Hajj is the sense of spiritual fulfilment and renewal. The Hajj is one of the five pillars of Islam, and it is considered a mandatory duty for every able-bodied and financially capable Muslim who can afford the journey. The rituals of the Hajj involve various acts of worship, such as circumambulating the Kaaba, standing at Arafat, and performing the symbolic stoning of Satan. These acts are meant to symbolize the submission, humility and devotion of the pilgrim to Allah (God) and to reinforce the bonds of brotherhood and sisterhood among Muslims from different backgrounds and nationalities.

Thus, the pilgrim's identity after the Hajj is marked by a deep sense of spiritual accomplishment and a renewed commitment to the Islamic faith. Another aspect of the identity of a pilgrim after the Hajj is the social and cultural significance of the pilgrimage. Hajj is a massive gathering of millions of Muslims from all over the world who converge on the holy city of Makkah to perform the rituals together. This experience creates a unique sense of unity and solidarity among the pilgrims, who often form lifelong friendships and connections with people from different cultures and backgrounds. Additionally, the Hajj has played a significant role in shaping Islamic culture and history, influencing art, literature and architecture, and fostering a sense of shared identity among Muslims worldwide.

The transformative nature of the Hajj (pilgrimage) is evident in the stories of many who have undertaken it. For some, the Hajj pilgrimage has been a turning point in their lives, inspiring them to become more devout Muslims and to live their lives in accordance with Islamic teachings. The significance of the Hajj and the impact of its rites assume great importance throughout the lives of pilgrims, which can be seen in numerous studies about the Hajj (Scupin, 1982; Peters, 1994; Wolfe, 1997; Bianchi, 2004; Raj, 2007). There are considerable physical challenges that pilgrims must overcome, including the heat, the crowds and the long distances that must be travelled. This physical exertion can be a transformative experience, as it requires pilgrims to push themselves beyond their comfort zones and develop resilience and determination resulting in personal growth and development.

The Hajj allows the pilgrim a new perspective on life, helping them to appreciate the things they have and to be more compassionate towards others. The Hajj is also seen as a transformative experience in terms of social and emotional aspects as well as increased compassion, empathy and social responsibility. Pilgrims are encouraged to reflect on their lives and their relationship with Allah, and to seek forgiveness for their sins (Fig.13.3). This introspection can lead to a greater sense of self-awareness and a deeper understanding of the purpose of life. For many pilgrims, the Hajj also has many positive impacts on their physical health due to the huge amount of walking that is involved. Whatever

Fig. 13.3. Pilgrims holding on to the entrance to the Kaaba, the House of God, as a form of seeking forgiveness and repenting past sins. Author's photo.

the outcome of a pilgrimage and the intentions to undertake it, true transformation or the test of their new-found inspiration or 'climax' (Camhi-Rayer *et al.*, 2012) occurs once the pilgrim returns home from their journey and how they intend to live the rest of their lives (i.e. pursuing a particular religious behaviour, driven by modesty, forgiveness, humility and by giving up a hedonistic lifestyle) (Bunyan, 1853; Fullam, 2006).

The Hajj: a Catalyst for Change

The Hajj, for those who attend and experience it, is an opportunity to bring about change not only in their everyday lives, but also in their understanding of their faith and how it should be applied in the world around them, their responsibilities to their families and communities they live in, and how best to serve humankind, regardless of colour, race and creed. Bigelow (2021) suggests that the Hajj can be a catalyst for social and political activism among Muslims.

The Hajj allows the returning pilgrim the chance to change a person's outlook on life, deepen their faith and help them become more compassionate and empathetic. To maximize spiritual benefit, pilgrims must strive to lead pious lives amid the ambivalences, contradictions and inconsistencies of their everyday lives (Al-Ajarma, 2020). Participants have reported feeling a sense of connectedness to their faith, a greater understanding of Islamic teachings, and a desire to live a more meaningful and compassionate life.

Throughout the whole Hajj, pilgrims are first-hand witnesses to the virtues of unity and togetherness. The whole environment and rituals of the Hajj places all pilgrims at the centre to witness, develop and practise a greater appreciation for the diversity of the Muslim ummah and to become more tolerant of different cultures and customs. The impact of the Hajj pilgrimage is not limited to the individual pilgrim but can also be seen in society. One significant impact of the Hajj is the promotion of unity and equality among Muslims. The Hajj pilgrimage brings together millions of Muslims from different countries, cultures and backgrounds. The shared experience of the Hajj can help to break down barriers and promote unity among Muslims. As stated by Ali (2021), the Hajj is the epitome of unity, equality and brother/sisterhood among Muslims. Bigelow (2021) found that pilgrimage can help to reinforce a sense of Muslim identity among pilgrims, as well as a sense of connection to the wider Muslim ummah (community). The Hajj pilgrimage is a journey that can be physically and mentally challenging.

The Hajj requires pilgrims to follow a strict set of rituals and to interact with people from different cultures and backgrounds. Additionally, the Hajj pilgrimage can also have economic benefits for the host country. It is a significant source of revenue for Saudi Arabia, the host country, and provides job opportunities for locals. Also it promotes cultural exchange, as pilgrims from different countries bring their unique customs and traditions to the host country.

On the world level, there are benefits associated with pilgrimage such as a good reputation that might impact a pilgrim's social interactions or commercial activities. Pilgrims are often invited as witnesses to marriages and to act as judges in cases of dispute as their opinions and ideas are highly respected in the community (Al-Ajarma, 2020). Returning pilgrims share community expectations of a morally elevated comportment in their daily lives after their return, while, at the same time, admitting a realistic sense of human imperfection.

Summary

Religious rituals are important contexts for making religion real. Belief is manifested in embodied practices by which the sacred and the divine are accessible to the senses. The sacred is present through the embodied practices (Orsi, 2005, pp. 73, 100). Experience in sacred spaces creates a lived religion that makes the invisible visible, the imagined realizable (Orsi, 1997; Ammerman, 2007; McGuire, 2008). Academic research suggests that pilgrimage has a positive effect on the pilgrim's well-being, including emotional, psychological and spiritual benefits. The transformative effects of pilgrimage can lead to positive changes in the pilgrims' attitudes, beliefs and values, and help them develop a greater sense of purpose and meaning in life. Overall, pilgrimage can have a positive impact on the psychological, emotional and spiritual well-being of the pilgrim. However, it is important to note that the effects of pilgrimage may vary depending on the individual and their specific beliefs, motivations and experiences during the pilgrimage. Pilgrims go to a place to fulfil their religious obligation.

Completion of the pilgrimage bestows upon the pilgrims many gifts that no other action could offer, such as receiving mercy from Allah as well as absolution from past sin. On successful completion of the Hajj, the pilgrim is also showered with honour and status (as explained previously, they are given the honorific title of al-hajj/al-hajja), that they will have during the remaining time on this Earth. The Hajj is significant in shaping individual moral conduct and bestowing an aura of religious merit on those who successfully completed it (Joll, 2011). Pilgrims have to negotiate their new status and the expectations that come with being al-hajj or al-hajja within the mundane and complex reality of everyday life. Even the title al-hajj or al-hajja is highly significant on both personal and social levels as a form of social, religious and moral capital (Bourdieu, 1986; Cooper, 1999, p. 93).

Pilgrimage has the potential to be a powerful and transformative experience for those who undertake it with an open mind and heart Moreover, we know from pilgrimage researchers that, after they return home, pilgrims tend to remain deeply affected by that experience in the context of their everyday life and practices (Frey, 2004; van der Beek, 2014, 2017). The Hajj pilgrimage is a transformative experience for pilgrims, providing them with a unique opportunity to deepen their faith, to develop a stronger sense of spirituality, seek forgiveness, and renew their commitment to Allah thus allowing them to get much closer to him. The Hajj rituals, practices, commitment and preparations can be seen as experiences of Muslim identities and as a type of religiosity that brings change to Muslims. Hajj is one of the obvious examples of the generation of moral change. It results ultimately in the cultivation of Islamic piety. Living as an authentic Muslim, behaving oneself in an Islamically appropriate way, is considered one of the aims to pursue after the Hajj. In fact, throughout the Hajj practices, a Muslim employs techniques of discipline that are adopted in their daily lives after they return from the holy sites. They open a door to pious subjects (Mahmood, 2005). It is a challenging journey that requires dedication and sacrifice, but for those who undertake it, the rewards are immeasurable.

References

Abrams, D. and Hogg, M.A. (1988) *Social Identifications: a Social Psychology of Intergroup Relations and Group Processes*. Routledge, London.

Al-Ajarma, K. (2020) Mecca in Morocco: articulations of the Muslim pilgrimage (Hajj) in Moroccan everyday life. PhD thesis, University of Groningen, Groningen, the Netherlands.

Al-Ajarma, K. (2021) After Hajj: Muslim pilgrims refashioning themselves. *Religions* 12: 36.

Alasmari, A.K., Edwards, P.J., Assiri, A.M., Behrens, R.H. and Bustinduy, A.L. (2020) Use of face masks and other personal preventive measures by Hajj pilgrims and their impact on health problems during the Hajj. *Journal of Travel Medicine* 27(8), 1–5. DOI: 10.1093/jtm/taaa155.

Ali, M.M. (2021) Hajj: Source of Human Unity and Universality. Available at: https://www.islamicity.org/53938/hajj-source-of-human-unity-and-universality/ (accessed 20 April 2023).

Alkhawaldeh, A.M. (2022) Religious tourism post-Covid-19 in the context of Muslim countries: destination image, perceived value, intention to revisit Makkah and health risk as moderator. *GeoJournal of Tourism and Geosites* 43(3), 858–865. https://doi.org/10.30892/gtg.43303-897

Ammerman, N.T. (2007) *Everyday Religion: Observing Modern Religious Lives*. Oxford University Press, Oxford.

Augé, M. (2008) *Non-places: an Introduction to an Anthropology of Supermodernity*. Verso, London.

Aziz, H. (2001) The journey: an overview of tourism and travel in the Arab/Islamic context. In: Harrison, D. (ed.) *Tourism and the Less Developed World: Issues and Case Studies*. CAB International, Wallingford, UK, pp. 151–160.

Badone, E. and Roseman, S. (eds) (2004) *Intersecting Journeys: the Anthropology of Pilgrimage and Tourism*. University of Illinois Press, Urbana, Illinois.

Barber, R. (1993) *Pilgrimages.* The Boydell Press, London.

Belhassen, Y., Caton, K. and Stewart, W.P. (2008) The search for authenticity in the pilgrim experience. *Annals of Tourism Research* 35(3), 668–689.

Bianchi, R.R. (2004) *Guests of God: Pilgrimage and Politics in the Islamic World.* Oxford University Press, New York.

Bigelow, A. (2021) *Islam Through Objects.* Bloomsbury Academic, London.

Bourdieu, P. (1986) The forms of capital, (trans. by Richard Nice). In: Richardson, J.G. (ed.) *Handbook of Theory and Research for the Sociology of Education.* Greenwood, New York, pp. 241–258.

Bremer, T.S. (2003) Il Genius Loci Ignotus di Eranos e la creazione di un luogo sacro. In: Barone, E., Riedl, M. and Tischel, A. (eds) *Eranos Monte Verità Ascona.* Edizioni ETS, Pisa, Italy, pp. 71–76. (in Italian)

Bunyan, J. (1853) *The Pilgrim's Progress.* Christian Classics Ethereal Library, Miami, Florida.

Caidi, N. (2019) Pilgrimage to the hajj: an information journey. *The International Journal of Information, Diversity and Inclusion* 3(1), 44–76.

Caidi, N. (2023) Curating post-hajj experiences of North American pilgrims: information practices as community-building rituals. In: Buitelaar, M. and van Leeuwen, R. (eds) *Narrating the Pilgrimage to Mecca: Historical and Contemporary Accounts.* Brill, Leiden, the Netherlands, pp. 369–390. https://doi.org/10.1163/9789004513174_017

Camhi-Rayer, B., Fregosi, F. and Hsun, C. (2012) Between religion and state: the Dajia pilgrimage in Taiwan. *Social Compass* 59(3), 298–310.

Carmichael, D.L., Hubert, J., Reeves, B. and Schanche, A. (eds) (1994) *Sacred Sites, Sacred Places.* Routledge, London.

Cazaux, F. (2011) To be a pilgrim: a contested identity on Saint James's Way. *Tourism: an International Interdisciplinary Journal* 59(3), 353–367.

Clingingsmith, D., Khwaja, A.I. and Kremer, M. (2009) Estimating the impact of the Hajj: religion and tolerance in Islam's global gathering. *The Quarterly Journal of Economics* 124(3), 1133–1170. https://doi.org/10.1162/qjec.2009.124.3.1133

Coleman, S. and Elsner, J. (1995) *Pilgrimage: Past and Present.* British Museum Press, London.

Cooper, B. (1999) The strength in the song: Muslim personhood, audible capital, and Hausa women's performance of the Hajj. *Social Text* 60, 87–110.

Cousineau, P. (1998) *The Art of Pilgrimage: the Seeker's Guide to Making Travel Sacred.* Conari Press, San Francisco, California.

Damari, C. and Mansfeld, Y. (2016) Reflections on pilgrims' identity, role and interplay with the pilgrimage environment. *Current Issues in Tourism* 19(3), 199–222. DOI: 10.1080/13683500.2014.957660.

del Alisal, R., Ackermann, M. and Martinez, D.P. (eds) (2007) *Pilgrimages and Spiritual Quests in Japan.* Routledge, Abingdon, UK.

Deng, F.M. (1995) *War of Visions: Conflict of Identities in the Sudan.* Brookings, Washington, DC.

Digance, J. (2003) Pilgrimage at contested sites. *Annals of Tourism Research* 30(1), 143–159.

Digance, J. (2006) Religious and secular pilgrimages: journeys redolent with meaning. In: Timothy, D. and Olsen, D.H. (eds) *Tourism, Religion and Spiritual Journeys.* Routledge, London, pp. 36–48.

Eickelman, D.F. and Piscatori, J. (1990) *Muslim Travellers: Pilgrimage, Migration, and the Religious Imagination.* University of California Press, Berkeley, California.

Eliade, M. (1959) *The Sacred and the Profane: the Nature of Religion.* Harper Torchbooks, New York.

Frey, N. (2004) Stories of the return: pilgrimage and its aftermaths. In: Badone, E. and Roseman, S. (eds) *Intersecting Journeys: the Anthropology of Pilgrimage and Tourism.* University of Illinois Press, Urbana, Illinois, pp. 89–109.

Fullam, L. (2006) Humility. A pilgrim's virtue. *New Theology Review* 19, 46–53.

Gatrad, A.R. and Sheikh, A. (2005) Hajj: journey of a lifetime. *British Medical Journal* 330, 133–137.

Gothóni, R. (1993) Pilgrimage = Transformation Journey. *Scripta Instituti Donneriani Aboensis* 15, 101–116. https://doi.org/10.30674/scripta.67208

Hassan, T., Carvache-Franco, M., Carvache-Franco, O.D. and Carvache-Franco, W. (2023) Sociodemographic relationships of motivations, satisfaction, and loyalty in religious tourism: a study of the pilgrimage to the city Mecca. *PLoS ONE* 18(3): e0283720.

Hervieu-Leger, D. (2001) *Religion as a Chain of Memory.* Rutgers University Press, New Brunswick, New Jersey.

Holm, J. and Bowker, J. (1994) *Sacred Place.* Pinter, London.

Inge, J. (2003) *A Christian Theology of Place.* Ashgate, Aldershot, UK.

Joll, C.M. (2011) *Muslim Merit-making in Thailand's Far-South. Muslims in Global Societies* Series 4. Berlin: Springer, Berlin. Available at: https://www.academia.edu/1975047/Muslim_Merit-making_in_Thailands_Far-south_Springer_2011 (accessed 15 April 2023).

Kenny, E. (2007) Gifting Mecca: importing spiritual capital to West Africa. *Mobilities* 2(3), 363–381.

Life of Muslim (2022) The Importance of the Hajj in the Quran and Hadith. Life of Muslim blog, 16 June. Available at: https://www.lifeofmuslim.com/blog/the-importance-of-the-hajj-in-the-quran-and-hadith (accessed 20 April 2023).

Liutikas, D. (ed.) (2021) *Pilgrims: Values and Identities*. CAB International, Wallingford, UK.

Lois-González, R.C. and Santos, X.M. (2015) Tourists and pilgrims on their way to Santiago. Motives, Caminos and final destinations. *Journal of Tourism and Cultural Change* 13(2), 149–164. https://doi.org/10.1080/14766825.2014.918985

Lois-González, R.C., Santos, X.M. and Romero, P.T.Z. (2018) The Camino de Santiago de Compostela: the most important historic pilgrimage way in Europe. In: Olsen, D.H. and Trono, A. (eds) *Religious Pilgrimage Routes and Trails: Sustainable Development and Management*. CAB International, Wallingford, UK, pp. 72–87.

Luz, N. (2020) Pilgrimage and religious tourism in Islam. *Annals of Tourism Research* 82: 102915. https://doi.org/10.1016/j.annals.2020.102915

Mahmood, S. (2005) *Politics of Piety: the Islamic Revival and the Feminist Subject*. Princeton University Press, Princeton, New Jersey.

Margry, P.J. (2008) *Shrines and Pilgrimage in the Modern World: New Itineraries into the Sacred*. Amsterdam University Press, Amsterdam.

Mawdudi, S.A. (1982) Origin and significance of the Hajj. Available at: http://www.islamicity.com/mosque/hajj/letsbemuslims.htm (accessed 10 April 2023).

McCleary, R.M. (2007) Salvation, damnation, and economic incentives. *Journal of Contemporary Religion* 22, 49–74.

McGuire, M.B. (2008) *Lived Religion: Faith and Practice in Everyday Life*. Oxford University Press, Oxford.

Morinis, A. (1984) *Pilgrimage in the Hindu Tradition: a Case Study of West Bengal*. Oxford University Press, Delhi.

Olsen, D.H. and Trono, A. (eds) (2018) *Religious Pilgrimage Routes and Trails: Sustainable Development and Management*. CAB International, Wallingford, UK.

Orsi, R. (1997) Everyday miracles: the study of lived religion. In: Hall, D. (ed.) *Lived Religion in America: Towards a History of Practice*. Princeton University Press, Princeton, New Jersey, pp. 3–21.

Orsi, R. (2005) *Between Heaven and Earth: the Religious Worlds People Make and the Scholars Who Study Them*. Princeton University Press, Princeton, New Jersey.

Osterrieth, A. (1989) Medieval pilgrimage: society and individual quest. *Social Compass* 36, 145–157.

Oviedo, L., de Courcier, S. and Farias, M. (2014) Rise of pilgrims on the Camino to Santiago: sign of change or religious revival? *Review of Religious Research* 56(3), 433–442. https://doi.org/10.1007/s13644-013-0131-4

Peters, F.E. (1994) *The Hajj: the Muslim Pilgrimage to Mecca and the Holy Places*. Princeton University Press, Princeton, New Jersey.

Quran Explorer (2021) Hajj pilgrimage and its significance. Quran Explorer blog, 7 September. Available at: https://www.quranexplorer.com/blog/events/hajj_pilgrimage_and_its_significance (accessed 13 April 2023).

Raj, R. (2007) The Festival of Sacrifice and travellers to the city of heaven (Makka). In: Raj, R. and Morpeth, N.D. (eds) *Religious Tourism and Pilgrimage Management: an International Perspective*. CAB International, Wallingford, UK, pp. 127–139.

Raj, R. and Griffin, K. (2015) *Religious Tourism and Pilgrimage Management: an International Perspective*, 2nd edn. CAB International, Wallingford, UK.

Reader, I. (2005) Making pilgrimages: meaning and practice in Shikoku. *The Journal of Asian Studies* 66(2), 249–253. DOI: 10.1017/S0021911807000721.

Schatzki, T.R. (1996) *Social Practices: a Wittgensteinian Approach to Human Activity and the Social*. Cambridge University Press, Cambridge.

Scupin, R. (1982) The social significance of the hajj for Thai Muslims. *Muslim World* 72, 25–33.

Sunnah.com (nd) Sahih al-Bukhari 1521, Book 25, Hadith 9. Available at: https://sunnah.com/bukhari:1521 (accessed 13 April 2023).

Timothy, D.J. (2018) Cultural routes: tourist destinations and tools for development. In: Olsen, D.H. and Trono, A. (eds) *Religious Pilgrimage Routes and Trails: Sustainable Development and Management*. CAB International, Wallingford, UK, pp. 27–37.

Timothy, D.J. and Olsen, D. (2006) Tourism and religious journeys. In: Timothy, D.J. and Olsen, D. (eds) *Tourism, Religion and Spiritual Journeys*. Routledge, New York, pp. 1–21.

Toguslu, E. (2017) The meaning of pilgrimage (Hajj): re-shaping the pious identity of Belgian Turkish Muslims. *Islam and Christian–Muslim Relations* 28(1), 19–32. DOI: 10.1080/09596410.2017.1282721.

Turner, V. and Turner, E. (1978) *Image and Pilgrimage in Christian Culture: Anthropological Perspectives*. Columbia University Press, New York.

van der Beek, S. (2014) Pilgrim narratives in dialogue. In: Farrelly, M. and Keely, V. (eds) *Pilgrim Paths: Journeys of Transformation*. Inter-Disciplinary Press, Oxford, pp. 45–53.

van der Beek, S. (2017) Ritual identity. *International Journal of Religious Tourism and Pilgrimage* 5(2), article 6.

Vukonić, B. (1996) *Tourism and Religion*. Elsevier Science, Oxford.

Wendt, A. (1992) Anarchy is what states make of it. *International Organization* 46(2), 391–425.

Westwood, J. (2003) *On Pilgrimage: Sacred Journeys Around the World*. Hidden Spring, Mahwah, New Jersey.

Wolfe, M. (1997) *One Thousand Roads to Mecca: Ten Centuries of Travelers Writing about the Muslim Pilgrimage*. Grove, New York.

Wu, C.H.-J. and Mursid, A. (2019) Loyalty motivations for religious tourism: Indonesian Muslim travelers umrah participating in umrah pilgrimage to Mecca, Saudi Arabia. *Tourism Review* 75(2), 466–478. https://doi.org/10.1108/TR-11-2018-0155

14 Selling Spirituality: Examples of a Mixture of Faith and Commercialization of Religious Symbols

Antonietta Ivona[1] and Donatella Privitera[2*]
[1]University of Bari Aldo Moro, Bari, Italy; [2]University of Catania, Catania, Italy

Abstract

This chapter discusses and analyses how globalization has played a crucial role within the religious tourism and pilgrimage management process. Pilgrims embark on sacred voyages to revered destinations, where they engage in shopping for sacred objects. Tourist actions have a profound impact in shaping the essence of a place, not only in religious tourism but also in the larger realm of human mobility. Further, the chapter offers insight into how spiritual symbols have become intertwined with the global market and the complexities this presents in the context of religious tourism. In the landscape of global commerce, Saint Nicholas, more commonly known as Santa Claus, stands as a quintessential example of a global figure. Operating within a global economic sphere, his reach extends to diverse destinations around the world. A defining moment in his global image is encapsulated in the iconic 1931 Coca-Cola advertisement, portraying Santa Claus as a ruddy-faced, red-clad, white-bearded figure travelling through the sky on a reindeer-drawn sleigh. This image of Santa Claus has permeated cultures and brought the veneration of this saint to new horizons.

Introduction

In this chapter we review the literature on religious symbols that have been commercialized by exploring images, social media and history to account for and understand how religious symbols emerged and became attractive to promote consumption and purchases.[1] The exploratory study shows the relationship between consumerism and religious symbols, demonstrating the centrality of the practice of using religious symbols to promote consumption to both faith communities and secular markets. Also, it highlights the importance of tourist actions and religiosity for place making in daily life, and in religious tourism (Iaquinto, 2020) as well as in mobility.

The order in which the instruments are discussed is thematic rather than temporal because what matters is not the story but the perspectives on this kind of use of religious symbols and territorial context. In some cases, then, the representation of religious symbols becomes a mark of territorial recognition precisely through souvenirs and images in a mixture of art, religion and global commerce. Globalization has had significant implications on icons, symbols and religious and cultural products used for religious tourism. There are some potential implications as well as a risk of the homogenization of symbols and icons. This means that distinct religious icons and symbols may become diluted or simplified to appeal to a broader audience.

*donatella.privitera@unict.it

DOI: 10.1079/9781800623675.0014

The topic of the commercialization of religious symbols in images has been explored in various fields, including religious studies, cultural and sociological studies, and media studies (Nardella, 2014). Carrette and King (2005) examined the commercialization of spirituality and religion in contemporary Western society and the ways in which religious symbols and practices are appropriated for commercial purposes in the global marketplace. They argue that this trend is indicative of the silent takeover of religion by contemporary capitalist ideologies. According to this view, the concept of spirituality is being used by companies and their consumers to pay 'lip service to the exotic' and historically significant religions of the world, while distancing themselves from their worldviews and forms of life (Carrette and King, 2005, p. 17). This trend implies that the use of spirituality or themes of religion are gradually being colonized by business and market logic. As a result, spirituality becomes a tool to justify the pursuit of materialistic goals, rather than a means to seek higher meaning and health of the soul. Here, a case study is explored: Saint Nicholas became Santa Claus as an example of commercialization of a religious symbol that became famous for secular worship.

Really, van Bühren (2017) confirmed that today the visual and verbal communications of Christian identity are still important. In fact, sacred places such as the Sistine Chapel can act as a means of communicating the faith to a large cross section of people that come from all four corners of the globe (Edwards, 2018): the mass of visitors, standing in the Sistine Chapel and looking at its unique artworks can effectively turn a 'tour' or 'visit' into a spiritual experience.

A strange combination is that of advertising and religious symbols, although the latter support the various religions through their recognizability on a wide variety of advertising channels. The values conveyed by advertising using religion have changed substantially over time. Values once associated with products advertised through religion, such as family, wisdom and practicality, have become marginal, while values such as status, high cost, freedom and vanity have grown. Religious symbols in advertising are used today in connection with value systems associated with the desire to stand out from others, complacency about oneself and one's body, collective participation, confidence and security.

Religious subject matter has thus proved to be highly adaptable to the cultural changes of the advertising audience. The advertising world has tried, successfully, to appropriate consensus on a symbolic system produced and firmly preserved by the religious world. If the advertiser can now turn the religious symbol into a tool for consumer identification with a brand, this means, as sociologist Nardella (2015) states, that a 'migration of symbols' is underway. The symbol no longer belongs exclusively to the original context, but becomes a door to be passed through, often or occasionally, to access a compatible and interchangeable religious or commercial system. The challenge of a religion that advertises seems great. Even greater is the challenge of an advertisement that does religion. Taking these issues into account, we will summarize the history of how Saint Nicholas became Santa Claus with an analysis of the evolution and some implications of what has been considered sacred and commercialized.

Religious Symbols

Symbols are powerful tools for expressing abstract concepts and conveying deeper meanings beyond their material form and language. Religious symbols are specific forms of communication and expression of sacred meanings and transcendence, which are essential in conveying the totality of religious belief and practice systematically. It is well affirmed that 'Religious symbols are bridges between consciousness, the world, and the sacred' (Kokosalakis, 2001, p. 15355).

Symbols are dynamic and primarily function as vehicles for salvation, revealing and concealing the sacred within the context of religious traditions and the social experience of the actors involved. Religious symbols are diverse, differing from one religious tradition to another and carrying direct social and political significance. An example of a religious symbol that has been commercialized for marketing is the Christian cross. The cross is an important symbol for Christians, but it has lost its original association with the crucifixion and resurrection of Jesus

Christ. Particularly over the past decade, the cross has become a popular fashion accessory, with many companies using it in their products to appeal to Christian consumers or to create a fashion trend (Einstein, 2008). For example, the cross has been used in clothing and jewellery. Some companies even use the cross in their branding or logo to make their products more appealing to Christian consumers and others.

Icons, which derive from the Greek word for picture or image, are a specific form of religious image and a religious symbol par excellence. While any image can be an icon, icons in a restricted sense convey substantive religious meaning and facilitate personal interaction between divine and human agents. Icons typically depict events and themes, but the human face is the focal point, endowing them with specific theological and cultural significance. In today's age of life aesthetics, there is an increasingly close relationship between religious faith and our way of life. As a result, the desires and actions of people have shifted from merely gazing at landscapes to seeking out a diverse range of experiences that match their personal preferences. Lifestyle plays a significant role in shaping the choices and habits of individuals. Religious faith and interior activity are often linked because religion often involves personal introspection and reflection, which can lead to increased self-awareness and understanding of one's thoughts, emotions and beliefs (Griffin and Raj, 2017).

In addition, some religious icons and symbols have cross-cultural appeal beyond their original religious context and increasing focus on commercialization and marketing become more popular where they are used primarily to generate revenue. For example, the Buddha is the founder of Buddhism and is venerated by Buddhists around the world. The Buddha statue is an important religious symbol for Buddhists, representing the Buddha's teachings and enlightenment. However, in recent years, the Buddha statue has become a popular decor item in Western countries, with many people purchasing Buddha statues as a decorative item without any connection to Buddhism. The commercialization of the Buddha statue has led to controversies among Buddhists who feel that the statue is being used inappropriately and disrespectfully (Oza, 2021).

Religious faith often involves a belief in a higher power or divine force that transcends human understanding. This belief can lead individuals to contemplate their own place in the world and their relationship with this higher power. Overall, religious faith and interior activity are often linked because religion can encourage personal introspection, cultivate virtues, and provide a sense of purpose and meaning in life.

Religious travelling and souvenir shopping are often interrelated movements because both are driven by a desire to connect with a particular religious tradition or place and to bring back a tangible reminder of that experience. It is of note that religion and tourism have much in common. It is hard to ignore the impression that in most places of pilgrimage the profane impacts of tourism are just as important, if not more so, than the religious (Vukonic, 1998, 2002).

When people engage in religious travel, they often visit significant religious sites such as temples, churches or holy sites. These sites may have cultural or historical significance, and they may also hold spiritual or religious importance for the individual. During the visit, individuals may engage in various activities such as prayer, meditation or participating in religious rituals. These activities are intended to deepen the individual's connection with the religious tradition and to help them feel closer to the divine (Moira *et al.*, 2019).

After the visit, individuals may want to bring back a tangible reminder of their experience. This is where souvenir shopping comes in. Souvenir shopping allows individuals to take a piece of the religious site or tradition with them, whether it is a piece of religious art, a religious text or a symbol of the religion. The souvenir serves as a reminder of the individual's spiritual journey and can help them to maintain a connection with the religious tradition even after they have returned home. In some cases, souvenir shopping may also be seen as a form of religious pilgrimage in and of itself. For example, in certain religious traditions, the act of acquiring a specific religious artefact or relic is seen as a sacred act that brings the individual closer to the divine.

Icons, saints, religious souvenirs and religious symbols are examples of the global economy because they are religious artefacts and objects that are produced, distributed and consumed on a global scale (Kim *et al.*, 2020). In some cases, the production and consumption of these religious artefacts can have economic implications for the communities that produce them. For example, the production of religious icons or symbols may be a major source of income for artisans or manufacturers in certain regions of the world. Overall, religious symbols (as well as souvenirs and icons) of pilgrimage tourism are a cultural product of globalization. It is of note that religion can evoke emotions, ranging from reverence and gratefulness to fear and anxiety. Moreover, the connotations that arise from religious icons are different and perform better than those of religious symbols, and advertisers pay more attention to various religious icons. In fact, both signs were found to positively impact brand evaluation and purchase intention, particularly among highly religious consumers (Agarwala *et al.*, 2021).

Saint Nicholas, a Historical Cult Without Borders

Saint Nicholas was born into a wealthy family in Asia Minor in a place called Patara (in present-day Turkey, south-west of Myra). He became bishop of the port city of Myra in the territory of Lycia. After his death on 6 December AD 343, his tomb in Myra soon became a pilgrimage destination and his relics were immediately considered miraculous due to a mysterious oily liquid, called the 'manna' of Saint Nicholas, which came out. When Lycia was occupied by the Turks in the 11th century, the Venetians tried to take it over, but were preceded by sailors from Bari who brought the relics to Apulia on 9 May 1087. The sailors wanted a church dedicated exclusively to the patron saint, while the archbishop wanted to keep the relics in the cathedral that had already been built. An armed confrontation was followed by the mediation of Abbot Elia, who in the meantime had taken delivery of the relics in his monastery of Saint Benedetto. Finally, the people and the archbishop agreed to entrust him with the demolition of some sacred and profane buildings and finally to raise the imposing and magnificent Basilica dedicated to Saint Nicholas.

Two years later the crypt of the new church was completed and Pope Urban II, escorted by the Norman knights of the lords of Apulia, placed the relics of the saint under the altar where they are still today. The translation of the relics of Saint Nicholas had an extraordinary echo throughout Europe, and in the Middle Ages the Apulian sanctuary became an important pilgrimage destination, with the result of the spread of the cult of Saint Nicholas of Bari (and not of Myra).

The feat of the sailors from Bari had various consequences on a religious and social level considering that Saint Nicholas, according to the preachers of the time, was at the forefront of popular devotion to the saints. With the theft of the bones of Saint Nicholas, Bari, while retaining its characteristics of a western city, consolidated its projection towards the East. The saint, in fact, lived in Asia Minor and was of Greek language and culture. Since then, the cult of Saint Nicholas has also spread widely in the West. The feat of the translation was the effect, and not the cause, of the exceptional cult of the Saint of Myra, already very much alive on the shores of the western Mediterranean (Cioffari, 2007).

The holiness of Saint Nicholas, known and venerated throughout the world, has never ceased to be the object of pilgrimages. In the 14th and 15th centuries there were many 'judicial pilgrimages' to Bari, imposed on sinners from Flanders and other European regions in lieu of serious sentences. The pilgrim, forced to take his leave by the authorities who had condemned him, wore the habit of the penitent, publicly acknowledging his faults with the contextual request for conciliation to the offended party. When the pilgrim arrived in the Basilica, he presented himself to the civil and religious authorities of Bari. The canons, custodians of the relics of Saint Nicholas, issued the condemned pilgrim a certificate attesting his arrival and said prayers in the sanctuary (i.e. the successful expiation of the sentence) (Bacci, 2009; Vantaggiato, 2010). This rite further favoured the diffusion and knowledge of the cult of Saint Nicholas towards Northern Europe.

What is most striking about this cult is the dimension of the phenomenon, with a location in Bari and its vast diffusion without borders, from East to West. The transfer of the relics to Bari, where the Bishop of Myra was already venerated in various churches, amplified his notoriety and favoured pilgrimage opportunities.

Saint Nicholas Religious Movements in the Modern World

The cult of Saint Nicholas is today almost widespread all over the world: the numerous churches dedicated to him bear witness to this. Evidence of the antiquity of worship of Saint Nicholas emerged, for example, during the archaeological excavations under the Palatine Hill in Rome, with the discovery of a fresco dedicated to him dated to around 760 AD; thus confirming that he was venerated in Italy already four centuries before the translation of his bones. Currently this image is considered the oldest existing image of Saint Nicholas in the world (Melchiorre, 1986).

Among the saints of the earliest Christian tradition, one can hardly identify another character who is the object of such a strongly rooted cult in both the East and the West and whose image, albeit through an almost uninterrupted series of local interpretations, has become so familiar to a multitude of people. He was honoured by the Byzantine church and soon revered and appropriated, even in the most literal sense of the term, by the Latin tradition and partly also by the Armenian one. The affirmation of the cult of Saint Nicholas can be considered definitive in the 6th century, a period to which an inscription found in Crete is dated in which his assistance is invoked, clearly demonstrating the fact that the fame of the saint had already spread widely along the routes of navigation of the eastern Mediterranean. By virtue of this process, the legendary physiognomy of the saint ends up assuming a different aspect, in some ways more conventional: from acts of generosity and moral integrity, the emphasis shifts more and more to the role of exceptional thaumaturge. Election to the episcopal role is no longer conceived as the elevation of status exceptionally granted to a layman by divine will, but as the final result of an adequate religious education and an ordinary ecclesiastical career (Ivona, 2021).

The testimonies on the universality of the cult of Saint Nicholas are numerous and consistent. The authoritative *The Oxford Dictionary of Saints* calls him one of the most universally revered saints in both East and West (Farmer, 2011). In *The Book of Saints* the Benedictine Monks of St Augustine's Abbey, Ramsgate (1989) call him one of the most popular saints in Christendom. Even more peremptory is *A Biographical Dictionary of the Saints* (Holweck, 1991), which defines him as the most popular saint in Christendom, highly celebrated by all nations, especially by the schismatic Russian church. The strong evangelical commitment of Saint Nicholas did not favour a narrative rich in miracles, as was the case with other saints; on the contrary, one gets the impression that the events narrated could exist even without the miracles and that the hagiographers wanted to add them to create a wonderful atmosphere. Certainly, the central element in the miraculous narrative of the figure of the saint of Myra is the sea, for which Nicholas, a new Poseidon, is presented as the saint capable of taming the force of the winds and the violence of the waves. Through miracles, the faithful increased their trust in the saint and he was suitably magnified. The stories of Saint Nicholas have not all been told in the same way. Each group of people reworked them according to their own sensitivity. Every medieval copyist put something of himself into it when he really didn't run into some translation or copying error (Charles, 1978; Bacci, 2009).

From Revered Saint to Globalized Icon

In the case of Saint Nicholas, it can be argued that he was a saint symbolic of an ante litteram globalization. From the moment he had begun to be celebrated in vast territories as the great thaumaturge par excellence, he had ceased to be a figure linked to his specific birthplace to become a large-scale saint who had visited the major centres of his cult, from Palestine and Cyprus to Rome via Syria and Armenia. He is a born globalist operating in a global business in several destinations (Harcourt, 2011).

Through the maritime communication routes from Lycia, Bishop Nicholas had reached the major centres of Christianity – Jerusalem, Rome, Constantinople – and from there his fame had spread everywhere. Towards the East he had joined the Byzantine-rite Christian communities on the Syrian and Palestinian coasts, and had often managed to gain the respect of the Islamic populations as well. Northwards there is evidence of his worship in the Caucasian regions: for example, in Orthodox Georgia he was frequently extolled in several pictorial works and a portrait of him from 1312 AD was found in a church in present-day Nagorno Karabakh. In Russian cities the saint had arrived from the Black Sea and from the Greek cities of Crimea up the great rivers until he soon reached Kiev. Soon the miraculous effigies of Saint Nicholas became second in number only to those of the Virgin Mary. This popularity also spread to the southern Slavic communities from which it was probably also transmitted to other groups such as the Vlachs and the Albanians (Ahrweiler, 1996; Bacci, 2009; Di Fazio, 2019).

The spread of the cult was also facilitated thanks to the strong veneration of the saint that was practised in Bari (Italy) since the translation of his bones, as previously mentioned. Hence the cult, given the geographical position of Bari that was central for land and sea pilgrimage routes, spread towards Mediterranean destinations as well as towards Rome, Santiago de Compostela, and the great sanctuaries of France and Germany. The cult of Saint Nicholas became associated with other important milestones in the international network of holy places, by expanding westwards into the Iberian Peninsula along the route leading to the tomb of the Apostle James in Galicia, and moving further northwards until it reached Iceland, where the holy bishop was celebrated like the ancient Viking heroes in the traditional literary form of saga.

It was precisely in Bari that, what we call today, the globalization of the image of Saint Nicholas took place, transformed and modified according to the secular and religious traditions where the cult landed as it went along. Through the multiple paths of worship, the images and legends that told of his life were mixed in a kind of secular and religious syncretism at the same time. Over the following centuries, for example, the special headdress of bishops, the mitre, was added to his image, and there was a tendency to make his face more in line with the appearance of Western prelates (i.e. rigorously shaved). Thus, it was that, especially north of the Alps, the great thaumaturge often found himself forced to adopt a look that made him almost unrecognizable, although this development never succeeded in supplanting the ancient far-more-popular bearded effigy. Western-style dress, in the case in point, became so strongly entrenched that it was exported even to the Levant territories that, following the Crusades, had become Latin possessions.

The more the cult and the notoriety of Saint Nicholas spread, the more the stories of his miracles spread, among which perhaps the best known is the one concerning the three young clerics saved from death, by the work of the saint, inflicted by an innkeeper where they had taken refuge to spend the night. The widespread diffusion of the episode of the three clerics was so popular that it occupied a prominent position in the large cycles of Saint Nicholas reproduced in stained-glass windows, in sculptures of cloisters and baptismal fonts, and in large fresco cycles from the 12th century onwards. Equally famous was the story of the three girls (or perhaps four) whose historicity, at least in its essential lines, is in any case guaranteed by the variety of ancient traditions that have survived to this day. The most extensive and known narrative is that of Michael the Archimandrite (8th century AD). He tells of a father who from a certain wealth had fallen into extreme poverty. He had some daughters of marriageable age, but due to the poverty into which the family had fallen they were not considered suitable marriage partners and were therefore marginalized, so he thought of solving the problem by making them prostitutes. Nicholas learned of this family drama and decided to intervene according to the method suggested in the Gospel; that is, he collected a sufficient sum of gold coins in a cloth and secretly threw it through the window into the man's house so that the sad matter could be solved, and so it was (Jones, 1983).

The 'Miracle of the Three Clerics' (transformed into three children in the narrative handed down over the centuries and which made him the patron saint of children) and the tale of the three maidens favoured a strange meeting of traditions and legends. This over time expressed itself mainly in a series of folk customs that are widespread in Northern Europe, very far from that Mediterranean Sea to which Saint Nicholas was deeply attached.

In this context, the special role of Saint Nicholas as bearer of gifts to children went on to be defined during the late Middle Ages, which was exalted in a whole series of folkloric customs that, albeit in a sweetened and commercialized form, have been passed down to the present day, especially in the German context.

Despite the emergence of new cults in the sphere of folk customs in the winter period that are widespread in Northern Europe, Saint Nicholas experienced ever-increasing success. Even in an environment such as the Calvinist Netherlands that was extremely hostile to saints, and especially to those who, like Nicholas, went dressed as members of the Roman ecclesiastical hierarchy, the popular cult of the 'gift-bearer' par excellence managed to perpetuate itself despite the persecution of Catholics and their state of hiding (Bacci, 2009).

But after the Protestant Reformation began in the 16th century, saints like Nicholas fell out of favour in much of Northern Europe. In many countries, including Germany, a tendency developed to concentrate the distribution of gifts on the night between 24 and 25 December rather than for the feast of Saint Nicholas on 6 December. To the figure of the child Jesus, the bearer of gifts to children, the tradition placed side by side the new figure of the 'Christmas Man' (*Weihnachtsmann*) to whom similar characteristics to those of the Saint of Myra were attributed.

Some of these Germanic figures were once again based on Nicholas, no longer as a saint but as a menacing sidekick such as Ru-klaus (Rough Nicholas), Aschenklas (Ashy Nicholas) and Pelznickel (Furry Nicholas). Dissimilar in appearance to the merry man in red, these colourful characters would later figure in the development of Santa Claus himself (Bowler, 2005).

Revival and Remaking of Saint Nicholas in the 19th century

During the first decades of the 19th century, everything changed thanks to a series of poets and writers who strove to make Christmas a family celebration, reviving and remaking Saint Nicholas.

For example, Washington Irving in his book *Knickerbocker's History of New York* (1809) recounted the successful venture of the enterprise of one of the first crews of colonizers of Lower Holland, who sailed from the city of Amsterdam and successfully arrived at the mouth of the Hudson thanks to the protection of Saint Nicholas whose effigy was drawn on the ship's prow. To commemorate the success of the enterprise and the foundation of the New Amsterdam, Irving recalls, albeit ironically, the 'pious ceremony' of hanging a stocking under the fireplace on the eve of Saint Nicholas, which in the morning is always found by the children of the home to be miraculously full. Jones (1983) states that without Irving there would be no Santa Claus (derived from the Dutch *Sinter Klaas*).

In 1821 an illustrated poem entitled 'The Children's Friend' (Stansbury, 1821) went much further in shaping the modern Santa Claus and associating him with Christmas. According to Bowler (2005) in this tale, the appearance of a Santa Claus is first outlined, who, while evoking the magic of Saint Nicholas, lost the religious characteristic and donned the robes of the 'Germanic gift-bearers'.

In 1823, Clement Clarke Moore wrote the poem 'A Visit from Saint Nicholas' (1862), better known today as 'The Night Before Christmas', for his six children, with no intention of adding to the burgeoning Santa Claus phenomenon. It was published anonymously the following year, and to this day the chubby, jolly Santa described in it rides a sleigh driven by eight tiny reindeer:

> He was dressed all in fur, from his head to his foot. [...]
> A bundle of toys he had flung on his back. [...]
> His eyes ... how they twinkled! His dimples how merry! His cheeks were like roses, his nose like a cherry!
> His droll little mouth was drawn up like a bow,
> And the beard of his chin was as white as the snow;
> The stump of a pipe he held tight in his teeth,
> And the smoke it encircled his head like a wreath;
> He had a broad face and a little round belly.
> (Moore, 1862, p. 8)

It was not until the late 19th century that the image of Santa Claus became standardized as a life-size adult, dressed in red with white fur trim, venturing from the North Pole on a reindeer-drawn sleigh and keeping an eye on children's behaviour.

The modern effigy of Santa Claus, developed in the second half of the century by German cartoonist Thomas Nast, who emigrated to the USA, made use of those features described by Moore to create the icon that still invades advertising today (Fig. 14.1).

Prior to Nast's creation of the character, the figure of Santa Claus/Saint Nicholas was quite different, tall and thin and wearing a green suit similar to an elf in Nordic tradition. Nast continued to draw and modify Santa Claus for 30 years in both appearance and the colours of his clothing, which varied from green to tan to the red for which he is known today.

Nast's creation turns out, then, to be a completely reimagined figure: he is pot-bellied, has a ruddy complexion, has a pipe in his mouth, is dressed in red and is laden with presents. Coca-Cola is to be credited with further humanizing the character holding the famous beverage (Fig. 14.2).

The icon created by Thomas Nast soon established himself as a collective symbol of American

Fig. 14.1. Thomas Nast's drawings of Santa Claus appeared in *Harper's Weekly* until 1886. From Widewalls, 2022, via Wikimedia Commons PDM 1.0 DEED (see Thomas Nast – a Cartoonist Who Created a Modern Image of Santa Claus).

Fig. 14.2. Haddon Sundblom's Coca-Cola Santa Claus. The postcard is from the personal archive of Antonella Ivona.

society, which imagined him as a true capitalist and captain of industry who owned, in the far north, a large toy factory on whose assembly line hundreds of worker elves worked day and night without ever going on strike. His popularity spread so widely that Coca-Cola advertisers thought it best to hire him as a testimonial to persuade the market to consume that refreshing drink even in the winter months. In order to circumvent a law that prohibited (due to the drink's caffeine content) the use of advertising images in which under-12s drank Coca-Cola, the Atlanta-based multinational decided to use Santa Claus as a testimonial, delegating the famous 'gift-giver' the task of winning over a key target of consumers (Lagioia, 2005).

The Coca-Cola Company wanted the campaign to show a healthy Santa Claus that was both realistic and symbolic, showing Santa Claus himself, not a man dressed as Santa. Illustrator Haddon Sundblom was commissioned to develop advertising images using Santa Claus; he, too, was inspired by Moore's poem. Sundblom's Santa Claus debuted in 1931 in Coca-Cola adverts in the *Saturday Evening Post* and appeared regularly in that magazine, as well as in *Ladies Home Journal, National Geographic, The New Yorker* and others. From 1931 to 1964, Coca-Cola advertisements showed Santa Claus delivering toys, stopping to read a letter and enjoy a Coke, and visiting children who had stayed up to greet him. The original oil paintings created by Sundblom were adapted for Coca-Cola advertising in magazines and on-store displays, billboards, posters, calendars and plush dolls. Many of these items today are popular collectibles (The Coca-Cola Company, 2023).

Through myths, legends, miracles, tales and traditions the figure of the Saint of Myra, revered for his thaumaturgical abilities, slowly gave way to a new figure less connoted in a confessional sense; old Klaus then transformed into a good-natured figure who reached from the frozen forests once a year to the interiors of urban homes to reward good children.

In short, if Saint Nicholas acts in a circumstantial way (apocryphal or not, his exploits are focused around very specific events), Santa Claus moves in the world of humans according to a nuanced and, at the same time, universal dynamic marked by recursiveness: he intervenes every year for the benefit of every child on the planet.

Summary

For many decades advertising has discovered religious symbols as a powerful vehicular system to spread promotional messages. The case study of the 'transfiguration' of Saint Nicholas to Santa Claus is an exact demonstration of this. Saint Nicholas and Santa Claus symbols, beliefs, narratives and disciplines have been commercialized and sold in the secular marketplace. The focus of this chapter was to examine how sacred images have been used to promote products and turned into commodities for consumption. This commodification can potentially weaken a brand (Hilpert, 2020) and diminish the sacred value of religious content.

In addition, Santa Claus became a tourism industry in Lapland; it became the Santa Claus Land Association, and it is now responsible for promoting Lapland's Santa Claus attractions in Finland where the Santa Claus Village was built purely to attract tourists. Saint Nicholas or Santa Claus is the history of the commodification of Christmas, as they are markers of the intangible sights of Christmas (Pretes, 1995).

The case study demonstrates that some images have become secular and these have usurped the original religious images required for a sacred experience (Bartunek and Do, 2011).

Religious tourism has caused the standardization of the tourist experience, presenting sites, symbols, souvenirs and rituals in a way that appeals to a broad audience and enhances a destination's appeal. This can result in a loss of authenticity, making the experience feel artificial. For instance, the commercialization of Christmas has made organized commercialism sacred, leading to a reduction in the sacred character of Saint Nicholas, as discussed by Bartunek and Do (2011).

Since ethical consumption has consistently grown in recent years, understanding how people with religious, spiritual or secular values behave in the marketplace could be of strategic importance (Spinelli *et al.*, 2019). In addition, the level of religiosity is relevant to an individual's attitudes, travels and values, as well as their participation in events and social life, and the expression of identities (Liutikas, 2021).

Overall, globalization has had both positive and negative implications for icons, symbols and religious and cultural products used for religious tourism. While it has opened new opportunities for exchange and understanding, it has also posed risks to the authenticity and integrity of these symbols and products.

Note

[1] While the chapter is the result of a joint work, sections 1, 2 and 7 are attributed to Donatella Privitera, and sections 3, 4, 5 and 6 to Antonietta Ivona.

References

Agarwala, R., Mishra, P. and Singh, R. (2021) Evaluating the impact of religious icons and symbols on consumer's brand evaluation: context of Hindu religion. *Journal of Advertising* 50(2), 1–19. DOI: 10.1080/00913367.2021.1940394.

Ahrweiler, H. (1996) Le culte de Saint Nicolas. *Transversalités* LVII/1, 147–153. (in French)

Bacci, M. (2009) *San Nicola il Grande Taumaturgo*. Ed. Gius. Laterza & Figli, Roma-Bari, Italy. (in Italian)

Bartunek, J.M. and Do, B. (2011) The sacralization of Christmas commerce. *Organization* 18(6), 795–806. DOI: 10.1177/1350508411416400.

Benedictine Monks of St Augustine's Abbey, Ramsgate (1989) *The Book of Saints: a Dictionary of Servants of God*. A & C Black, London.

Bowler, G. (2005) *Santa Claus: a Biography*. McClelland & Stewart, Toronto, Canada.

Carrette, J. and King, R. (2005) *Selling Spirituality: the Silent Takeover of Religion*. Routledge, London.

Charles, W.J. (1978) *Saint Nicholas of Myra, Bari, and Manhattan: Biography of a Legend*. University of Chicago Press, Chicago, Illinois.

Cioffari, G. (2007) *Pellegrini a San Nicola di Bari nella storia.* Centro Studi Nicolaiani, Bari, Italy. (in Italian)

Di Fazio, A. (2019) *Pellegrini di San Nicola. Sul tratto pugliese della via Francigena*. Adda Editore, Bari, Italy. (in Italian)

Edwards, E. (2018) Survey of audience reception in the Sistine Chapel: decoding the message of sacred art. *Church, Communication and Culture* 3(3), 260–282. DOI: 10.1080/23753234.2018.1539625.

Einstein, M. (2008) *Brands of Faith: Marketing Religion in a Commercial Age*. Routledge, London.

Farmer, D. (2011) *The Oxford Dictionary of Saints*. Oxford University Press, Oxford.

Griffin, K. and Raj, R. (2017) The importance of religious tourism and pilgrimage: reflecting on definitions, motives, and data. *International Journal of Religious Tourism and Pilgrimage* 5(3). DOI: 10 21427/D7242Z.

Harcourt, T. (2011) Santa Claus as a symbol of the global economy. *The Globalist*, 23 December. Available at: https://www.theglobalist.com/santa-claus-as-a-symbol-of-the-global-economy (accessed 20 March 2023).

Hilpert, M. (2020) Branding the sacral landscape: religious content in regional tourism marketing, set in a post-secular society. *International Journal of Religious Tourism and Pilgrimage* 8(4), 50–58. DOI: arrow.tudublin.ie/ijrtp/vol8/iss4/6/.

Holweck, F.G. (1991) *A Biographical Dictionary of the Saints: With a General Introduction on Hagiology*. Omnigraphics Inc, Detroit, Michigan.

Iaquinto, B.L. (2020) Understanding the place-making practices of backpackers. *Tourist Studies* 20(3), 336–353. DOI: 10.1177/1468797620927308.

Irving, W. (1809) *Knickerbocker's History of New York*. W.B. Conkey Company Publishers, Chicago, Illinois.

Ivona, A. (2021) La Peregrinacion de San Nicolás. Geografía de una tradición centenaria. *Sémata: Ciencias Sociais e Humanidades* (33), 1–20. DOI: 10.15304/semata.33.7878. (in Spanish)

Jones, C.W. (1983) *San Nicola. Biografia di una Leggenda*. Laterza, Bari, Italy. (in Italian)

Kim, B., Kim, S. and King, B. (2020) Religious tourism studies: evolution, progress, and future prospects. *Tourism Recreation Research* 45(2), 185–203. DOI: 10.1080/02508281.2019.1664084.

Kokosalakis, N. (2001) Symbolism (religious) and icon. In: Smelser, N.J. and Baltes, P.B. (eds) *International Encyclopedia of the Social & Behavioral Sciences*. Elsevier, Oxford, pp. 15354–15357.

Lagioia, N. (2005) *Babbo Natale. Dove si racconta come la Coca-Cola ha plasmato il nostro immaginario*. Fazi Editore, Roma. (in Italian)

Liutikas, D. (2021) The expression of identities in pilgrim journeys. In: Liutikas, D. (ed.) *Pilgrims: Values and Identities*. CAB International, Wallingford, UK, pp. 17–34.

Melchiorre, V.A. (1986) *Bari e San Nicola*. Edipuglia, Bari, Italy. (in Italian)

Moira, P., Mylonopoulos, D. and Parasxi, A. (2019) Preserving traveller's memory in religious destinations. *International Journal of Religious Tourism and Pilgrimage* 7(2). DOI: 10.21427/ercj-j216.

Moore, C.C. (1862) *A Visit From Saint Nicholas*. James G. Gregory Publisher, New York.

Nardella, C. (2014) Studying religion and marketing: an introduction. *Sociologica* 3, 1–15.

Nardella, C. (2015) *La Migrazione dei Simboli. Pubblicità e Religione*. Guerini Scientifica, Milano, Italy. (in Italian)

Oza, P. (2021) Buddhist iconography and religious symbolism in different Buddhist statues. *SSRN* [Social Science Research Network], 12 May. Available at: https://ssrn.com/abstract=3842904 (accessed 17 March 2023).

Pretes, M. (1995) Postmodern tourism: the Santa Claus industry. *Annals of Tourism Research* 22(1), 1–15.
Spinelli, G., Nelson-Becker, H. and Ligossi, R. (2019) Consumer competence strategies, spiritually inspired core values and locus of control: what are the links? *Sustainability* 11(17): 4787. DOI: 10.3390/su11174787.
Stansbury, A.J. (1821) *The Children's Friend*. William Gilley Publisher, New York.
The Coca-Cola Company (2023) Haddon Sundblom and the Coca-Cola Santas. Available at: https://www.coca-colacompany.com (accessed 10 March 2023).
van Bühren, R. (2017) Caravaggio's 'Seven Works of Mercy' in Naples. The relevance of art history to cultural journalism. *Church, Communication and Culture* 2(1), 63–87. DOI: 10.1080/23753234.2017.1287283.
Vantaggiato, L. (2010) *Pellegrinaggi giudiziali*. Ed. Compostelliane, Pomigliano d'Arco, Italy. (in Italian)
Vukonic, B. (1998) Religious tourism: economic value or an empty box? *International Review of Economics and Business* 1(1), 92–100.
Vukonic, B. (2002) Religion, tourism and economics: a convenient symbiosis. *Tourism Recreation Research* 27(2), 59–64, DOI: 10.1080/02508281.2002.11081221.
Widewalls (2022) Thomas Nast – a cartoonist who created a modern image of Santa Claus. Widewalls, 3 December. Available at: https://www.widewalls.ch/magazine/thomas-nast-santa-claus (accessed 30 July 2023).

15 Preparing Students for Healing: Intersectional Approaches to Pilgrimage and Travel

Alison T. Smith*
College of Charleston, Charleston, South Carolina, USA

Abstract
This chapter explores the vast potential of pilgrimage and transformative travel as powerful tools for nurturing identity development and fostering self-care among university students. It uncovers the transformative possibilities these journeys offer, delving into how they can serve as catalysts for personal healing, recovery from traumas, and the cultivation of positive and authentic identities. It includes observations of how others have used transformative travel to shape their sense of identity in positive and authentic ways. As a culminating project, students design and embark upon personal microjourneys that are relevant to their own self-care needs. These microjourneys are inexpensive and locally based, thereby expanding the healing potential of transformative travel and pilgrimage to more participants. This chapter includes further research into how to empower students with the tools and knowledge to embrace their own transformative journeys, enabling them to explore the intersection of identity development and self-care through pilgrimage and travel. It underscores the universality of these concepts and the potential for them to positively impact the lives of young scholars as they navigate the complexities of university life and beyond.

Introduction

Questions of identity are complex, and embarking on a path of personal exploration can lead to grappling with uncomfortable truths. It is important for young people to realize that they are not alone in their unease, and it can be instructive to read about others who have found positive ways to deal with their demons. Pilgrimage scholars know that walking and the quest for identity often go hand in hand, much as exercise scientists understand the physiological and psychological benefits of walking as exercise. My research explored 'purposeful, contemplative walking with a mind toward peace' in a previous article (Smith, 2018a), and that practice is well-suited for anyone facing an array of stressors, including university students and professors. Students in my courses read a variety of works that frame walking long distances as a transformative experience. They explore shorter microjourneys that are transformative as well. As a final project, they design their own transformative microjourney, which can take place either in the area surrounding our university in Charleston, South Carolina, or in their hometown. Research examined the incorporation of micro-pilgrimage as a high-impact learning practice for first-year students in a previous article on pedagogy and mapping tools (Smith,

*smitha@cofc.edu

DOI: 10.1079/9781800623675.0015

2018b). Having students develop a microjourney of their own design beyond the space of the traditional classroom is consistent with the spatial–temporal extension of the classroom into what are called **learningscapes** among practitioners of engaged pedagogy. When the traditional features of the classroom are changed, room is made for 'new learning opportunities to emerge' (Kinloch *et al.*, 2021, p. 18). Engaged pedagogies encourage students, teachers and community members to learn together; following 'students' sites and sights' can lead us 'to think and move in dynamic, connected ways' (Kinloch *et al.*, 2021, p. 113). The scope and variety of final projects in my courses are an impressive indicator of the kinds of challenges students face and the ways they seek to engage with them. Matters of faith, identity and oppression are among the themes of their projects, in addition to the need to relieve stress and to care for oneself.

In this chapter I discuss how a university professor who works with students in their first year of study has a front-row seat from which to witness the myriad trials our young people face. Their struggles include depression, anxiety, substance abuse, questions of gender and sexuality, racism, xenophobia, and scores of other challenges to their well-being. The transition from high school to university can be academically daunting, with personal and social issues compounding the problems. Questions of identity are at the forefront of many of these challenges as it is the first time many students find themselves living independently and away from familial support and pressures. At the threshold of adulthood, first-year university students are just beginning to understand who they are as they seek to find their place in the world. As one of a group of faculty members working to incorporate self-care and wellness into university courses, I aim to integrate research and practice into the curriculum, infusing first-year students with readings and assignments that promote mental and physical wellness. The readings for the course emphasize walking and transformative travel as tools that are available to students to help them navigate their 4 years at the university and indeed their life beyond academia.

The institutionalized forms of oppression add to the burdens and challenges faced by our students, especially those who are struggling with issues of gender identity or coping with established systems of racial and sexual oppression. Intersectionality 'illuminates how gender, class, nation, sexuality, religion, and other categories of identity, power, privilege, and oppression interconnect to affect the lives of individuals and groups ... at community, societal, and global levels' (Kirk and Emerita, 2012, p. 3). In our post-pandemic world, students seem less prepared than ever to meet the demands of academic life, and at-risk groups often find that their needs are neglected as universities fall behind in their ability to meet students' mental health issues. Recently a small, interdisciplinary faculty group was created at the College of Charleston to focus on meeting first-year students' mental wellness needs within the context of our courses. FLASC (Faculty Liberal Arts and Sciences Colloquium) groups at our university are faculty-led reading and discussion groups on a specific topic, with the anticipated outcome of creating new courses or infusing existing courses with updated content. A colloquium on mental wellness was facilitated by Dr Lisa Ross, a professor of psychology. With representatives from various disciplines and backgrounds, we met each week for the duration of one semester to discuss readings related to students' mental health. We also evaluated existing strategies and developed new ones for promoting mental wellness and self-care in our first-year seminars and learning communities. Building upon the work of this faculty focus group, a first-year seminar course called 'Travel for Transformation' was developed. My goal is to introduce students to the benefits that scholars recognize in pilgrimage while also integrating strategies to promote self-care and mental wellness. An additional positive outcome is that by encouraging our students to incorporate regular practices of self-compassion in their daily lives, a space opens to allow for showing compassion to others as well.

To help students equip themselves with some essential tools for self-care, we considered examples from works of non-fiction, fiction, films and videos, while also attending lectures and visiting local sites. Non-fiction readings included Ben Montgomery's (2014) *Grandma Gatewood's Walk* and Cheryl Strayed's (2012) *Wild.* Among the films viewed were *The Way* by Emilio Estevez (2010) and *The Last Tourist* by Tyson

Sadler (2022). *The Unlikely Pilgrimage of Harold Fry* by Rachel Joyce (2013) was the only novel included in the readings, although students cited others as examples in class discussions. As one model for the kinds of microjourneys the students might explore, they read Clint Smith's (2021) *How the Word is Passed: a Reckoning with the History of Slavery Across America*. These longer works and films were supplemented by articles on pilgrimage, walking, transformative travel and mindfulness. *Grandma Gatewood's Walk* and *Wild* were selected for their relevance to the discipline of Women's and Gender Studies, the academic programme that sponsors the seminars. *The Way* fits within my current areas of research, which include Pilgrimage Studies and the Camino de Santiago. *How the Word is Passed* was the common read for all incoming students at our university for the 2022–2023 academic year, and students had the opportunity to meet with author Clint Smith and hear him speak as they engaged with his work. Other course readings offered context for the ideas explored in these works, such as George Greenia's (2014) article 'What is Pilgrimage?' Finally, short videos featuring pilgrimage scholars Kip Redick (2019) and John Shultz (2021), among others, served to strengthen the connections students made between pilgrimage, transformative travel, and the microjourneys they might choose to pursue (Fig. 15.1).

One of the objectives of my course is to reinforce the links between teaching and research, a key component of inquiry-guided learning (IGL), described by Virginia S. Lee (2012) as teaching that makes the most of faculty who are actively engaged in research. Presented with problem solving rather than lecture-driven learning, 'IGL students become increasingly comfortable with and able to make good decisions and judgements under conditions of uncertainty, a hallmark of intellectual growth and maturity' (Lee, 2012, p. 6). These outcomes align with the goal of fostering self-care and self-compassion in our students, and for this reason students are encouraged to practise positive

Fig. 15.1. Transformative books and the film *The Way* (Estevez, 2010). Author's photo.

cognitive restructuring techniques within the IGL process. 'Cognitive restructuring includes lower-order actions such as being optimistic, engaging in positive thinking, and playing down negative consequences ...' (Allen and Leary, 2010, p. 109). Using the problem-solving situations to which they are exposed in combination with the positive cognitive restructuring techniques they are encouraged to practise, students design a transformative microjourney in the spirit of the transformative experiences they read about or view. Through this pedagogical practice student learning needs to be at the forefront of research.

Transformative Films and Narratives

Emma Gatewood: a survivor and model of self-reliance

The story of Emma Gatewood is one of perseverance and determination despite the odds, and her accomplishments are legendary. Emma 'Grandma' Gatewood was the first woman to hike the entire Appalachian Trail as a 'thru hike' (i.e. the whole length of the trail is walked from end to end in a single trip), and she was the first person to complete the thru hike twice. Ben Montgomery (2014) tells Emma's compelling story against the background of the political and social climate of the historic moment when she completed the 2100-mile journey for the first time in 1955, while also setting the stage with a history of long-distance walking in the USA as well as of the Appalachian Trail itself. This exhaustive background includes a history of important women hikers in the USA, and Montgomery's portrayal of Emma Gatewood also serves to increase our knowledge of that aspect of women's history. Growing up in a farming community in rural Ohio, Emma was married at a youthful age to an abusive husband, with whom she had 11 children. To cope with the mistreatment that she received from her husband, Emma sought refuge in nature, taking long walks in the woods to escape her abuser. When she finally managed to divorce her tormentor, she made the decision to walk the Appalachian Trail after reading about it in *National Geographic* magazine. Emma Gatewood offers an example of a woman who thrived and even became a folk hero after she removed herself from a situation of domestic violence. The transformative journey of walking the Appalachian Trail led Emma Gatewood to a new life in which she was in charge. It was unimaginable to many in the 1950s USA that a 67-year-old grandmother could complete the extreme physical and psychological challenge presented by the Appalachian Trail.

Emma Gatewood offers a powerful example of resilience and courage, and several students were inspired by her to take on physical challenges they had previously believed themselves incapable of doing, such as running in long-distance endurance competitions ranging from 5Ks (5 km) to a marathon.

Cheryl Strayed: walking one's way to healing

Like Emma Gatewood, Cheryl Strayed turns to nature and walking to transition out of a demanding situation in life. In many ways, Strayed's (2012) memoire *Wild* has become a classic feminist outdoor adventure story. The author's brutal honesty about the wreck her life had become and her lack of preparation for the journey she undertakes on the Pacific Crest Trail stand in direct contrast to the iconic masculine stories of bravado that are more typical of the genre established by Ernest Hemingway. Like Emma's *National Geographic* impetus, Cheryl's decision to walk the Pacific Crest Trail results from a chance encounter with a book on the topic. A blizzard causes Cheryl to go to the REI (Recreational Equipment, Inc.) store in Minneapolis to purchase a shovel; while there, she sees a guidebook for the Pacific Crest Trail and reads the back cover as she waits in line. Cheryl is at a particularly vulnerable point in her life. At 26, she has lost her mother to cancer, been married and divorced, has begun engaging in unprotected sex with multiple partners and is experimenting with heroine. Moreover, she learns a few days after the visit to REI that she is pregnant. Knowing that she will terminate the unwanted pregnancy, she returns to the REI store, buys the book, and formulates a plan to walk the Pacific Crest Trail as a means of changing her out-of-control life:

> *I had to change* was the thought that drove me in those months of planning. Not into a different person, but back to the person I used to be—strong and responsible, clear-eyed and driven, ethical and good. And the PCT [Pacific Crest Trail] would make me that way. There, I would walk and think about my entire life. I would find my strength again, far from everything that had made my life ridiculous.
>
> (Strayed, 2012, p. 57)

Prior to her mother's cancer diagnosis, Cheryl had been an excellent student with a promising future. Her childhood had not been easy, however. Like Emma Gatewood, Cheryl's mother had also been a victim of domestic violence at the hands of Cheryl's father. During her 1100-mile walk from the Mojave Desert through California and Oregon to the Bridge of the Gods at the Washington State border, she has ample time to reflect on her difficult childhood, the absence of her father, and the brave optimism of her mother despite the countless challenges she faced in her short life of 45 years. Cheryl reflects on her own mistakes, and despite numerous setbacks she completes the arduous journey she set out to do. Walking the Pacific Crest Trail gave Cheryl self-confidence and a renewed sense of self-worth. She leaves drug abuse and irresponsible sexual encounters behind her as she moves forward to embrace the adult life she had dreamed of as a professional writer.

Like Emma, Cheryl inspired students with her story of physical achievement and psychological well-being. That she was closer to their age at the time of her walk made the story more relevant for some. Several students remarked that they found her ability to walk away from drug use inspiring.

The Way: walking to friendships through grief

Emilio Estevez's (2010) film *The Way* is yet another story of how people deal with trauma by walking, this time on the ancient pilgrimage path, the Camino de Santiago. At the story's centre is a troubled father–son relationship. Although they love one another, they obviously do not understand each other. Approaching the age of 40, Daniel (the son) has decided to abandon his pursuit of a doctoral degree in anthropology and to walk the Camino de Santiago instead. His father Tom, a successful physician, disapproves of this decision. When Daniel dies accidentally on his first day of walking the Camino, Tom goes to Saint-Jean-Pied-de-Port, France, to retrieve his son's body. Once in Saint-Jean, often understood to be the beginning of the Camino Francés walk across Spain to Santiago de Compostela, Tom decides to walk the Camino himself, carrying Daniel's ashes and scattering them along his journey.

While healing from grief is the primary theme of the film, Tom encounters several other pilgrims with different troubles as he makes his way on the 800-km trek to Santiago de Compostela. Unlike the long-distance walks of Emma Gatewood and Cheryl Strayed on the Appalachian Trail and the Pacific Crest Trail, respectively, the Camino de Santiago is historically a religious pilgrimage route. Although most of Tom's fellow pilgrims do not embark on the journey for religious reasons, there is nevertheless an aspect of faith that informs some of their journeys at different points along the way. Tom describes himself as a 'lapsed Catholic', but he accepts a rosary from a priest he meets on the trail. The pilgrims Tom meets and befriends are grappling with a host of issues – domestic violence, grief following the termination of a pregnancy, marital troubles related to body image and shaming, addiction to nicotine, cancer and writer's block are among the problems weighing upon the pilgrims. As he learns to process his own grief, Tom opens up to the other walkers and gains their support and trust. From this film, students learn the importance of community when faced with difficult issues. Certain characters in the film utilize the previously mentioned practice of positive cognitive restructuring. While *Grandma Gatewood's Walk* and *Wild* offer stories of individual triumph, *The Way* suggests that Tom's transformation and healing are only possible within the context of the camaraderie and mutual support of his new-found friends. By opening himself up to learn about other people and their suffering, Tom learns something about himself and can cope with his own suffering as well.

Students who took this lesson to heart decided to seek out friendships in structured ways. One student, a self-described loner, began to take evening walks with a friend when that friend finished working. Instead of simply walking back to the dorm together, they selected a destination and prolonged the walk to allow time for relaxation and appreciation of the beauty of the Charleston, South Carolina area at nighttime. Another student who had had a challenging time making friends joined a cycling group, where she gained a supportive community while also challenging herself physically.

Harold Fry: walking for love and self-acceptance

The Unlikely Pilgrimage of Harold Fry (Joyce, 2013) appears to be a story of a very ordinary retired English man. During his spontaneous walking journey, we learn that his life has been far from ordinary; rather, Harold Fry has been dealt an extraordinary amount of tragedy in his life. Harold's adventure begins when he receives an unexpected letter from a former co-worker named Queenie Hennessy. Although Harold and Queenie have not been in touch for over 20 years, she says that she is writing to thank him for the kindness he once showed her and to let him know that she is dying of untreatable cancer. Harold and his wife Maureen have a stagnant relationship that consists of little more than inhabiting the same home, stuck in a life of mindless routines that keep them from truly connecting with one another. The surprising letter from Queenie is like a jolt from the past. Harold dashes off a reply and heads out the door to put the letter in the nearest post box (mailbox). When he passes several post boxes without depositing his reply letter, he realizes: 'He had started something and he didn't know what it was, but now that he was doing it, he wasn't ready to finish' (Joyce, 2013, p. 11). A chance conversation with an employee at a petrol (gas) station leads him to the conclusion that he must walk to Queenie Hennessy, and that by walking it will help her have the faith she needs to live. The daunting aspect of this walk is that Harold begins where he lives in Kingsbridge, located in the south-west corner of England. Queenie is in hospice care in Berwick-upon-Tweed, located in the north-east corner of England near the border with Scotland, a total distance of 627 miles. Harold is not accustomed to walking long distances, and he sets off on this long journey with absolutely no gear and shod in a pair of yachting shoes.

Harold's completely out-of-character decision to walk the length of England with no preparation and no gear throws both Harold and Maureen into new territory. They are forced to face traumatic events that they have not had the courage to address, either individually or as a couple. Having been abandoned by his mother in his teens and left to be raised by an alcoholic father suffering from post-traumatic stress disorder, Harold lacks the emotional skills necessary for marriage and parenting. Maureen had erroneously believed that she could fill every void in Harold's life, only to become critical and bitter upon discovering that this was not possible. During Harold's journey, he has encounters and forms brief relationships with a host of people. Most importantly, he learns to listen, and bearing witness to their stories allows him to see that suffering is an inherent part of the human condition. He becomes a more caring and compassionate person for the stories he hears. Among the traumas that become known in his conversations with an array of interlocutors are repressed sexualities, marital infidelity, homelessness, terminal illness, abandonment, betrayal of trust and suicide. Harold gains an ability to deal with the greatest challenges life presents us with by facing them directly and through compassionately listening to others. He also learns that there is beauty in the simplicity of each moment. During his unplanned journey, he learns to accept the kindness of strangers with humility and dignity. Most importantly, Harold learns to open himself up to intimacy, recognizing the friendship with one woman and rekindling his love for another as he pays his final respects to Queenie Hennessy:

> "She was a good woman," said Maureen at last. "She was a true friend."
>
> She felt something warm against the tip of her fingers, and then she felt the pressure of his hand gripping her own.
>
> (Joyce, 2013, p. 316)

On his long walk from Kingsbridge to Berwick-upon-Tweed, Harold has interactions with people from all walks of life. He learns to listen without judgement, and for the first time in his life, he feels that he is doing something important. Although he is not a religious man in the traditional sense of the word, he comes to understand the significant role that faith plays in human interactions. Harold's journey helps him gain closure in some of the most difficult challenges he has ever faced, and in the end, it brings new life to his marriage to Maureen.

Although some students had difficulty relating to this story of an ordinary retiree, others found it to be the most compelling work we read or viewed during the semester. The breadth and difficulty of the traumas witnessed and experienced by such an ordinary person allowed students to understand that external appearances cannot be relied upon to tell us who a person is and what they have lived.

Transformative essays by Clint Smith: a personal and collective reckoning with slavery

Not every student is dealing with a personal trauma or understands that they may be dealing with one; for this and other reasons, it was important to include in the course information about how we can approach collective traumas like the legacy of the international slave trade in ways that foster responsible reflection. Clint Smith's (2021) *How the Word is Passed: a Reckoning with the History of Slavery Across America* proved to be an ideal work to both encourage student reflection and discussion while also serving as a model for the microjourneys in the Charleston area that they would undertake for their final course project. Originally selected to be the common read for the entire incoming class of first-year students, that designation was abruptly shifted to an opt-in for students, despite the plan for Smith to visit our campus and engage with students in both formal lecture formats and more casual in-class conversations. While the downside is that not all of our students had the opportunity to meet with Smith and discover his landmark work, the upside is that those students whose professors did decide to opt in, and require the book for class reading and discussion, had more direct access to the author and had the ability to ask him questions and interact with him in small-group settings. In his book, Smith describes site visits to places both familiar and unknown. His journeys represent his desire to understand 'how slavery is remembered and reckoned with' with the idea of 'teaching myself all of the things I wish someone had taught me long ago' (Smith, 2021, p. 5). Although most of the sites he visits to write his book are in the southern states that once belonged to the Confederacy, he also visits New York City and Dakar, Senegal, to emphasize that the ramifications of the global slave trade extend beyond the geographical region of the south-eastern USA. Using Smith's work in the class aligned well with the goal of engaged pedagogies to promote interconnectivity by 'residing in the local and finding resonance in the global ...' (Kinloch, 2021, p. 113).

Although I ask my students to read the entire book, in our classroom discussions we focus on three specific chapters, each of which has an evocative subtitle: (i) 'Monticello Plantation' ('There's a difference between history and nostalgia'); (ii) 'New York City' ('We were the good guys, right?'); and (iii) 'Gorée Island' ('One slave is too much').

Monticello was the home of Thomas Jefferson, the third President of the USA and author of the Declaration of Independence, and a visit to Monticello is a secular pilgrimage that borders on sacred for many Americans. Niya Bates, who was Monticello's director of African American history at the time of Clint Smith's visit, explained to Smith that Monticello is central to Americans' understanding of who they are in the scope of US history. She tells him, 'I think people come to us because they're grappling with their own identity' (Smith, 2021, p. 46). Because of Jefferson's complicated legacy as the author who penned the words 'all men are created equal', so central to the foundational values that most Americans hold to be true, while simultaneously being the owner and indeed father of enslaved humans, many Americans come to his home to try to understand not only Jefferson's place in history but also their own.

In the chapter on New York City, Smith goes on a tour centred on uncovering the history of slavery and the Underground Railroad. The tour guide affirms at the beginning of the visit that what they will learn is not New York or US history, but in fact world history 'that has been completely whitewashed and wiped out of our education systems globally ...', acknowledging that her presentation will contradict the history the group has been taught, perhaps even making them uncomfortable and causing them to question their personal beliefs (Smith, 2021, p. 208). On the tour, Smith learns that places like Wall Street and Central Park are tied to the oppression of and violence against both the native Lenape people and enslaved and freed Africans who were brought to what is now New York as early as 1626.

Finally, the chapter on Gorée Island explores the history of a place described by Clint Smith himself as a place of 'pilgrimage for those attempting to reckon with the history of chattel slavery' (Smith, 2021, p. 243). This is the final reading for my course, the one we discuss just prior to the student presentations of their own final projects or microjourneys. Under the colonial administration of first the Portuguese and later the French, Gorée Island's history has been linked with that of the transatlantic slave trade since the 16th century. A UNESCO (United Nations Educational, Scientific and Cultural Organization) World Heritage Site, Gorée Island has been visited by US Presidents Bill Clinton, George W. Bush and Barack Obama, as well as Pope John Paul II and activist Angela Davis, among countless other prominent world figures. While the full history of Gorée Island's place in the transatlantic slave trade is the subject of ongoing research, it serves as an important site of memory as we collectively try to fill the gaps. Clint Smith uses the existence of these lacunae to underscore the connections between history, place and the individual, in the exploration of personal identity:

> There are the gaps that sit at the edge of Gorée Island, a place attempting to tell a story that some feel has been forgotten, grappling with the tension that exists at the nexus of fact and truth, a place that still holds the ghosts of thousands and remains a symbol for the plight of millions.
>
> There are the gaps that exist inside me, a Black man in America unable to trace my roots past a certain point in history. Whose lineage beyond the plantations where my ancestors were held remains obscured by the smog of displacement. They are the gaps that I am trying to understand, the gaps I am trying to fill.
>
> (Smith, 2021, p. 269)

Students Filling the Gaps

Students, who are trying to fill their own gaps, both those created by personal and collective trauma, had the opportunity to speak with Smith and ask him questions directly when he visited College of Charleston's campus in October 2022 in a smaller group discussion prior to his formal talk to a larger audience. When asked about his own transformative experiences during his university years, he told of going to a poetry slam during the summer after his first year of college. A talented soccer player, Smith had dreams of playing professionally one day. The night at the poetry slam changed everything for him and set him on the path of becoming a poet and a teacher. He urged the students to try new things and to take courses on subjects that appealed to them rather than limiting themselves to things they deem to be practical. Smith's advice to the students expressed exactly what was hoped for them as they selected their personal microjourney or micro-adventure for their final projects. It was my hope that they might choose an excursion, a site visit, or a walk that would be as meaningful and as life-changing for them as the poetry slam was for Smith, which put him on a path to become a poet and writer and ultimately led him to take a pilgrimage to Gorée Island. Opening new worlds can heal wounds and fill gaps, and the journey always begins with an initial step in a well-chosen direction.

Students' microjourneys and projects have been as varied as the students are themselves. Many elected to visit historic homes, forts, plantations and museums in the Charleston area, but they did so with a critical eye, eager to seek out the stories that have been left untold, either fully or partially. Others elected to take on physical challenges, such as running a long distance for the first time. Examples of interesting site visits included Hampton Park, 9 College Way, the Charleston Museum and Laurel Hill County Park. Today a beautiful urban park and garden, Hampton Park was once part of a plantation and is named after a Confederate general. The student who selected Hampton Park was surprised to learn that it had also served as a prison site for Union soldiers during the Confederacy, and that upon being freed at the end of the Civil War, the formerly enslaved people held what some consider to have been the very first Memorial Day service to honour those Union soldiers who had died while there. A visit to 9 College Way, located on College of Charleston's campus, allowed students to explore the history of one of Charleston's prominent LGBTQ+ (lesbian, gay, bisexual, transgender, queer/questioning and more) couples from the mid-20th century, who lived and owned a bookstore at that address. Although a plaque in front of the building states that the two men were business partners, the truth of their love story is now being told (Greene, 2022). One student visited the Charleston Museum, the oldest museum in the USA, and noted with disappointment that the history of slavery was downplayed there. Other students walked trails they had been familiar with since childhood but did so with a new perspective. Some students completed these walks while at home over breaks, and others walked on Charleston area trails such as the ones in Laurel Hill County Park. Students brought a sense of curiosity about the history of familiar trails, often doing solo walks that allowed them to reflect on the different people who may have walked those same paths across the centuries. Being alone gave them the space to reflect, not only on the history of the places through which they moved, but also on the course of their own lives and their experience of the same walk over the course of several years.

One student initiated a practice of taking her singing bowls out to Sullivan's Island, where she would sit in contemplation, melding the sounds of her bowls with those of the ocean. She used this time for physical relaxation and as a spiritual practice, offering the music of her bowls as a form of prayer. The student s microjourneys mimicked ones made by Emma Gatewood prior to her separation from her husband and embarking on the Appalachian Trail. To escape her abusive husband, Emma sought refuge in nature. She would head into the woods, where she would take long walks and write poetry. My student's song bowl practice might also be interpreted as one of collective healing on the site that was once the nexus of the transatlantic slave trade. Prior to being taken to Charleston to be sold, the enslaved Africans who survived the Middle Passage spent a time in quarantine on Sullivan's Island. The Toni Morrison Society installed a bench for reflection on Sullivan's Island in 2007, and the bench was inaugurated by Toni Morrison herself. The evocative, healing vibrations of the sound bowls might serve as an aid for some in the reflective and healing process. Her intention to add a sacred element by using the bowls fits in here.

While these microjourneys may not qualify as pilgrimages in the strictest sense, many of them do share elements that have been used to define pilgrimage. Among those attributes cited by George Greenia (2014) in his article 'What is Pilgrimage?' we find that at least two were present in most of the students' projects:

> Being present at a site that others have designated as significant for non-material reasons: being there is transformative and enriching, and connects one to values beyond the normal reach of the individual ...
>
> The experience of pilgrimage is seen *in advance* as capable of creating an enduring memory one returns to in later life ...
>
> (Greenia, 2014, p. 10)

Summary

Students enrolled on the 'Travel for Transformation' course read about transformative travel and pilgrimage experiences with the goal of designing their own microjourney. In formulating their final project, they put IGL into practice. By undertaking a microjourney of their own design and choosing, they practise self-care and learn to use positive cognitive restructuring techniques. The works we read and view in class offer multiple examples of how people undertake journeys both large and small, allowing them to discover how to reframe negative experiences in a positive light. The students also explore how others have used travel and pilgrimage to grapple with personal and collective traumas as they seek to understand their place in the world and work through their own questions of identity. Gender, sexuality and race are at the forefront of my students' inquiries, but they are not the only arenas of discovery. Pilgrimage to a place of personal, historical or religious interest is encouraged, but students are also welcome to pursue adventures of their own design in consultation with me and following guidelines established at the beginning of the course. Ideally, the course is a safe space in which I hope that students will feel free to pursue interests that may have previously seemed inaccessible, as long as that pursuit comes from a place of self-care. Not all students will feel personally transformed after the course, but they will have been exposed to readings that help fill the gaps in what they have previously learned or that speak to topics on which discussion may have been restricted, as in the areas of gender and racial identity. Women like Emma Gatewood and Cheryl Strayed are role models who display independence and resilience. Clint Smith shows students how they might fight systems of oppression by grappling with difficult historical topics and asking the necessary questions so that untold stories become known. The fictional characters of Tom Avery (*The Way*) and Harold Fry show how the physical act of walking through one's grief can help us to open up to others. Our sense of personal identity is forged at the intersections of race, gender, age, nationality, ethnicity, social class and ability, as well as numerous additional categories depending on where and how one lives. Through the journey of self-discovery that students elaborate both in my course and in others of similar design and intent, they learn to equip themselves with some of the tools and aids needed to navigate the many overlapping systems of oppression and individual challenges they face in their daily lives, both on campus and beyond.

References

Allen, A.B. and Leary, M.L. (2010) Self-compassion, stress, and coping. *Social and Personality Psychology Compass* 4, 107–118.

Estevez, E. (dir.) (2010) *The Way*. Film. Filmax/Elixir Films, USA.

Greene, H. (2022) 9 College Way: Site of the Book Basement, 1946–71. Available at: https://discovering.cofc.edu/items/show/2?tour=1&index=1 (accessed 3 December 2023).

Greenia, G. (2014) What is pilgrimage? In: Harman, L.D. (ed.) *A Sociology of Pilgrimage: Embodiment, Identity, Transformation*. Ursus Press, London, pp. 8–27.

Joyce, R. (2013) *The Unlikely Pilgrimage of Harold Fry*. Random House, New York.

Kinloch, V., Nemeth, E.A., Butler, T.T. and Player, G.D. (2021) *Where is the Justice: Engaged Pedagogies in Schools and Communities*. Teachers College, Columbia University, New York.

Kirk, G. and Emerita, O.M. (2012) *Women's Lives: Multicultural Perspectives*, 6th edn. McGraw-Hill, New York.

Lee, V.S. (2012) What is inquiry-guided learning? *New Directions for Teaching and Learning* 129, 5–14.

Montgomery, B. (2014) *Grandma Gatewood's Walk: the Inspiring Story of the Woman Who Saved the Appalachian Trail*. Chicago Review Press, Chicago.

Redick, K. (2019) Pilgrimage on the Appalachian Trail. YouTube video. Available at: https://www.youtube.com/watch?v=9ztDurfg3Wk (accessed 3 December 2023).

Sadler, T. (dir.) (2022) *The Last Tourist*. Film. Utopia, USA.

Shultz, J. (2021) Vertical Pilgrimage: Japanese Mountain Religious Experience and American Big Wall Climbing. Video presented at Sacred Journeys 7th Global Conference. Available at: https://youtu.be/z_gT6LCs8dQ (accessed 3 December 2023).

Smith, A. (2018a) Pilgrimage, place, pedagogy: how mapping tools can enhance first year experience abroad. *International Journal of Humanities and Arts Computing* 12(1), 81–90.

Smith, A. (2018b) Walking meditation: being present and being pilgrim on the Camino de Santiago. *Religions* 9(3), 82. Available at: https://www.mdpi.com/2077-1444/9/3/82 (accessed 3 December 2023).

Smith, C. (2021) *How the Word is Passed: a Reckoning with the History of Slavery Across America*. Little, Brown and Company, New York.

Strayed, C. (2012) *Wild: From Lost to Found on the Pacific Crest Trail*. Vintage, New York.

Index

Note: The page numbers in italics and bold represents figures and tables respectively.

Printed and bound by CPI Group (UK) Ltd, Croydon, CR0 4YY
13/09/2026
14961111-0001